ALLE ZEIT WACH
1842

Reinhard Selten (Ed.)

Game Equilibrium Models II

Methods, Morals, and Markets

With Contributions by
D. Abreu, W. Albers, K. Binmore, R. Gardner
W. Güth, R. M. Harstad, H. Kliemt, W. Leininger
A. Okada, E. Ostrom, D. Pearce, L. Phlips
S. H. Schanuel, L. K. Simon, J. Sobel, G. Stephan
E. van Damme, J. W. Walker, F. Weissing, W. R. Zame

With 79 Figures

Springer-Verlag Berlin Heidelberg GmbH

Professor Dr. Reinhard Selten
Institut für Gesellschaft- und
Wirtschaftswissenschaften
der Universität Bonn
Wirtschaftstheoretische Abteilung I
Adenauerallee 24-42
D-5300 Bonn 1, FRG

ISBN 978-3-642-08109-5 ISBN 978-3-662-07365-0 (eBook)
DOI 10.1007/978-3-662-07365-0

Originally published by Springer-Verlag Berlin Heidelberg New York Tokyo in 1991

2142/7130-543210

Preface to the Series "Game Equilibrium Models"

The four volumes of the series "Game Equilibrium Models" are the result of a research year at the Center for Interdisciplinary Research of the University of Bielefeld, Germany. The German name of this center is Zentrum für interdisziplinäre Forschung, but everybody who is familiar with this unique institution refers to it by the official acronym *ZiF*.

In the time from October 1, 1987, to September 30, 1988, the ZiF was the home of the interdisciplinary research group which produced the papers in the four volumes of this series. Participants coming from many parts of the world lived in the guest apartments of the ZiF for the whole time or part of it and worked on a common project. The name of the project was "Game Theory in the Behavioral Sciences". It occurred to me only later that "Game Equilibrium Models" - the title of the series - even more appropriately indicates the unifying theme of the research group.

Among the participants were economists, biologists, mathematicians, political scientists, psychologists and a philosopher. A lively interaction resulted from the mix of disciplines. The common methodological basis of non-cooperative theory was the shared culture which facilitated communication across disciplines. The intense exchange of ideas coming from different fields had a profound influence on the thinking of many among the participants.

It was not easy to find a coherent way to group the papers into the four volumes and to find appropriate titles for the books. These and other difficult decisions have been made by an editorial committee consisting of Wulf Albers, Rudolf Avenhaus, Eric van Damme, Werner Güth, Peter Hammerstein, Ronald Harstad, Franz Weissing, and myself.

In the behalf of the whole research group I want to thank all those who helped to make the research year possible. We owe special thanks to the staff of the ZiF and in particular to Mrs. Lilo Jegerlehner for her technical help in the preparation of the four volumes.

Finally, I want to express my gratitude to all those who assisted me in the organizational and editorial work, especially to Franz Weissing whose efforts were indispensable.

Bielefeld/Bonn, January 1991 Reinhard Selten

Contents

Contributors

Dilip Abreu, Economics Department, Harvard University, Cambridge, Mass., 02138, USA

Wulf Albers, Institut für Mathematische Wirtschaftsforschung der Universität Bielefeld, Universitätsstr. 25, D-4800 Bielefeld 1, FRG

Ken Binmore Economics Department, University of Michigan, Ann Arbor, MI 48109, USA

Eric van Damme, CentER for Economic Research, Hogeschoollaan 225, P.O. box 90153, 5000 LE Tilburg, The Netherlands

Roy Gardner, Department of Economics, Indiana University, Ballantine Hall, Bloomington, IN 47405, USA

Werner Güth, Johann Wolfgang Goethe-Universität Frankfurt/M., Fachbereich Wirtschaftswissenschaften, Mertonstr. 17, D-6000 Frankfurt am Main 11, FRG

Ronald M. Harstad, Economics Department, Virginia Commonwealth University, Richmond, VA 23284, USA

Hartmut Kliemt, Universität Duisburg, Fachbereich 1: Philosophie, Postfach 101503, D-4100 Duisburg 1, FRG

Wolfgang Leininger, Universität Bonn, Wirtschaftstheoretische Abteilung III, Adenauerallee 24-46, D-5300 Bonn 1, FRG

Akira Okada, Graduate School of Policy Science, Saitama University, Urawa 338, Japan

Elinor Ostrom, Department of Political Science, Workshop in Political Theory and Policy Analysis, Indiana University, 513 North Park, Bloomington, IN 47405, USA

David Pearce, Economics Department, Yale University, New Haven, CON 06520, USA

Louis Phlips, European Institute, Department of Economics, Badia Fiesolona, Via dei Roccettini 5, I-50016 S. Domenico die Fiesole (FI), Italy

Stephen H. Shanuel, Department of Mathematics, SUNY at Buffalo, Buffalo, NY 14214, USA

Leo K. Simon, University of California, Berkeley, CA 94720, USA

Joel Sobel, Economics Department of California at San Diego, La Jolla, CA 92093, USA

Gunter Stephan, Volkswirtschaftliches Institut der Universität Bern, Abtl. Angewandte Mikroökonomie, Gesellschaftsstraße 27, CH-3012 Bern, Switzerland

James M. Walker, Department of Economics, Workshop in Political Theory and Policy Analysis, Indiana University, 513 North Park, Bloomington, IN 47405, USA

Franz Weissing, Rijksuniversiteit Groningen, Biologisch Centrum, Kerklaan 30, Postbus 14, 9751-NN Haren, The Netherlands

William R. Zame, Department of Mathematics, SUNY at Buffalo, Buffalo, NY 14214, USA

Introduction to the Series "Game Equilibrium Models"

Game equilibrium models are descriptions of interactive decision situations by games in extensive or normal form. The analysis of such models is based on the equilibrium point concept, often refined by additional requirements like subgame perfectness. The series consists of four volumes:

I: Evolution and Game Dynamics
II: Methods, Morals and Markets
III: Strategic Bargaining
IV: Social and Political Interaction.

The game equilibrium models presented in these books deal with a wide variety of topics. Just one example from each of the volumes may provide an illustration: Egg trading in hermaphrodite fish (*Friedman and Hammerstein* in Volume I), the social organization of irrigation systems (*Weissing and Ostrom* in Volume II), wage bargaining (*Haller* in Volume III), and beheading games in mediaeval literature (*O'Neill* in Volume IV).

Non-cooperative game theory is a useful research tool not only in economics and the social sciences, but also in biology. Game theory has been created as a theory of conflict and cooperation among rational individuals. For a long time strong rationality assumptions seemed to be indispensable at the foundations of game theory. In this respect, biological applications have changed our perspectives. Game equilibrium may be reached as the result of processes of natural selection without any rational deliberation. Therefore, game equilibrium models can contribute to the explanation of behavioral and structural features of animals and plants.

The interpretation of game equilibrium as the result of evolution raises the question of dynamic stability with respect to evolutionary processes. Similar problems also arise in theories of game learning. The first volume contains three papers on game dynamics. Two of them are concerned with the dynamic foundations of evolutionary game theory and the third one explores stability in a model of anticipatory learning. The remaining papers in the first volume present evolutionary game equilibrium models ranging from abstract investigations of phenomena like bluffing or group-based altruism to the examination of concrete systems observed in nature like "competition avoidance in a dragonfly mating system". Not only theoretical clarifications of the foundations of evolutionary game theory and related research can be found in **Evolution and Game Dynamics**, but also exciting new biological applications.

The title of the second volume, **Methods, Morals, and Markets**, points to several areas of research which attract the interest mainly of economists, but also of political scientists, mathematicians and philosophers. The first paper is a sophisticated mathematical contribution which applies new tools to basic questions of non-cooperative game theory. The word "method" mainly refers to this paper, but to some extent also to the next three contributions, which discuss basic conceptual problems in the interpretation of game equilibrium. Two papers relate to the philosophical notion of the social contract and its exploration with the help of game theoretical models. This work in concerned with "morals", a theme which is also touched by a paper on irrigation institutions. The remaining four papers of the second volume explore game equilibrium models of markets; two of these contributions are experimental and compare theoretical solutions with experimental data.

The third volume on **Strategic Bargaining** collects ten papers on game equilibrium models of bargaining. All these papers look at bargaining situations as non-cooperative games. Unlike in cooperative game theory, cooperation is not taken for granted, but explained as an outcome of equilibrium analysis. General models of two-person and n-person bargaining are explored, sometimes enriched by institutional detail like the availability of long-term contracts. Other papers explore bargaining in special contexts like wage negotiations. Two contributions concern spatial games; one of these contributions is experimental.

The exploration of strategic models of bargaining is an active field of research which attracts the attention of many game theorists and economists. The ten papers in the third volume contribute to the progress in this field.

The fourth volume on **Social and Political Interaction** mainly presents game equilibrium models in the area of political science. Three of the papers concern topics in other fields: the distribution of foreign language skills, altruism as a social dilemma (an experimental paper) and beheading games in mediaeval literature. Five contributions to the area of international relations deal with game theoretical models of the balance of power, of alliance formation, and of an issue in armament policy. An investigation of inspection problems like those arising in connection with the non-proliferation treaty also touches the area of international relations. Other papers on problems of political science deal with the game theoretical resolution of the Condorcet paradox by equilibrium selection, the modelling of political pressure exerted by firms on the government and the draft resistance problem.

The main emphasis is on biology in Volume I, on economics in Volumes II and III, and on political science in Volume IV. This is the result of an attempt to group the great variety of papers resulting from a year long interdisciplinary research project in a reasonably coherent way. However, not only biologists, but also economists and a psychologist have contributed to Volume I. Similarly, not only economists and mathematicians, but also political scientists as well as a biologist and a psychologist are among the authors of Volumes II and III. All four volumes are the result of the cooperation of researchers from many disciplines united by a common interest in game equilibrium models within and beyond the borders of their fields.

Bielefeld/Bonn, January 1991 — Reinhard Selten

Introduction to Volume II: "Methods, Morals and Markets"

This volume contains contributions to the methodology of analysis of games, game-theoretic contributions to the fundamental ethical questions facing societies, and game-theoretic analyses of market environments. The analysis throughout is strictly noncooperative, in the usual sense that players are expected to abrogate an agreement unless the nature of the agreement provides incentives to comply. A recurrent theme is the importance of timing and incentives in extensive-form games. Indeed, it would not be far off the mark to consider the volume as being subtitled "On the necessity, sufficiency and predictive power of subgame-perfect equilibria." *Güth, Leininger and Stephan* find that restriction to subgame-perfect equilibria and to an analogue, truncation-consistent equilibria leaves only stationary equilibria in supergames. *Abreu and Pearce* ask when subgame-perfection may not pay appropriate attention to whether the group wishes to deviate from plans made, that is, to renegotiate.

Questions of the influence of actions and information flows upon the further play of the game are also at the heart of analyses of ethics and societal formation and organization considered in this volume. *Binmore* presumes, somewhat more palatably than in analyses of societal organization by Rawls and Harsanyi, that agreements will only be maintained when no party sees a benefit to renegotiation; notions of justice and fairness are considerably influenced when this rough analogue to subgame perfectness is introduced. Subgame perfectness is even more apparent in *Okada and Kliemt*, where agreements to enforce cooperative behavior in the face of free-rider problems are recognized to arise only if there are incentives for voluntary participation in collectives which may reach these agreements.

The market applications concluding the volume tend also to focus on questions of order of and predictability of moves in extensive form games. In considering how multiple equilibria of the Spence job market signaling game might be reduced to more specific predictions, *van Damme and Guth* contrast the way in which the refinements literature examines beliefs following unanticipated moves with the way in which the equilibrium selection theory of Harsanyi and Selten (1988) analyzes uniform perturbations of the game which reach every node. *Phlips and Harstad* consider subgame-perfect equilibria of futures markets in which participants anticipate that, after the futures market closes prior to maturity, oligopolistic producers of the resource traded on the futures market will alter their time paths of extraction in response to the futures positions they have taken. *Albers and Harstad* uncover a type of overbidding in English common-value auctions with independent information: the price at which the first bidder quits exceeds the symmetric equilibrium predictions in laboratory observations. They then consider how the remaining bidders (do and ought to) adjust following this overbidding.

So variations on the theme of the role of subgame perfection keep arising in this volume. The scope of topics covered, though, is vast.

Often new insights become available when problems previously regarded as distinct are shown to be analyzable by a common logical technique. Quite a powerful combination arises in the presentation of the underlying algebraic geometry of games by *Schanuel, Simon and Zame*. From an elementary, accessible beginning, they take the reader into an exploration of the usefulness of the Tarski-Seidenberg theorem for games. This theorem shows that any first-order formula is equivalent to some formula without quantifiers; its fundamental usage in game theory is to provide straightforward equations which are equivalent to assertions of the existence of such characterizations as equilibrium points, and various refined equilibrium points. The principal illustration of the power of the algebraic geometry approach is of independent interest: they explore whether the tracing procedure at the heart of the equilibrium selection theory of Harsanyi and Selten (1988) is necessarily capable of producing a unique prediction, as is its function. They also explore whether the computationally simpler linear tracing procedure, when it converges, converges to the same prediction as the more complex logarithmic tracing procedure, as claimed. Perhaps the major contribution of Schanuel, Simon and Zame, though, is pointing to a fruitful methodology for future insights.

The term "supergame" refers to a situation in which an ordinary game, called the "stage game" for clarity, is repeatedly played by the same set of players, countably many times. For all players to play an equilibrium point of the stage game in every repitition constitutes an equilibrium point of the supergame, in particular a "stationary" equilibrium. But the interplay between stage games that results when players adopt strategies dependent upon what has happened in previous stage games yields a much wider collection of equilibrium outcomes. The so-called "Folk Theorem" of repeated games is a fable that any outcome which each player prefers to some stationary equilibrium can be obtained by history-dependent equilibrium play in an infinitely repeated game. Conditions yielding precise variants of this fable, as well as precise approximations to it for game where stage games are repeated finitely many times, appear in the literature. *Güth, Leininger and Stephan* offer a conceptual discussion of the Folk Theorem, and of axiomatic restrictions on the class of equilibria sufficient to restrict to stationary outcomes, for finitely and infinitely repeated play.

Abreu and Pearce take a different tack, while still asking what outcomes are to be predicted when games are repeated. Rather than introducing axioms to be satisfied, they are concerned with the meaning of self-enforcingness of agreements, the notion behind equilibrium concepts. Agreements which are history-dependent may allow for higher payoffs to be "cooperatively" reached in view of the repetition, which may make adopting such agreements rational. But self-enforcingness has traditionally been viewed as a condition on individual incentives. In repeated games for which history-dependent agreements are being considered, one also has to ask when an agreement fails to be self-enforcing after some histories because the group of players as a whole would like to renegotiate the agreement. If cooperation is to be possible, agreements will have to call for history-dependent play, and at times history may lead the group to a position wherein they arrive at incentives to re-forge the agreement. Abreu and Pearce argue that agreements once reached can depend on history in ways that credible deviations from agreements cannot, so credible deviations can be expected to be stationary. Agreements can then be called renegotiation-proof if no credible deviation would be agreed to by the group following any history reachable by the agreement. They discuss how the resulting set of renegotiation-proof equilibria depends upon the assumptions made about when the group might agree to deviate.

What makes a game-player good (or, for that matter, effective)? "It is not what you know, it is who you play" is an answer offered by *Sobel*. He presents a natural argument that intuitive increases in ability to play games are necessarily helpful in two-player, zero-sum games. Sobel presents striking examples to show that ability is not necessarily more helpful in games that are not purely contests. For a significant class of games, increased ability need not enhance payoffs unless a player's rivals' abilities increase as well.

The basic nature of social organization, its relation to the consent of members of society, and the implications for distribution of society's economic surplus, for millennia subjects of philosophers' writings, have in the past century become multidisciplinary, with economists, political scientists, anthropologists and mathematicians contributing proposals, justifications, and critiques. Game-theoretic methods can clearly contribute to a surer understanding of the assumptions underlying justifications. *Binmore*'s paper is a substantial effort to isolate the role of commitment in such analyses. If the hypothetical technique of bargaining "in the original position" is to be applied in a society without perfect commitment to agreements reached in the prior bargaining, sensible agreements must be consistent with a hypothetical opportunity to re-enter original position and renogiate. Interestingly, this approach, thorough in a sense that is original, supports an approach to societal assessment of inequality that is recognizably in the spirit of Rawls' well-known analysis, with the added benefit that Rawls' denial of orthodox decision theory as a relevant analytical tool for bargaining is no longer needed.

Binmore discusses briefly the question of how to determine who enters the hypothetical bargaining, in the context of determining who is a "citizen." Much of moral philosophy, particularly in discussing "contractarianism", expects any active or consensual role of a citizen to be endogenous: someone becomes involved in a social order, becoming a citizen or a member of some collective, only out of their own free will. Nonetheless, a straightforward inference of game theory, that incentive problems arise in the original formation of any such collective, as well as in deciding what the collective behavior is to be, has escaped careful attention in widely cited discussions of contractarianism. *Okada and Kliemt* point to fundamental difficulties by building and analyzing an illustrative model, in which players decide whether to enter into a collective group, and collective group members decide whether to utilize an enforcement mechanism to overcome a known free-rider problem. To maximize opportunities for success of collective action, the model is simplified to ignore costs of an enforcement mechanism, and to ignore any incentive problems that may arise with respect to the agents who operate an enforcement mechanism. Nonetheless, in their illustration, the Pareto-efficient outcome wherein all players join the collective and enforce cooperative behavior cannot be supported as a subgame-perfect equilibrium. Their paper closes with an indication of how this illustration fits into an understanding of contractarian literature, in particular Nozick's position.

Social organization also involves questions of how society organizes usage of interrelated resources. Perhaps the most telling example is water usage in irration systems. The fundamental problem arising in irrigation institutions is establishment of a system of rules which effectively utilize the water, and enforcement of such rules in the presence of individual incentives to undermine them. *Weissing and Ostrom* point to an extensive literature showing widespread failures of irrigation systems, often with abuses rendering huge investments worthless. The story of failed institutions is not universal, as several examples exist of institutions where rules are largely followed, and irrigation proceeds effectively. Extant literature, however, has not systematically examined the inter-related incentives for cheating and for costly monitoring of other irrigators' behavior. Here game theory can help isolate these incentives, and indicate relationships between structural and institutional parameters and strength of incentives. Weissing and Ostrom begin such an exercise, simplifying by assuming symmetry in both incentives and behavior (ultimately, their analysis will need to be extended to take account of such fundamental asymmetries as unidirectional gravitational flow of water). The essence of such situations has a turntaker who is allotted a certain amount of water, but decides how much actually to take, and a turnwaiter who will become a turntaker in due course, and in the meantime must decide whether to incur resource costs monitoring the current turntaker's behavior. All scenarios studied suggest a fundamental difficulty in attaining perfect enforcement as equilibrium behavior. How frequently the rules of an irrigation system are violated at equilibrium depends crucially on structural parameters (like the size of the system), the incentives of the players and established "norms of behavior". Weissing and Ostrom close their paper with a discussion of the relative importance of "primary" and "secondary" incentives for equilibrium behavior in a positive context.

The recent Harsanyi and Selten (1988) book takes on the daunting task of supplying a unique (normative) prediction of the outcome of any game (in agent normal form). Such an effort is necessarily a laborious specification of an order of relative importance in applying tools which restrict a set of equilibrium points to a more plausible subset. *Van Damme and Güth* offer an example of Harsanyi-Selten equilibrium selection theory in a simple game for which there is in the literature a benchmark method of narrowing predictions. A benchmark is not readily available for games with multiple strict equilibria. It is available for games where multiple equilibria are in the main caused by behavior at unreached information sets, which are treated in the large literature on equilibrium refinements. Van Damme and Güth compare the equilibrium selection method with the refinements approach for the example of the labor market signaling game due to Spence. In it, a worker has private information as to whether his productivity is high or low; the worker chooses an education level which two competing potential employers observe before making wage offers.

The traditional literature on futures market activity assumes competitive behavior, ignoring the potential impact on futures trading of an oligopolistic structure in the industry producing the underlying commodity. Several results in the literature suggest a futures market will be completely inactive, as trades based upon differential information face a fundamental hurdle: if a trader wishes to sell July oil to you at \$p, he must possess some information which would lead you to refuse to buy July oil at \$p if you knew that information. *Phlips and Harstad* offer an alternative: modeling a game of inconsistent incomplete information in which differences in beliefs are fundamental differences in opinion, rather than in information. Their model has duopoly producers altering the time path of extraction of a natural resource to influence the profitability of futures market positions. Nonetheless, the futures market is active in subgame-perfect equilibrium.

An important example of a type of economic market where behavior often appears at odds to game-theoretic predictions is what is called a "common-value" auction. In such a market, an asset is auctioned to bidders, with asset value the same for all bidders, but each faces uncertainty over what the asset is worth. A standard example is an offshore oil lease, where the amount of oil extractable will not depend on the identity of the winning bidder, but each bidder knows only a geologic estimate, not the actual amount of oil. *Albers and Harstad* report on controlled laboratory experiments with common-value auctions, simplified by enforcing statistical independence of bidders' estimates (this makes their experiments less useful as a model of such markets as oil leases, but adds insight on divergences between theory and observation). They conducted "English" auctions, in which an auctioneer calls out prices ascendingly until only one bidder remains in competition. In addition, they provide evidence on the impact of a framing effect, as two theoretically identical decision treatments are presented to subjects in different ways: the asset is worth either the sum of, or the mean of, bidders' estimates.

Markets with a common pool resource (such as a fishery) have generally been presumed to lead, in the absence of well-defined property rights, to complete dissipation of economic rents. This presumption is at odds with game-theoretic predictions of economic rents at socially suboptimal, but nonzero levels. *Walker, Gardner and Ostrom* report on laboratory tests. In each of their experiments, 8 subjects simultaneously decide how much of an endowment of tokens to convert to cash, and how much to invest in a market where marginal returns are diminishing in the total amount invested. This setup is repeated 20-25 times. They find decidely suboptimal appropriation, but clearly at variance with the usual presumption: whenever marginal returns fall near zero, subjects notably reduce appropriation in the next period. Nash equilibrium is a poor predictor of individual subject choices, but the deviations from Nash predictions approximately balance, so that aggregate appropriation is observed surprisingly close to the levels predicted by game theory.

We trust several insights found here will lead other researchers usefully to employ game-theoretic analyses of topics related to the many discussed in this volume.

Bielefeld/Richmond, VA, January 1991 Ronald M. Harstad

Reference:

Harsanyi, J. and R. Selten (1988) A General Theory of Equilibrium Selection in Games. Cambridge, Mass.: MIT Press

THE ALGEBRAIC GEOMETRY OF GAMES AND THE TRACING PROCEDURE

Stephen H. Schanuel

Leo K. Simon

William R. Zame

---

1. INTRODUCTION

This paper has two purposes. The immediate purpose is to point out some difficulties with the tracing procedure of Harsanyi and Selten, and show how they can be dealt with. The other purpose is to describe the theory of semi-algebraic sets and a few of its applications in game theory.

The *tracing procedure* is the heart of the extensive theory of equilibrium selection in games which has been developed by Harsanyi and Selten (Harsanyi [1975], Harsanyi-Selten [1988]). For each (normal or extensive form) game, the theory of Harsanyi and Selten prescribes (on the basis of Bayesian and risk analysis) a prior probability distribution over strategies for this game. Given this prior probability distribution, the *logarithmic tracing procedure* identifies a unique Nash equilibrium (the

---

We thank Nicolas Goodman and Reinhard Selten for helpful conversations. The third author was supported by NSF Grants DMS-8602839 and SES-8720966.

logarithmic solution) from the set of all Nash equilibria. In part because the logarithmic solution is difficult to compute in practice, Harsanyi and Selten also use another procedure, which they call the *linear tracing procedure* . In contrast to the logarithmic tracing procedure, the linear tracing procedure is relatively easy to compute, but does not always lead to a unique solution. However, the linear tracing procedure leads to a unique solution for "most" games, and when it does lead to a unique solution, it leads to the logarithmic solution. These two properties of the linear tracing procedure make it useful in applications.

Unfortunately, there are some difficulties with the descriptions and constructions that Harsanyi and Selten give for the tracing procedures. In particular, the arguments for the crucial properties of the tracing procedure (that the logarithmic tracing procedure always leads to a unique solution, and that the linear solution - when it is unique - coincides with the logarithmic solution) are not rigorous. (We discuss the tracing procedure and the difficulties in Section 4.)

In this paper we clarify the definitions of the logarithmic and linear tracing procedures and give rigorous proofs for the crucial properties. Our methods also show that, viewed as a function of the the initial data (i.e., the payoffs of the game and the prior probability distribution), the logarithmic solution is continuous on a dense open set of full measure. (It could not possibly be continuous everywhere.)

We believe that our methods are of interest in themselves, and will have wide applicability in game theory. Primarily, we make use of *real algebraic geometry* , which is the study of algebraic and semi-algebraic sets. (An *algebraic set* is defined by polynomial equalities; a *semi-algebraic* set is defined by (conjunctions and disjunctions of) polynomial inequalities.) It has been observed by Kohlberg and Mertens [1986] (and perhaps by many others) that certain of the constructions of game theory lead to semi-algebraic sets, and that semi-algebraic sets have a very special structure which is relevant for game theory. In particular, the set of Nash equilibria of any game is a semi-algebraic set, and hence has a finite number of connected components. This fact plays a significant role in the theory of stable equilibrium.

In this paper we show that virtually *all* of the constructions of game theory give rise to semi-algebraic sets. This is a consequence of a deep and remarkable result from mathematical logic, the Tarski-Seidenberg theorem (Tarski [1931], Seidenberg [1954]). The Tarski-Seidenberg theorem asserts that any first order formula in the language of the real numbers (i.e., a formula which does not involve quantification over sets) is equivalent to a first order formula which involves no quantifiers at all. (Indeed, the Tarski-Seidenberg theorem actually gives an explicit procedure for this "elimination of quantifiers.") A first order formula which involves no quantifiers is simply a conjunction and disjunction of polynomial inequalities, and hence defines a semi-algebraic set. The import of the Tarski-Seidenberg theorem for game theory is that virtually all of the usual game-theoretic constructions are (or are equivalent to) first order constructions, and hence give rise to semi-algebraic sets. (The Kohlberg-Mertens [1986] notion of *stability* may be an exception here. Since stability is a set-valued notion, it does not seem clear whether it has a first order formulation.)

As a consequence of the Tarski-Seidenberg theorem, we show (Theorem 1) that virtually all of the usual game-theoretic equilibrium correspondences (Nash, subgame perfect, sequential, perfect, etc.) have semi-algebraic graphs. It follows (Corollary 1.1) that each of these correspondences is continuous at every point of a dense open set of full measure (previously, it had not been known that the perfect equilibrium correspondence had any points of continuity at all) and admits a selection which is continuous at every point of the same set. In fact, the logarithmic solution provides such a selection (previously, the generic continuity properties of the logarithmic solution were unknown). It also follows (Corollary 1.2) that for every game, the set of Nash (respectively, subgame perfect, sequential, perfect) equilibria has a finite number of connected components, and that there is a bound for this number which depends only on the game form. Finally (Corollary 1.3), for every game, the set of Nash (respectively, subgame perfect, sequential, perfect) equilibria is the finite union of connected real-analytic manifolds (of various dimensions).

We stress that the above conclusions are immediate applications of the definitions and known facts about semi-algebraic sets. For other, less immediate, applications, see Simon [1987] and Blume and Zame [1989]. Blume and Zame use the theory of semi-algrebraic sets to show that, for generic games, all sequential equilibria are (trembling hand) perfect. Simon uses a generalization (due to van den Dries [1986]) of the theory of semi-algebraic sets to establish the existence of mixed-strategy equilibria for continuous time games. The only other application (of which we are aware) of the theory of semi-algebraic sets in game theory is the previously mentioned observation of Kohlberg-Mertens [1986] that the set of Nash equilibria of a game is a semi-algebraic set, and hence has only a finite number of connected components. A different aspect of the Tarski-Seidenberg theorem (that first order formulas true in one real closed field are true in all real closed fields) was used by Bewley and Kohlberg [1976] in their work on stochastic games.

The remainder of the paper is organized in the following way. Section 2 describes the basics of the theory of semi-algebraic sets and the Tarski-Seidenberg theorem, and gives some simple examples. Section 3 details the general applications to game theory, and in particular gives the results about equilibrium correspondences described above. Section 4 briefly reviews the tracing procedure, isolates what appear to be the crucial difficulties, and gives rigorous arguments to circumvent them. Finally, Section 5 briefly explicates the relationship of the tracing procedure to the entire Harsanyi-Selten equilibrium selection procedure.

2. SEMI-ALGEBRAIC SETS

In this section, we describe the basics of the theory of algebraic and semi-algebraic sets and the Tarski-Seidenberg theorem. Excellent general references are Bochnak-Coste-Roy [1988] and Delfs-Knebusch [1981a, b]. For a very brief, but readable, synopsis, we recommend van den Dries [1986].

By definition, an *algebraic set* in $\mathbb{R}^N$ is defined by (a finite number of) polynomial *equalities*. More precisely, an algebraic set is a set of the form:

$$A = \{ x = (x_1,\ldots,x_N) \in \mathbb{R}^N : p_1(x) = \ldots = p_n(x) = 0 \},$$

where $p_1, \ldots p_n$ are polynomials (with real coefficients). Note that the vanishing of an arbitrary collection of polynomials is equivalent to the vanishing of a finite number of polynomials because the ring of real polynomials in N variables is *Noetherian*. Note also that the vanishing of at least one of the polynomials $q_1, \ldots q_n$ is equivalent to the vanishing of the product $\prod q_i$, and that the simultaneous vanishing of the polynomials $p_1, \ldots p_n$ is equivalent to the vanishing of the sum $\sum p_i^2$. (The reader might note that this elementary fact already draws a sharp distinction between real polynomials and complex polynomials.) Hence, any arbitrary conjunction and/or finite disjunction of polynomial equalities is equivalent to a single polynomial equality. In particular, the family of algebraic sets is closed under arbitrary intersections and finite unions. Note that, aside from $\mathbb{R}$ itself, the algebraic subsets of $\mathbb{R}$ are precisely the finite sets.

By definition, a *semi-algebraic set* in $\mathbb{R}^N$ is the union of a finite number of sets, each defined by a finite number of polynomial *inequalities* . More precisely, a semi-algebraic set is a finite union of sets of the form:

$$B = \{x \in \mathbb{R}^N : p(x) = 0 , q_1(x) < 0 , \ldots , q_n(x) < 0\} .$$

where $p , q_1, \ldots q_n$ are polynomials (with real coefficients). (In contrast with the case of polynomial equalities, a finite conjunction and/or disjunction of polynomial inequalities need not be equivalent to a single polynomial inequality.) Note that the complement of a semi-algebraic set is a semi-algebraic set and that the union and intersection of (a finite number of) semi-algebraic sets is a semi-algebraic set. In other words, the family of semi-algebraic sets forms a *Boolean algebra* of sets. We may also describe the semi-algebraic sets as the smallest Boolean algebra of sets containing all those defined by a single polynomial inequality $q(x) \leq 0$. The semi-algebraic subsets of $\mathbb{R}$ consist of the all finite unions of intervals (closed or open or half open, finite or infinite).

Every algebraic set is semi-algebraic, and that every semi-algebraic set is the union of (relatively) open subsets of algebraic sets.

If A , B are semi-algebraic sets, a function $f : A \to B$ (or more generally, a correspondence $F : A \twoheadrightarrow B$) is *semi-algebraic* if its graph is a semi-algebraic set. Note that we do not require a semi-algebraic function to be continuous, nor do we require a semi-algebraic correspondence to be upper or lower hemi-continuous, or have closed values, or even to have non-empty values.

Two aspects of the theory of semi-algebraic sets are of primary interest to us. The first is that semi-algebraic sets admit an alternate description, which allows us to recognize as semi-algebraic many sets which are not presented as the solution sets of polynomial inequalities. The second is that semi-algebraic sets (and hence semi-algebraic functions and correspondences) have a very special structure.

To explain the first aspect, it is convenient to discuss in a very informal way the first order theory of the real numbers. We begin with the first order language, which is built up from the usual logical symbols (and, or, not, such that, implies, $\forall$, $\exists$), the real numbers (as constants), (real) variables, the algebraic operations (+, $\cdot$), equality (=), and the order relation (<). A *first order formula* is any formula in this language in which all quantifiers are extended only over elements of $\mathbb{R}$ and not over *sets* . (Keep in mind that many substructures of $\mathbb{R}$ do not have names in this language. In particular, there is no name for the integers, and there is no predicate for "is an integer".) The following are examples of first order formulas:

(F1) $x > 0$

(F2) $v^2 - 4uw > 0$

(F3) $\exists y$ such that $x = y^2$ and $y \neq 0$

(F4) $\exists y$ and $\exists z$ such that $y \neq z$ and $uy^2 + vy + w = 0$ and $uz^2 + vz + w = 0$

In these formulas we have followed the usual convention that "unbound variables are free." If a formula contains n free variables $x_1,\ldots,x_n$, substituting a particular real number r_i for each free variable x_i yields a *sentence* (i.e., a formula with no free variables) which may be true or false. If this sentence is true, we say that $r_1,\ldots,r_n$ *satisfy* the formula. The set of all n-tuples $(r_1,\ldots,r_n)$ satisfying the formula is the set *defined* by the formula.

Thus, formulas (F1) and (F3) each define a set of real numbers, while (F2) and (F4) each define a subset of $\mathbb{R}^3$. Clearly (F1) and (F3) define the same set, namely the set of positive real numbers. (F4) defines the set of triples $(u,v,w) \in \mathbb{R}^3$ such that the polynomial $ut^2 + vt + w$ has two distinct real roots. Since a quadratic polynomial has two distinct real roots exactly when its discriminant is positive, we see that the formulas (F2) and (F4) define the same set.

If $\Phi(x_1,..x_n,y)$ is a first order formula involving the free variables $x_1,\ldots,x_n$ and y , then we obtain first order formulas whenever we specialize or bound the variable y ; i.e., if r is any real number then the following are also first order formulas in which only the variables $x_1,\ldots,x_n$ are free:

(F5) $\Phi(x_1,\ldots,x_n,r)$

(F6) $\exists y\ ,\ \Phi(x_1,\ldots x_n,y)$

(F7) $\forall y\ ,\ \Phi(x_1,\ldots x_n,y)$

As we have noted, this language contains no name for the set of integers and no predicate for "is an integer" and there is no first order formula (in this language) which is satisfied precisely by the integers (or the positive integers, or the rational numbers). The restriction to first order formulas is crucial here; the formula

(F8) $\forall X \subset \mathbb{R}\ ,\ \{\ 0 \in X \text{ and } (y \in X \Rightarrow y+1 \in X\ )\ \} \Rightarrow y \in X$

(with the single free variable y) is satisfied precisely by the positive integers. Of course it is not a first order formula, since it involves quantification over a set.

Note that a first order formula cannot involve a polynomial of unspecified degree. However, it may certainly involve a polynomial of a particular, pre-specified degree (in a particular, pre-specified number of variables), since a polynomial of pre-specified degree in a pre-specified number of variables is determined entirely by its (finite, pre-specified mumber of) coefficients.

As we have noted, the two formulas (F1) and (F3) above are satisfied by the same values of the variable x , and define the same subset of $\mathbb{R}$; i.e., they are *equivalent* . (In view of this simple observation, it might seem that the order relation $<$ is redundant, since it can be expressed in terms of multiplication. However, doing so *requires* the use of quantifiers, while it is precisely the *elimination* of quantifiers which provides the powerful tool we shall use.) Similarly, the two formulas (F2) and (F4) are equivalent; they are satisfied by the same values of the variables u , v , w , and define the same subset of $\mathbb{R}^3$. The forms of these formulas are notable; (F3) and (F4) involve quantifiers, while (F1) and (F2) do not. That is, (F3) and (F4) are equivalent to formulas from which *the quantifiers have been eliminated* . The following theorem of Tarski and Seidenberg (Tarski [1931], Seidenberg [1954]) says that this is always possible. (The Tarski-Seidenberg theorem is frequently phrased as a statement about real closed fields, and it is in this form that it was used by Bewley-Kohlberg [1976]. We have phrased it as a statement about the real numbers because that is more convenient for our purposes. For a more formal discussion of the relationship between the two formulations, see Blume-Zame [1989].)

TARSKI-SEIDENBERG THEOREM (version I): Every first order formula is equivalent to a first order formula with no quantifiers.

A first order formula with no quantifiers is just a conjunction and disjunction of polynomial inequalities, and hence defines a semi-algebraic set. Thus, the Tarski-Seidenberg theorem can be phrased in the following (equivalent) way, which is more convenient for us:

TARSKI-SEIDENBERG THEOREM (version II): A subset of $\mathbb{R}^N$ is semi-algebraic if and only if it can be defined by a first order formula.

As we shall see, the Tarski-Seidenberg theorem is remarkably powerful and at the same time remarkably easy to apply. A few examples may help to suggest the kind of logical manipulations involved.

PROPOSITION 1: The image of a semi-algebraic set under a semi-algebraic map is a semi-algebraic set.

PROOF: If $f : A \to B$ is a semi-algebraic map between semi-algebraic sets, then graph(f) is a semi-algebraic subset of $A \times B$. The image of f is defined by the formula:

(*) $b \in B$ and $\exists a \in A$ such that $(a,b) \in$ graph(f)

Since A, B and graph(f) are semi-algebraic sets and hence are defined by first order formulas, (*) "is" also a first order formula. A little more precisely: if Φ_A, Φ_B, Ψ are the first order formulas which define A, B, graph(f), then (*) is shorthand for the first order formula

(*') $\Phi_B(b)$ and $\exists a$ such that $\Phi_A(a)$ and $\Psi(a,b)$

Hence, the Tarski-Seidenberg theorem implies that image(f) is semi-algebraic. ∎

PROPOSITION 2: The closure of a semi-algebraic set is a semi-algebraic set.

PROOF: If A is a semi-algebraic set in $\mathbb{R}^N$, then its closure $\bar{A}$ is just the set of points $x = (x_1,\ldots,x_N) \in \mathbb{R}^N$ satisfying the formula:

$$(**) \qquad \forall \varepsilon > 0, \exists y \in A \text{ such that } |x - y|^2 < \varepsilon$$

Since $|x - y|^2$ is a polynomial and A is defined by a first order formula, (**) is (shorthand for) a first order formula. Hence, the Tarski-Seidenberg theorem implies that $\bar{A}$ is semi-algebraic. ∎

As a final example of the sort of manipulations which are sometimes useful, we give an extension of Proposition 1 which will be needed in Section 4. By Proposition 1, if A is a semi-algebraic subset of $\mathbb{R}^n$ and $\varphi : \mathbb{R}^n \to \mathbb{R}^m$ is a semi-algebraic mapping, then $\varphi(A)$ is a semi-algebraic set. Hence $\varphi(A)$ can be defined by polynomial inequalities. We would like to have bounds for the number of polynomials required, and their degrees. One suspects that it should be possible to find such bounds if we constrain A to lie in a "semi-algebraic family of sets" and constrain φ to lie in a "semi-algebraic family of mappings" and this is what we shall demonstrate. (Precise statements of this sort can be formulated in many different ways; the formulation we choose is simply one that is convenient for the application we need.) We formalize the idea of a semi-algebraic family of sets by beginning with a semi-algebraic subset B of $\mathbb{R}^{n+1}$ and considering the family sets B_η obtained by intersecting B with the hyperplane $x_{n+1} = \eta$. Rather than formalizing the idea of a semi-algebraic family of mappings, we simply consider the family of all linear transformation from $\mathbb{R}^n$ to $\mathbb{R}^m$; identifying a linear transformation with its matrix yields a natural "semi-algebraic parametrization" of this family.

To state our result precisely, we first collect some notation. A family of subsets of $\mathbb{R}^N$ is a *Boolean algebra of sets* if it is closed under the formation of complements, finite unions and finite intersections. If E is a family of subsets of $\mathbb{R}^N$, then by $B(E)$ we mean the Boolean algebra generated by E; i.e., the smallest Boolean algebra of sets containing E. Now, $B(E)$ can be constructed from E closing under complements, then finite intersections, and finally, finite unions. In particular, if E is a finite family, then $B(E)$ is also a finite family of sets. (It is not difficult to see that the cardinality of $B(E)$ is at most 2^k, where

$k = 2^{\text{cardinality}(E)}$.) If $f_1,\ldots,f_n$ are polynomials, then we shall abuse notation to write $B(f_1,\ldots,f_n)$ for the Boolean algebra generated by the sets $\{x : f_i(x) < 0\}$ and $\{x : f_i(x) \le 0\}$.

The following proposition contains the precise result we need in Section 4. The reader should keep in mind that it is merely a particular example of the way in which the Tarski-Seidenberg theorem can be "bootstrapped" to obtain bounds on the number and degrees of polynomials required to express certain sets.

PROPOSITION 3: Let B be a semi-algebraic subset of $\mathbb{R}^{n+1}$, and write $\pi : \mathbb{R}^{n+1} \to \mathbb{R}$ for the projection on the last coordinate. For each $\eta \in \mathbb{R}$, set $B_\eta = B \cap \{x : \pi(x) = \eta\}$. Let m be a fixed integer, and let $L = L(\mathbb{R}^{n+1},\mathbb{R}^m)$ be the family of all linear transformations $\varphi : \mathbb{R}^{n+1} \to \mathbb{R}^m$. Then there are integers K and D such that for every linear transformation $\varphi \in L$ and every $\eta \in \mathbb{R}$, there are polynomials $f_1,\ldots,f_K$ (where the degree of each f_k is bounded by D), such that $\varphi(B_\eta) \in B(f_1,\ldots,f_K)$.

PROOF: We identify a linear transformation $\varphi : \mathbb{R}^{n+1} \to \mathbb{R}^m$ with its matrix (in the standard bases), and hence view L as $\mathbb{R}^{m(n+1)}$. Define the function $\Lambda : \mathbb{R}^{n+1} \times L \to \mathbb{R}^m \times L \times \mathbb{R}$ by $\Lambda(x,\varphi) = (\varphi\cdot x,\varphi,\pi(x))$; this is a semi-algebraic (in fact, polynomial) mapping, and $B \times L$ is clearly a semi-algebraic subset of $\mathbb{R}^{n+1} \times L$. Hence by Proposition 1, there are polynomials $F_1,\ldots,F_r$ in $m + m(n+1) + 1$ variables such that $\Lambda(B \times L) \in B(F_1,\ldots,F_r)$. Writing y for the variables in $\mathbb{R}^m$, φ for the variables in $L = \mathbb{R}^{m(n+1)}$ and η for the last variable, observe that for fixed $\varphi^* \in L$, and fixed $\eta^* \in \mathbb{R}$, each $F_i(y,\varphi^*,\eta^*)$ is a polynomial in m variables. It is easily seen that

$$\varphi^*(B_{\eta^*}) \in B(F_1(y,\varphi^*,\eta^*),\ldots,F_r(y,\varphi^*,\eta^*))$$

which yields the desired result. ∎

We turn now to the second aspect of the theory of semi-algebraic sets which is of importance to us: semi-algebraic sets (and semi-algebraic functions and correspondences) have a very special structure. The most important consequences of this special structure (at least for our purposes) are given below.

Recall that a *finite simplicial complex* in $\mathbb{R}^N$ is a finite disjoint collection $\{K_j\}$ of open simplices (of various dimensions), having the property that if K_ℓ meets the closure of K_i, then K_ℓ is an open lower dimensional face of K_i.

TRIANGULABILITY (Lojasiewicz [1964, 1965], Hironaka [1975]): Every semi-algebraic set can be semi-algebraically triangulated. Indeed, for every semi-algebraic subset A of $\mathbb{R}^N$, there is a finite simplicial complex $\{K_j\}$ in $\mathbb{R}^N$ and a semi-algebraic homeomorphism $h : \mathbb{R}^N \to \mathbb{R}^N$ such that $h(K) = A$.

Since simplices are connected, it of course follows immediately that semi-algebraic sets have only a finite number of connected components.

STRATIFIABILITY (Whitney [1957], Bochnak-Coste-Roy [1988, pp. 188-189]): Every semi-algebraic set is the disjoint union of a finite number of semi-algebraic subsets, each of which is a real-analytic manifold.

In view of this, we may speak unambiguously of the *dimension* of a semi-algebraic set ($\dim A$ = maximum of the dimension of smooth submanifolds of A), and the *dimension* of a semi-algebraic set *at a point* ($\dim_p A$ = maximum of the dimension of smooth submanifolds of A whose closures contain p). Note that $\dim A = \max\{\dim_p A : p \in A\}$. (By convention, the dimension of the empty set is -1.) Note that the map $p \to \dim_p A$ is a semi-algebraic function.

If A is any set, we write $\overline{A}$ for its topological closure. By its *Zariski closure* $\mathrm{Zar}(A)$ we mean the smallest algebraic set containing A; equivalently, $\mathrm{Zar}(A)$ is the set of common zeroes of all polynomials which vanish on A. Note that $\mathrm{Zar}(A) \supset \overline{A}$. As we have noted before, the set of common zeroes of all polynomials which vanish on A is in fact the set of common zeroes of a single polynomial. In order that A be Zariski closed (i.e., that $A = \mathrm{Zar}(A)$) it is necessary and sufficient that there be a polynomial vanishing at every point of A and nowhere else; i.e., that A be an algebraic set.

DIMENSION (Bochnak-Coste-Roy [1988, pp. 47, 237]): If A is a non-empty semi-algebraic set, then $\dim \mathrm{Zar}(A) = \dim \overline{A} = \dim A$ and $\dim (\overline{A} - A) < \dim A$. If $f : A \to \mathbb{R}^k$ is a semi-algebraic mapping, then $\dim f(A) \le \dim A$; if f is one-to-one, then $\dim f(A) = \dim A$.

It should of course be kept in mind that no such result is true for arbitrary sets A (even for sets defined by smooth inequalities) or for arbitrary mappings f (even for mappings that are continuous and differentiable almost everywhere.) For example, set $A = \{(m/n, 1/n) \in \mathbb{R}^2 : m , n$ positive integers$\}$. It is easily seen that A is a discrete subset of the upper half plane, and in particular is a differentiable submanifold of dimension 0 . On the other hand , $\overline{A}$ contains the line $\{(r,0) : r \in \mathbb{R}\}$, and so has dimension 1 .

PIECEWISE MONOTONICITY (van den Dries [1986]): Every semi-algebraic function $f : (a,b) \to \mathbb{R}$ is piecewise monotone. That is, there exist points $c_i \in (a,b)$ with $a = c_0 < c_1 < \ldots < c_k = b$ such that the restriction of f to the subinterval (c_i,c_{i+1}) is either constant, or continuous and strictly monotone. In particular, the one-sided limits:

$$\lim_{x \to a+} f(x) \qquad \text{and} \qquad \lim_{x \to b-} f(x)$$

both exist (as extended real numbers).

Note that the analogous statement is false for functions of two variables; the function $f(x,y) = xy/(x^2 + y^2)$ has no limit as (x,y) approaches $(0,0)$.

GENERIC LOCAL TRIVIALITY (Hardt [1980], Bochnak-Coste-Roy [1988, p. 195]): Let A , B be semi-algebraic sets, and let $\varphi : A \to B$ be a continuous semi-algebraic function. Then there are is a (relatively) closed semi-algebraic subset $B' \subset B$ with $\dim B' < \dim B$ such that for each of the (finite number of) connected components B_i of $B \setminus B'$ there is a semi-algebraic set C_i and a semi-algebraic homeomorphism $h_i : B_i \times C_i \to \varphi^{-1}(B_i)$ with the property that $\varphi \circ h_i(b,c) = b$ for each $b \in B_i$, $c \in C_i$.

Informally: except for a small subset of the range, every continuous semi-algebraic function is locally a product. (Caution: this is false without the requirement that φ be continuous; see Bochnak-Coste-Roy [1988, p. 196].)

Generic local triviality has many striking consequences. For our purposes, the most important are that semi-algebraic correspondences and semi-algebraic functions are generically continuous, that semi-algebraic correspondences admit generically continuous semi-algebraic selections, and that semi-algebraic functions are generically real-analytic.

GENERIC CONTINUITY: If X, Y are semi-algebraic sets and $F: X \twoheadrightarrow Y$ is a semi-algebraic correspondence with closed values, then there is a closed semi-algebraic set $X' \subset X$ of lower dimension such that F is continuous at each point of $X \setminus X'$. In particular, a semi-algebraic function is continuous at each point of the complement of a lower dimensional sem-algebraic set.

PROOF: Set $B = X$, $A = \text{graph}(F) \subset X \times Y$, and let $\varphi : A \to B$ be the projection onto the first factor and $\psi: A \to Y$ the projection onto the second factor. Generic triviality yields a semi-algebraic set $B' \subset B$, and connected components B_i of $B \setminus B'$, semi-algebraic sets C_i, and semi-algebraic homeomorphisms $h_i : B_i \times C_i \to \varphi^{-1}(B_i)$ as above. It is evident that the restriction of F to B_i is continuous. Moreover, each B_i is a relatively open subset of B (being one of the finite number of connected components of the open set $B \setminus B'$), and $\cup B_i = B \setminus B'$. We conclude that F is continuous at each point of $B \setminus B'$. The second statement follows from the observation that a single-valued correspondence is a function. ∎

SELECTIONS: If X, Y are semi-algebraic sets and $F: X \twoheadrightarrow Y$ is a semi-algebraic correspondence with non-empty (not necessarily closed) values, then there is a semi-algebraic function $f: X \to Y$ and a closed semi-algebraic set $X' \subset X$ of lower dimension such that $f(x) \in F(x)$ for each $x \in X$ and f is continuous at each point of $X \setminus X'$.

PROOF: Set $B = X$, $A = \text{graph}(F) \subset X \times Y$, and let $\varphi : A \to B$ be the projection onto the first factor and $\psi: A \to Y$ the projection onto the second factor. Generic triviality yields a semi-algebraic set $B' \subset B$, and connected components B_i of $B \setminus B'$, semi-algebraic sets C_i, and semi-algebraic homeomorphisms $h_i : B_i \times C_i \to \psi^{-1}(B_i)$ as above. Choose (for each i) a point $c_i \in C_i$ and define f_1 on $\cup B_i$ by $f_1(b) = \psi \circ h_i(b,c_i)$; f_1 is evidently a continuous, semi-algebraic selection, defined on $X \setminus X_1$. We may now apply this procedure to the restriction of F to X_1 , obtaining a lower dimensional semi-aklgebraic subset $X_2 \subset X_1$ and a continuous semi-algebraic selection f_2 defined on $X_1 \setminus X_2$. Since the dimensions of the sets X_i are strictly decreasing, this is a finite process. Finally, we define the selection $f : X \to Y$ by $f(x) = f_1(x)$ for $x \in X \setminus X_1$ and $f(x) = f_{i+1}(x)$ for $x \in X_i \setminus X_{i+1}$. ∎

We have already noted that semi-algebraic functions are generically continuous; in fact we can say much more.

GENERIC REAL ANALYTICITY: If X , Y are semi-algebraic sets, X is a real-analytic manifold, and $f : X \to Y$ is a semi-algebraic function, then there is a closed semi-algebraic set $X' \subset X$ of lower dimension such that f is real analytic at each point of $X \setminus X'$.

PROOF: As noted above, there is a closed semi-algebraic subset $X'' \subset X$ such that f is continuous at each point of $X \setminus X''$. Let $Z \subset (X \setminus X'') \times Y$ be the graph of the restriction $f|(X \setminus X'')$, and let $\pi_1 : (X \setminus X'') \times Y \to (X \setminus X'')$ and $\pi_2 : (X \setminus X'') \times Y \to Y$ be the projections. Since Z is a semi-algebraic set, it is the union of real-analytic manifolds M_i . By the semi-algebraic version of Sard's theorem (Bochnak-Coste-Roy [1988, p. 205]), the set C_i of critical values of $\pi_1|M_i$ is a semi-algebraic subset of $\pi_1(M_i)$ of lower dimension. Since $f|(\pi_1(M_i) \setminus C_i) = \pi_2 \circ \pi_1{}^{-1}$, f is real-analytic on $\pi_1(M_i) \setminus C_i$. Taking the union over all M_i and replacing the lower dimensional semi-algebraic set $\cup C_i$ by its closure yields the desired result. ∎

3. EQUILIBRIUM

As we have noted in the Introduction, the general relevance of the Tarski-Seidenberg theorem to game theory is that virtually all of the constructions of game theory have - or can be given - first order descriptions, and hence define semi-algebraic sets. In particular, this is the case for almost all of the usual equilibrium correspondences, so the results of the preceeding section lead very easily to extremely strong conclusions about these equilibrium correspondences.

To make this precise, fix an extensive form Γ ; i.e., a finite set of players, a game tree, and information sets for each player). An extensive form game is obtained from Γ by specifying a probability distribution over initial nodes, and payoffs (for each player) at each terminal node. If there are N players, I initial nodes, and Z terminal nodes, then we may parametrize the set of all such games by $\Pi(I) \times \mathbb{R}^{NZ}$, where $\Pi(I)$ is the probability simplex of dimension $I-1$; the dimension of this set is $NZ+I-1$. (Alternatively, we could view the probability distribution π_0 over initial nodes as fixed and the payoffs u as variable, or vice versa.) If $\pi \in \Pi(I)$ and $u \in \mathbb{R}^{NZ}$, we denote the corresponding game by $\Gamma_{\pi,u}$. A (behavioral) strategy for a player in this game is a function from his information sets to probability distributions on available actions at these information sets. We write Δ^i for the set of behavioral strategies of player i , and $\Delta = \Delta^1 \times \ldots \times \Delta^N$ for the set of behavioral strategy profiles. Note that $\Pi(I) \times \mathbb{R}^{NZ}$ and Δ are naturally identified with subsets of Euclidean space, defined by the appropriate linear equalities and inequalities; in particular, these sets are semi-algebraic. For each $\pi \in \Pi(I)$ and $u \in \mathbb{R}^{NZ}$, a Nash (or subgame perfect or sequential or (trembling hand) perfect) equilibrium for the corresponding game $\Gamma_{\pi,u}$ is an element of Δ , so each of these equilibrium notions yields a correspondence $\Pi(I) \times \mathbb{R}^{NZ} \twoheadrightarrow \Delta$. (If we view the probability distribution π_0 over initial nodes as fixed, and the payoffs u as variable, each of these equilibrium notions yields a correspondence $\{\pi_0\} \times \mathbb{R}^{NZ} \twoheadrightarrow \Delta$.)

THEOREM 1: For every game form Γ , the Nash, subgame perfect, sequential, and (trembling hand) perfect equilibrium correspondences are all semi-algebraic.

PROOF: To see that the Nash equilibrium correspondence $NE : \Pi(I) \times \mathbb{R}^{NZ} \twoheadrightarrow \Delta$ is semi-algebraic, we write the set

$$\text{graph}(NE) = \{(\pi,u,\sigma) \in \Pi(I) \times \mathbb{R}^{NZ} \times \Delta : \sigma \text{ is a Nash equilibrium for } \Gamma_{\pi,u}\}$$

in terms of polynomial inequalities. This is elementary. Write $E_i(\sigma_i,\sigma_{-i})$ for the expected payoff to player i if he follows the strategy σ_i and everyone else follows follows the strategy profile σ_{-i} . For σ to be a

Nash equilibrium it is necessary and sufficient that $E_i(\sigma_i,\sigma_{-i}) \geq E_i(s_i,\sigma_{-i})$ for every i and for every pure strategy s_i of player i. Hence

$$\text{graph(NE)} = \{(\pi,u,\sigma) : \forall i \, , \, \forall s_i \, , \, E_i(\sigma_i,\sigma_{-i}) \geq E_i(s_i,\sigma_{-i}) \}$$

Since each $E_i(\sigma_i,\sigma_{-i})$ and $E_i(s_i,\sigma_{-i})$ is a polynomial in σ_i, σ_{-i}, π, and u, we conclude that graph(NE) is a semi-algebraic set, as desired.

The argument for the subgame perfect equilibrium correspondence is similar, and equally elementary.

The argument that the sequential equilibrium correspondence is semi-algebraic is no longer elementary. We must show that

$$\text{graph(SE)} = \{(\pi,u,\sigma) : \sigma \text{ is a sequential equilibrium for } \Gamma_{\pi,u}\}$$

can be written in terms of polynomial inequalities. In view of the Tarski-Seidenberg theorem, it will suffice to show that graph(SE) can be defined by a first order formula. By definition, σ is a sequential equilibrium for the game $\Gamma_{\pi,u}$ if there exist beliefs θ such that the assessment (σ,θ) is consistent and sequentially rational. Since beliefs are probability distributions over actions, they can be represented as elements of $\mathbb{R}^{\ell}$ for sufficiently large ℓ. Hence assessments lie in $\Delta \times \mathbb{R}^{\ell}$. Following Kreps and Wilson [1982], write $\Psi^0 \subset \Delta \times \mathbb{R}^{\ell}$ for the set of assessments (σ,θ) with the property that each action of each player is given positive probability, and beliefs are derived by Bayes' rule. It is evident that Ψ^0 is defined by a polynomial equalities and inequalities. Since the set of consistent assessments Ψ is the closure of Ψ^0, it follows from Proposition 1 that Ψ is a semi-algebraic set. Sequential rationality means that, for each player i and each information set h for that player, the strategy σ_i is optimal against σ_{-i}, starting from h, given the beliefs θ. Write $E_i{}^h(\sigma_i,\sigma_{-i} \mid \theta)$ for the expected payoff to player i starting from the information set h, if he has the beliefs θ, follows the strategy σ_i, and everyone else follows the strategy profile σ_{-i}; note that this is a polynomial in σ_i, σ_{-i}, π, and u. Putting everything together, we obtain:

$$\begin{aligned}\text{graph(SE)} = \{(\pi,u,\sigma) : \exists\, \theta \in \mathbb{R}^{\ell} \text{ such that } (\sigma,\theta) \in \Psi \text{ and} \\ \forall i \, , \, \forall \tau_i \, , \, E_i{}^h(\sigma_i,\sigma_{-i} \mid \theta) \geq E_i{}^h(\sigma_i,\sigma_{-i} \mid \theta)\}\end{aligned}$$

The Tarski-Seidenberg theorem now guarantees that graph(SE) is a semi-algebraic set, as desired.

The proof that the perfect equilibrium correspondence is semi-algebraic follows the same outline.

Finally, to see that each of these correspondences remains semi-algebraic when we view the probability distribution π_0 over initial nodes as fixed, we need only observe that $\{\pi_0\} \times \mathbb{R}^{NZ}$ is a semi-algebraic subset of $\Pi(I) \times \mathbb{R}^{NZ}$, and that the restriction of a semi-algebraic correspondence to a semi-algebraic subset of its domain is again a semi-algebraic correspondence. ∎

With this observation in hand, we can derive some striking consequences. The first follows immediately from Theorem 1, Generic Continuity and Selection.

COROLLARY 1.1: For every game form Γ, there is a closed, semi-algebraic subset $X \subset \Pi(I) \times \mathbb{R}^{NZ}$ of dimension $NZ + I - 1$ such that the Nash, subgame perfect, sequential, and perfect equilibrium correspondences are continuous at every point of $(\Pi(I) \times \mathbb{R}^{NZ}) \setminus X$. Moreover, each of these correspondences admits a semi-algebraic selection which is continuous at each point of $(\Pi(I) \times \mathbb{R}^{NZ}) \setminus X$.

If we view the probability distribution π_0 over initial nodes as fixed, then we obtain a closed, semi-algebraic subset $X_{\pi_0} \subset \mathbb{R}^{NZ}$ of dimension $NZ - 1$ such that the Nash, subgame perfect, sequential, and perfect equilibrium correspondences are continuous at every point of $(\{\pi_0\} \times \mathbb{R}^{NZ}) \setminus X_{\pi_0}$. Moreover, each of these correspondences admits a semi-algebraic selection which is continuous at each point of $(\{\pi_0\} \times \mathbb{R}^{NZ}) \setminus X_{\pi_0}$.

It is instructive to compare the conclusions of Corollary 1.1 with those available from general facts about correspondences. Recall (Hildenbrand [1974]) that if A, B are complete metric spaces, and $F : A \twoheadrightarrow B$ is an upper hemi-continuous correspondence with compact values, then F is continuous at each point of a residual set (and hence admits a selection which is continuous at each point of the same residual set). Applying this result to the correspondences above leads to the conclusion that the Nash, subgame perfect, and sequential equilibrium correspondences are continuous at each point of a residual set (and admits a selection which is continuous at each point of the same residual set). It should be kept in mind, however, that a residual set might fail to be open (indeed, its complement could be dense) and might have (Lebesgue) measure zero. On the other hand, from Corollary 1.1 and the fact that a closed lower dimensional semi-algebraic subset of $\mathbb{R}^{\ell}$ has no interior and is of (Lebesgue) measure zero, we conclude that the Nash, subgame perfect, and sequential equilibrium correspondences are continuous at each point of a dense open set whose complement has measure zero (and admit selections which are continuous at each point of the same dense open set). This is clearly a substantial improvement over the conclusions available from general facts about correspondences. In the case of the perfect equilibrium correspondence, the improvement is much more dramatic, since the perfect equilibrium correspondence is not upper hemi-continuous, so might *a priori* have no points of continuity whatsoever.

Recall that Kohlberg and Mertens [1986] have used the fact that the set of Nash equilibria of any game is semi-algebraic to conclude that the set of Nash equilibria has only a finite number of connected components. (This fact plays an important role in the theor of stable equilibria.) The following Corollary sharpens this observation: the finiteness statement remains valid for various refinements, and there are uniform bounds on the number of connected componenets. The proof involves a useful theme: generic local triviality gives information off a lower dimensional set; restricting to that lower dimensional set yields a situation to which generic local triviality can be applied again, yielding finer information, etc.

COROLLARY 1.2: For any game form Γ, there is an integer r (depending only on Γ) such that the number of connected components of the set of Nash equilibria (respectively, subgame perfect equilibria, sequential equilibria, perfect equilibria) of any game $\Gamma_{\pi,u}$ is finite and bounded by r.

PROOF: We give the argument only for Nash equilibria. Applying generic local triviality to the projection

$$\text{proj}:\text{graph(NE)} \to \Pi(I) \times \mathbb{R}^{NZ}$$

implies that there is a lower dimensional set $X_1 \subset \Pi(I) \times \mathbb{R}^{NZ}$ and a finite number of semi-algebraic sets $Y_{11}, Y_{12}, \ldots, Y_{1k_1}$ with the property that, for every $(\pi,u) \in (\Pi(I) \times \mathbb{R}^{NZ} \setminus X_1)$, the set of Nash equilibria of the game $\Gamma_{\pi,u}$ is (semi-algebraically) homeomorphic with one of the sets Y_{1i}. Applying generic local triviality to the restriction

$$\mathrm{proj}\,|\,\mathrm{proj}^{-1}(X_1) \to X_1$$

implies that there is a lower dimensional set $X_2 \subset X_1$ and a finite number of semi-algebraic sets $Y_{21}, Y_{22}, \ldots, Y_{2k_2}$ with the property that, for every $(\pi,u) \in (X_1 \setminus X_2)$, the set of Nash equilibria of the game $\Gamma_{\pi,u}$ is (semi-algebraically) homeomorphic with one of the sets Y_{2j}. Continuing in this way, we obtain a finite collection $\{Y_{ij}\}$ of semi-algebraic sets with the property that for every $(\pi,u) \in (\Pi(I) \times \mathbb{R}^{NZ}$, the set of Nash equilibria of the game $\Gamma_{\pi,u}$ is (semi-algebraically) homeomorphic with one of the sets Y_{ij}. Triangulability now yields the desired conclusion. ∎

The last Corollary follows immediately from Theorem 1 and Stratifiability.

COROLLARY 1.3: For every game, the set of Nash equilibria (respectively, subgame perfect equilibria, sequential equilibria, perfect equilibria) is a finite disjoint union of connected real-analytic manifolds.

(Compare the discussion in Kreps and Wilson [1982] about the set of sequential equilibria.)

4. THE TRACING PROCEDURE

In this section, we use the general theory of semi-algebraic sets (as described above) and some other ideas from real algebraic geometry to repair some gaps in the description of the tracing procedures and the verifications of the crucial properties. We begin with a review of the tracing procedure, and a discussion of the difficulties.

We first collect some notation. In what follows, we fix an N-player normal form game Γ (the argument for extensive form games requires only minor modifications). Let $\Delta \subset \mathbb{R}^{\ell}$ be the set of profiles of mixed strategies and let Δ^* be the set of profiles of completely mixed strategies. Write $W = (0,1) \times (0,1) \times \Delta^*$, $W^* = (0,1] \times (0,1] \times \Delta^*$ and

$\overline{W} = [0,1]\times[0,1]\times\Delta$; these are subsets of $\mathbb{R}^{2+\ell} = \mathbb{R}\times\mathbb{R}\times\mathbb{R}^{\ell}$; W is open and $\overline{W}$ is its closure. Let $\pi_1, \pi_2 : \mathbb{R}\times\mathbb{R}\times\mathbb{R}^{\ell} \to [0,1]$, be the projections into the first two variables, and let $\sigma : \mathbb{R}\times\mathbb{R}\times\mathbb{R}^{\ell} \to \Delta$ be the projection into the last variable. We usually write $w = (\eta,t,q)$ for a typical element of $\mathbb{R}^{2+\ell}$, so that $\eta = \pi_1(w)$, $t = \pi_2(w)$ and $q = \sigma(w)$; we frequently refer to q as the *strategy part* of w. For $A \subset \mathbb{R}\times\mathbb{R}\times\mathbb{R}^{\ell}$ and $\eta \in [0,1]$, write $A(\eta) = \{w \in A : \pi_1(w) = \eta\}$.

Now, let H_i be the payoff function for player i in the game Γ ; it is convenient to view H_i as a (affine) function from Δ to $\mathbb{R}$. Fix a probability distribution $p \in \Delta^*$. For $\eta \in [0,1]$, $t \in [0,1]$, we define a game $\Gamma(\eta,t)$ with the same set of players and the same strategies as Γ, but with payoffs $H_{i,t,\eta}$ defined by:

$$H_{i,t,\eta}(q_i,q_{-i}) = (1-t)H_i(q_i,q_{-i}) + tH_i(q_i,p_{-i})$$
$$+ \eta t\alpha_i \sum \log(q_{ik})$$

where α_i is a suitable positive constant. (Here we have written q_{ik} for the k-th component of the strategy vector q_i for player i. Note that if $t = 0$ the logarithmic term disappears and the game $\Gamma(\eta,0)$ coincides with Γ. However, when $0 < t \le 1$ and $0 < \eta$ the logarithmic term is very important. Indeed, as Harsanyi and Selten show, for these values of t and η, every equilibrium of the game $\Gamma(\eta,t)$ is in completely mixed strategies. (A little care must be exercised here. If $0 < t \le 1$ and $0 < \eta$, the payoff functions $H_{i,t,\eta}(q_i,q_{-i})$ are, strictly speaking, only defined if q_i is a completely mixed strategy; otherwise, $H_{i,t,\eta}(q_i,q_{-i}) = -\infty$. However, as Harsanyi and Selten show, this causes no difficulties. In particular, each of the games $\Gamma(\eta,t)$ does indeed have an equilibrium, in completely mixed strategies.) As a consequence, for these values of t and η, a completely mixed strategy profile $q \in \Delta^*$ is an equilibrium exactly when it satisfies the first order conditions. Because the derivative of $\log(t)$ is $1/t$ (a rational function), these first order conditions can be written as the conjunction of a finite number of polynomial identities, and hence as the vanishing of a single polynomial $P(\eta,t,q)$.

Write

$$L = \{(\eta,t,q) \in W^* : q \text{ is an equilibrium of } \Gamma(\eta,t)\},$$

the graph of the (Nash) equilibrium correspondence. In view of the above,

L is the zero set of the polynomial $P(\eta,t,q)$, and so is an algebraic set. For each $\eta > 0$, the game $\Gamma(\eta,1)$ is separable and has a unique equilibrium $w(\eta,1)$. It follows that there is a t_0 sufficiently close to 1 such that the game $\Gamma(\eta,t)$ has a unique equilibrium for $t_0 < t < 1$. Equivalently, the restriction $\pi_2 | L(\eta) \cap \pi_2^{-1}((t_0,1))$ is a homeomorphism of the smoth curve $L(\eta) \cap \pi_2^{-1}((t_0,1))$ onto the interval $(t_0,1)$.

The logarithmic tracing procedure may now be described in the following way: For each $\eta > 0$, $L(\eta)$ is an algebraic set which contains the point $w(\eta,1)$, and near this point, $L(\eta)$ is a smooth one dimensional curve. Hence, by Puiseaux's theorem, we may analytically continue this curve until it first leaves W^*. Harsanyi and Selten prove that every limit point w of $L(\eta)$ in $\overline{W}$ has the property that the strategy part $\sigma(w)$ is a Nash equilibrium of the game $\Gamma(\pi_1(w),\pi_2(w))$. Since all equilibria of the games $\Gamma(\eta,t)$ are in completely mixed strategies when $\eta > 0$ and $t > 0$, this means that $L(\eta)$ can only leave W^* through a point $w(\eta)$ such that $\pi_2(w(\eta)) = 0$, so the strategy profile $\sigma(w(\eta))$ is an equilibrium of the game $\Gamma(\eta,0)$. Taking limits as $\eta \to 0$ yields the logarithmic solution.

Unfortunately, there are some difficulties with this construction. The first of these is that Puiseaux's theorem applies only to algebraic *curves* (i.e., one dimensional algebraic sets), and the algebraic set $L(\eta)$ is generally *not* a curve. (It is an algebraic set and a portion of it is one dimensional, but it may also have higher dimensional portions.) Hence Puiseaux's theorem does *not* imply that $L(\eta)$ can be analytically continued *as a curve*, and in such a case the recipe of following $L(\eta)$ is not well-defined. (This difficulty cannot be remedied by the "obvious" means of passing to the "one dimensional branch" of $L(\eta)$, since it is entirely possible that $L(\eta)$ has both one dimensional portions and higher dimensional portions, but only one branch. See Bochnak-Coste-Roy [1988, pp. 53-54].) The second difficulty is more subtle (and more difficult to deal with). Harsanyi and Selten assert that $\sigma(w(\eta))$ has a limit as $\eta \to 0$ because $L(\eta)$ is a curve and depends algebraically on η. As we have already noted, $L(\eta)$ need not be a curve. Even if $L(\eta)$ is a curve, however, it may branch, in which case the procedure of following $L(\eta)$ will be well-defined, but *not* an algebraic procedure. (See Figure 1.) To put it another way: If $L(\eta)$ is an algebraic curve, then following $L(\eta)$ until it first leaves W^* amounts to finding the intersection with $\{t = 0\}$ of the irreducible branch $L^*(\eta)$ of $L(\eta)$ that contains the portion of $L(\eta)$ near $\{t = 1\}$; but the irreducible branch $L^*(\eta)$ does *not* depend algebraically on η. Thus, there is no reason to suppose that the limit point $w(\eta)$ depends nicely on η, and in particular, there is no reason to suppose that $\sigma(w(\eta))$ has a limit as $\eta \to 0$.

Our approach has the same intuition but avoids the pitfalls identified above. After a preliminary construction (Lemma 1), we show (Lemma 2) that there is a finite set $E \in (0,1)$ with the property that for all $\eta \notin E$ there is a unique irreducible analytic curve $C(\eta) \subset W$ containing the portion of $L(\eta)$ near $\{t = 1\}$. (Recall that an *analytic subset* of W is a

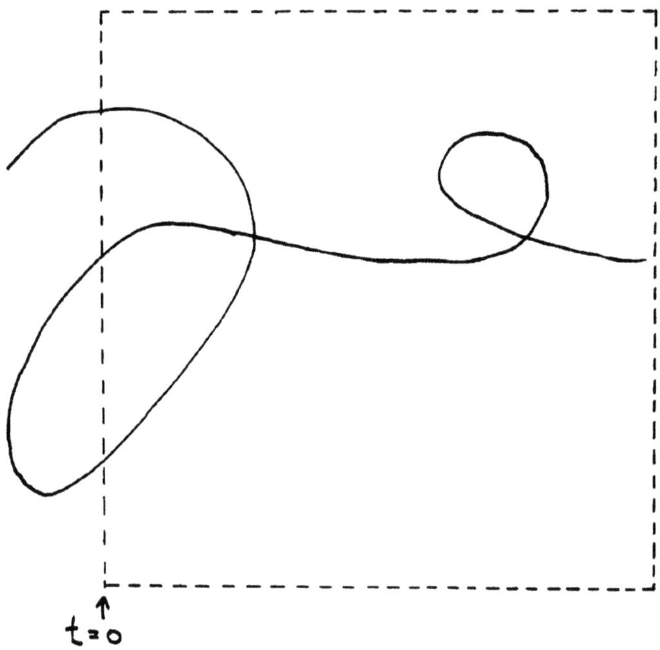

FIGURE 1

closed set which is locally the set of zeroes of a finite number of real analytic functions; an analytic curve is a one dimensional analytic set. See Milnor [1968].) Although $C(\eta)$ is not an algebraic curve, it is a semi-algebraic curve, and depends semi-algebraically on η . We then show (Lemma 3) that for each $\eta \notin E$, and for t sufficiently close to 0 , there is a unique point $z(\eta,t) \in C(\eta)$ with $\pi_2(z(\eta,t)) = t$. Because $C(\eta)$ depends semi-algebraically on η , The point $z(\eta,t)$ depends semi-algebraically on η and t . Hence, for each fixed η , $z(\eta,t)$ has a limit $z(\eta)$ as t approaches 0 , and $z(\eta)$ in turn depends semi-algebraically on η. Hence $z(\eta)$ in turn has has a limit as η approaches 0 (Lemma 4). This limit is the logarithmic solution.

To make these ideas precise, fix $\eta > 0$. As noted above, for t sufficiently close to 1, there is a unique point $z(\eta,t) \in L(\eta)$ such that $\pi_2(z(\eta,t)) = t$. Hence there is a $t \in [0,1)$ such that the restriction $\pi_2 | L(\eta) \cap \pi_2^{-1}((t,1))$ is one to one. We write t_η for the smallest such t. Evidently, t_η is defined by a first order formula and hence is a semi-algebraic function of η. Set $L(\eta,t) = L(\eta) \cap \pi_2^{-1}((t_\eta,1))$, and note that $L(\eta,t)$ is a connected one dimensional manifold (since it is homeomorphic to the interval $(t_\eta,1)$). Our first task is isolate the relevant portion of $L(\eta)$.

LEMMA 1: There is a two dimensional algebraic surface K and a finite subset E of $(0,1]$ such that:

(1) $K^* = K \cap W^* \subset L$;

(2) for all η, $K(\eta) \cap \pi_2^{-1}((t_\eta,1)) = L(\eta,t)$;

(3) if $\eta \notin E$ then $K(\eta)$ is a one dimensional algebraic curve.

PROOF: Set $L^0 = \{w \in L : \pi_2(w) > t_{\pi_1(w)}\}$. Since t_η is a semi-algebraic function of η, L^0 is a semi-algebraic set. Note that for each η,

$$L^0(\eta) = \{w \in L^0 : \pi_1(w) = \eta\} = L(\eta,t)$$

and that this is a one dimensional curve. In particular, L^0 is a two dimensional semi-algebraic set, and its Zariski closure $K = \mathrm{Zar}(L^0)$ is a two dimensional algebraic surface. Since $L^0 \subset L$ and L is the intersection of W^* with an algebraic set, we conclude that $K^* = K \cap W^* \subset L$. In particular, $K(\eta) \cap \pi_2^{-1}((t_\eta,1)) = L(\eta,t)$.

Since K is two dimensional and $K(\eta) \cap \pi_2^{-1}((t_\eta,1)) = L(\eta,t)$, which is one dimensional, we conclude in particular that $K(\eta)$ has dimension at least one and at most two (for every η). Moreover, the set of η for which $K(\eta)$ has dimension two is a semi-algebraic set; hence either it is finite or it contains an interval. However, if it contains an interval, K would necessarily have dimension at least three, a contradiction. Hence, for all but a finite number of η the set $K(\eta)$ is an algebraic curve. We write E for the (possibly empty) set of η for which $K(\eta)$ is not an algebraic curve. ∎

For each $\eta \notin E$, $K(\eta)$ is an algebraic curve so $K(\eta) \cap W$ is a one dimensional analytic curve. Let $C(\eta)$ be the analytically irreducible branch of $K(\eta) \cap W$ that contains $L(\eta,t)$. (That is, $C(\eta)$ is the smallest analytic subset of $K(\eta) \cap W$ containing $L(\eta,t)$.) Set $C = \{w \in W : \pi_1(w) \notin E \text{ and } w \in C(\pi_1(w))\}$.

It is important to note that the set $C(\eta)$ is uniquely defined, and independent of the choices made in this construction. Indeed, it follows immediately from uniqueness of analytic continuation that for $\eta \notin E$ and $t_\eta < t < 1$, $C(\eta)$ is the unique irreducible analytic curve in W that contains $L(\eta) \cap \pi_2^{-1}((t,1))$.

LEMMA 2: C is a semi-algebraic set, and, for each $\eta \notin E$, $C(\eta)$ is a semi-algebraic set.

PROOF: We are going to provide a first order description of $C(\eta)$; to do so, we use some of the structure theory of algebraic curves. (See Bochnak-Coste-Roy [1988] and Milnor [1968].) By a *local algebraic curve* (in $\mathbb{R}^n$) we mean the intersection of an algebraic curve with an open semi-algebraic subset of $\mathbb{R}^n$. For J a local algebraic curve and $z \in J$, we say that z is a *(topological) regular point* if some neighborhood of z in J is an arc; otherwise, z is a *branch point*. (A word of caution: this is a topological notion of regularity, not a smooth notion. A point may be be regular in this sense and still be a cusp.) An alternate description of regular and branch points can be given in the following way. Write $B(z,\varepsilon)$ for the open ball with center z and radius ε, and $\partial B(z,\varepsilon)$ for its boundary. For any point z in the algebraic curve J, and for all $\varepsilon > 0$, we consider the number of points in the intersection $\partial B(z,\varepsilon) \cap J$. If ε is sufficiently small, then this number of points becomes independent of ε, and is an even integer $2\beta(z)$. The integer $\beta(z)$ is called the *branching order* of J at z; z is a branch point exactly if the branching order is greater than 1.

We write J_b for the (necessarily finite) set of branch points of J and $J_r = J \setminus J_b$ for the set of regular points; note that J_b and J_r are semi-algebraic sets. Let $J_r = \cup A_i$ be the decomposition of J_r into (a finite number of) connected components; A_i is an arc and is a semi-algebraic set. We usually refer to the sets A_i as the *regular branches* of J.

We say that connected components A_i, A_j of J_r (or more generally, any pair of disjoint connected subsets of J) are *continuations of each other at the branch point* z if z is a limit point of both A_i and A_j and there are a neighborhood U of z, and connected components $\hat{A}_i$ and $\hat{A}_j$ of $A_i \cap U$ and $A_j \cap U$ respectively such that $\hat{A}_i \cup \hat{A}_j \cup \{z\}$ is an arc and is an analytic subset of U. (We cannot insist that $A_i \cup A_j \cup \{z\}$ itself be an analytic set, because of the possibility of loops. Note that this definition is entirely local in nature. See Figure 2.)

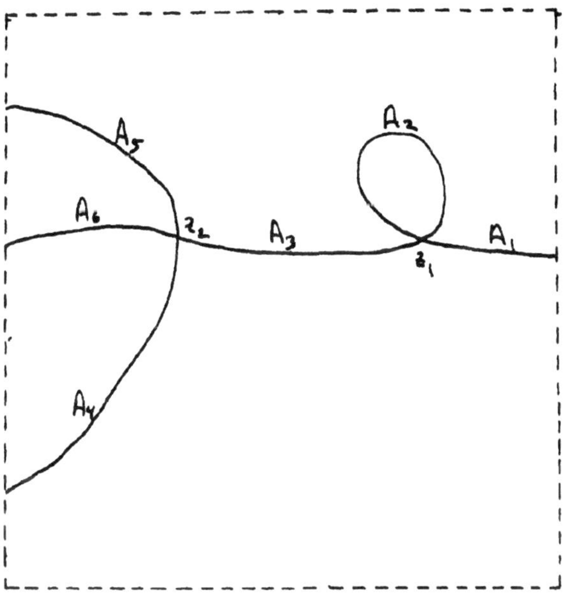

FIGURE 2

A_1, A_2 and A_2, A_3 are continuations of each other at z_1.
A_3, A_6 and A_5, A_6 are continuations of each other at z_2.
A_3, A_5 are not continuations of each other at z_2.

We will show that $C(\eta)$ is the union of regular branches of $K(\eta) \cap W$, together with some branch points. Hence to show that $C(\eta)$ is semi-algebraic (and depends semi-algebraically on η), we must deal with the problem of recognizing when two regular branches of a local algebraic curve are continuations of each othe at a point z. We show that this analytic problem can in fact be reduced to a semi-algebraic problem. In essence, this is because the algebraic nature of the curve implies that two regular branches are continuations of each other at z if they agree to sufficiently high order at z . The first step in this program is to reduce to a planar problem (i.e., a problem in $\mathbb{R}^2$.)

Consider a linear mapping $\theta : \mathbb{R}^n \to \mathbb{R}^2$ with the property that there is a finite set $J' \subset J$ such that $\theta|(J \setminus J')$ is one to one. (Such linear maps always exist (Bochnak-Coste-Roy [1988, p. 203]). The image of J is a semi-algebraic set and so belongs to the Boolean algebra generated by a finite number of polynomials; without loss, we may assume that none of these polynomials is identically zero. The set M of common zeroes of these polynomials is a plane algebraic curve contining $\theta(J)$. If the connected components A_i , A_j of J_r are continuations of each other at the branch point z , then the images $\theta(A_i)$, $\theta(A_j)$ are continuations of each other at $\theta(z)$. On the other hand, if z is a branch point of J which is in the closure of both A_i and A_j and A_i , A_j are not continuations of each other at z , then there is another component A_k which is a continuation of A_i at z . It follows that $\theta(A_i)$ and $\theta(A_k)$ are continuations of each other at $\theta(z)$. Since $\theta|J$ is one to one on the complement of a finite set, uniqueness of analytic continuation implies that $\theta(A_j)$ cannot also be a continuation of $\theta(A_i)$ at $\theta(z)$. Conclusion: A_i and A_j are continuations of each other at z if and only if $\theta(A_i)$ and $\theta(A_j)$ are continuations of each other at $\theta(z)$.

We have now reduced our original problem to a two dimensional problem; the next step is to reduce it to the solution of a power series equation. The image $\theta(J)$ is a semi-algebraic set and is contained in the algebraic curve M , which is the set of common zeroes of a finite number of polynomials, and hence of a single polynomial f . In view of Proposition 3, there is a bound on the degree d of this polynomial that depends only on J , and not on the choice of linear mapping θ . Fix a point $p = (x_0, y_0) \in M$. If V is a sufficiently small neighborhood of p , then $M \cap V$ is the union of a finite number of analytic arcs, each containing p ; these are the *local analytic branches* of M at p . Puiseaux's theorem (see Milnor [1968] for example), implies that, (for V sufficiently small) each local analytic branch of M at p can be parametrized by a pair of functions:

$$x(t) = x_0 + t^{\mu} \quad , \quad y(t) = y_0 + \sum_{i=1}^{\infty} c_i t^i$$

where μ is an integer not exceeding the degree of f and the power series $\sum c_i t^i$ is convergent on some interval $(-\varepsilon, +\varepsilon)$. (Or else such a parametrization exists with the roles of x, y interchanged, a possibility we shall henceforth not mention.) Given disjoint connected subsets M_1, M_2 of M, it follows (using uniqueness of analytic continuation) that M_1 and M_2 are continuations of each other at p if and only if there exist an integer μ and a convergent power series $\sum c_i t^i$ such that for arbitarily small $\varepsilon > 0$, the image of the interval $(-\varepsilon,+\varepsilon)$ under the mapping $x(t) = x_0 + t^\mu$, $y(t) = y_0 + \sum c_i t^i$ is contained in M and meets both $M_1 \cup M_2$. Note that, if this is the case, then the image of $(-\varepsilon,0)$ must be contained in one of M_1, M_2 and the image of $(0,+\varepsilon)$ must be contained in the other.

We now show that this power series problem can be reduced to a semi-algebraic problem. To do this we make the following observation. Suppose we are given an integer μ and a polynomial $g(t)$ of degree ℓ and no constant term; we ask: when it is possible to find a convergent power series $\sum c_i t^i$ whose first ℓ terms agree with $g(t)$, such that the functions $x(t) = x_0 + t^\mu$, $y(t) = y_0 + \sum c_i t^i$ parametrize one of the local analytic branches of M at p? Since M is the zero set of the polynomial f, this will be possible if and only if it is possible simply to find a convergent power series $\sum c_i t^i$ whose first ℓ terms agree with $g(t)$ such that $f(x(t),y(t)) = 0$. An elementary calculation with coefficients (or an appeal to a result of E. Cartan) shows that this will be possible if $\ell \geq \mu d + 1$ (where d is the degree of the polynomial f) and (1) $f(x_0 + t^\mu, y_0 + g(t))$, which is a polynomial in t, has no terms of order less than $\mu d + 1$. Given such a polynomial $g(t)$, Taylor's theorem and the observation at the end of the last paragraph imply that, in order that the parametrization $(x(t),y(t))$ meet both M_1 and M_2 and be contained in $M_1 \cup M_2$, we must also have, in addition to the above (reversing the roles of M_1 and M_2 if necessary): (2) for all sufficiently small $\varepsilon > 0$, and all $t \in (-\varepsilon,0)$, there is a point $(x_1,y_1) \in M_1$ such that:

$$|x_1 - x_0 - t^\mu| < |t|^{(\mu d+(1/2))} \text{ and } |y_1 - y_0 - g(t)| < |t|^{(\mu d+(1/2))}$$

(3) for all sufficiently small $\varepsilon > 0$, and all $t \in (0,+\varepsilon)$, there is a point $(x_2,y_2) \in M_2$ such that

$$|x_2 - x_0 - t^\mu| < |t|^{(\mu d+(1/2))}$$

$$|y_2 - y_0 - g(t)| < |t|^{(\mu d+(1/2))}$$

That is, M_1 and M_2 are continuations of each other at $p = (x_0,y_0)$ if and only if there is an integer $\mu \le d$ and a polynomial $g(t)$ of degree $\mu d+1$ and no constant term satisfying the conditions (1), (2), (3). The existence of such an integer and polynomial is a first order statement. (Because it amounts to the the assertion that there is a polynomial of degree $d+1$ or a polynomial of degree $2d+1$, . . . , or a polynomial of degree d^2+1 , and because asserting the existence of a polynomial of degree ℓ with no constant term really amounts to asserting the existence of ℓ real numbers - the coefficents of the polynomial.)

Putting all of this together, we see that we have reduced the analytic problem of determining when two regular branches of the local algebraic curve J are continuations of each other to the existence of a linear mapping θ , an integer $\mu \le d$ and a polynomial $g(t)$ of degree at most d . This is a first order - hence semi-algebraic - problem, as desired.

Specializing to the local algebraic curve $K(\eta) \cap W$, we write $K_b(\eta)$ for the set of branch points, $K_r(\eta)$ for the set of regular points, and $K_r(\eta) = \cup A_i(\eta)$ for the decomposition into regular branches. We may number these connected components so that $A_1(\eta)$ is the regular branch containing $L(\eta,t)$. We can describe $C(\eta)$ as the closure (in W) of the union of all those connected components $A_i(\eta)$ of $K_r(\eta)$ which are either equal to $A_1(\eta)$, or are continuations of $A_1(\eta)$ at some branch point, or are continuations of a regular branch which is a continuation of $A_1(\eta)$, etc. Since each of the regular branches $A_i(\eta)$ is semi-algebraic, the set $C(\eta)$ is the closure of the union of (a finite number of) semi-algebraic sets and hence is semi-algebraic.

What remains is to show that the set C is semi-algebraic. To do this, we first appeal to Proposition 3 to guarantee the existence of an integer D such that, for every $\eta \notin E$ and every linear mapping $\theta : \mathbb{R}^{2+\ell} \to \mathbb{R}^2$, the image $\theta(K(\eta) \cap W)$ is contained in an algebraic curve defined by a single polynomial of degree at most D . We then appeal to generic triviality to conclude that there are a finite number of $\eta_i \in (0,1)$, with $0 = \eta_1 < \eta_2 \ldots < \eta_k = 1$, semi-algebraic curves J_i and semi-algebraic homeomorphisms

$$\Phi_i : (\eta_i,\eta_{i+1}) \times J_i \to C \cap \pi_1^{-1}((\eta_i,\eta_{i+1}))$$

which map $\{\eta\} \times J_i$ onto $C(\eta)$ for each $\eta \in (\eta_i,\eta_{i+1})$. In particular, for each i and each $\eta \in (\eta_i,\eta_{i+1})$, the curves J_i and $C(\eta)$ have the same number of regular branches; we let $J_{i1}, \ldots, J_{im}$ be the regular branches of J_i . Since $A_1(\eta)$ is the regular branch of $C(\eta)$ that contains

$L(\eta,t)$, we may number so that $\Phi_i(\eta,J_{i1}) = A_1(\eta)$. We can now write a first order description of the set C : it is the set of all points $z = (\eta,t,q)$ in $K(\eta) \cap W$ satisfying one of the following statements:

(a) $\eta = \eta_i$ for some i , and $z \in C(\eta)$;

(b) $\eta \in (\eta_i,\eta_{i+1})$, and $z \in \Phi_i(\eta,J_{i1})$ or there is an index j such that $z \in \Phi_i(\eta,J_{ij})$ and $\Phi_i(\eta,J_{ij})$ and $\Phi_i(\eta,J_{i1})$ are continuations of each other, or there is an index j and an index k such that $z \in \Phi_i(\eta,J_{ik})$ and $\Phi_i(\eta,J_{ik})$ and $\Phi_i(\eta,J_{ij})$ are continuations of each other and $\Phi_i(\eta,J_{ij})$ and $\Phi_i(\eta,J_{i1})$ are continuations of each other, or . . . ;

(c) $\eta \in (\eta_i,\eta_{i+1})$, and $z \in K_b(\eta)$ and for every $\delta > 0$ there is a point $z' \in K(\eta)$ such that $|z - z'| < \delta$ and z' satisfies (b) above.

Since we have reduced the problem of determining when regular branches of $C(\eta)$ are continuations of each other to a first order problem this provides a first order description of C . The Tarski-Seidenberg theorem therefore guarantees that C is a semi-algebraic set. ∎

With this hard work out of the way, the remainder is relatively straightforward.

LEMMA 3: For each $\eta \notin E$, there is an $s \in (0,1)$ such that the restriction $\pi_2 | C(\eta) \cap \pi_2^{-1}((0,s))$ is a one to one map onto the interval $(0,s)$.

PROOF: Since π_2 is a semi-algebraic map and $C(\eta)$ is a semi-algebraic set, it follows that for each $t \in (0,1)$ the set $C(\eta) \cap \pi_2^{-1}(t)$ is semi-algebraic and of dimension zero or one. If $C(\eta) \cap \pi_2^{-1}(t)$ were of dimension one then it would contain a relatively open subset of $C(\eta)$, which would imply that all of $C(\eta)$ lies entirely in the hyperplane $\pi_2^{-1}(t)$, an absurdity. We conclude that for each $t \in (0,1)$, $C(\eta) \cap \pi_2^{-1}(t)$ is a finite set.

Generic triviality implies that there is a $\delta > 0$ and an integer r (possibly 0) such that $C(\eta) \cap \pi_2^{-1}((0,\delta))$ is the union of arcs $A_1,\ldots,A_r$ with the following properties: (1) each A_i is a relatively open subset of $C(\eta) \cap \pi_2^{-1}((0,\delta))$; (2) $A_i \cap A_j = \emptyset$ if $i \neq j$; (3) the restriction $\pi_2 | A_i$ is a homeomorphism of A_i onto the interval $(0,\delta)$. The desired conclusion is that $r = 1$.

That $r \neq 0$ follows as in Harsanyi-Selten [1988]. Because $C(\eta)$ is an analytic curve in W , every point of $C(\eta)$ has a neighborhood in $C(\eta)$ which is the union of a finite number of open (analytic) arcs. Hence if we delete from $C(\eta)$ a small neighborhhood of $z(0,1)$, what remains must have a limit point ζ on the boundary of W ; i.e., $\zeta \in \overline{W} \setminus W$. Moreover, the strategy part $\sigma(\zeta)$ must be an equilibrium for the game $\Gamma(\pi_1(\zeta),\pi_2(\zeta))$. Since $\pi_1(\zeta) = \eta \in (0,1)$, this can only be the case if $\pi_2(\zeta) = 0$, or $\pi_2(\zeta) = 1$, or $0 < \pi_2(\zeta) < 1$ and $\sigma(\zeta) \in \Delta \setminus \Delta^*$. The last of these is impossible, since for $0 < \eta < 1$ and $0 < t < 1$, every equilibrium of the game $\Gamma(\eta,t)$ is in completely mixed strategies. The second of these is impossible because it would imply that, for t arbitrarily close to 1 , the game $\Gamma(\eta,t)$ had two distinct equilibria. We conclude that $\pi_2(\zeta) = 0$ and hence that $r \neq 0$.

It remains to show that r cannot exceed 1 . Since $C(\eta)$ is an irreducible analytic curve, there is a real analytic function $\varphi : (0,1) \to C(\eta)$ and a finite set $B \subset (0,1)$ such that $\varphi | (0,1) \setminus B$ is a homeomorphism onto $C(\eta) \setminus \varphi(B)$. Shrinking δ if necessary, there is no loss of generality in assuming that $\varphi^{-1}(A_i) \cap B = \emptyset$ for each i . It follows that the sets $\varphi^{-1}(A_i)$ are disjoint open intervals in $(0,1)$; say that $\varphi^{-1}(A_i) = (\alpha_i,\beta_i)$. For each i , $\varphi | (\alpha_i,\beta_i)$ is a homeomorphism onto A_i and $\pi_2 | A_i$ is a homeomorphism onto $(0,\delta)$, so it follows that $\pi_2 \circ \varphi | (\alpha_i,\beta_i)$ is a homeomorphism onto $(0,\delta)$. Now if $r \geq 2$ then we can find indices j , k such that $\beta_j < \alpha_k$. Since $\pi_2 \circ \varphi$ maps (α_k,β_k) homeomorphically onto the interval $(0,\delta)$, and $\pi_2 \circ \varphi$ maps all of $(0,1)$ onto $(0,1)$, it can only be that $\pi_2 \circ \varphi(\alpha_k) = \delta$ and $\beta_k = 1$. Similarly, we conclude that $\pi_2 \; \varphi(\beta_j) = \delta$ and that $\alpha_j = 0$. But this means that $(0,1)$ is the union of the sets $\pi_2 \circ \varphi((0,\beta_j))$, $\pi_2 \circ \varphi((\alpha_k,1))$ and $\pi_2 \circ \varphi([\beta_j,\alpha_k])$. However, the first two of these coincide with the interval $(0,\delta)$, and the third is compact,which contradicts the fact that the image of $\pi_2 \circ \varphi$ is the entire interval $(0,1)$. We conclude that $r = 1$, as desired. ∎

Now, for each $\eta \notin E$, we define s_η to be the largest s so that the restriction $\pi_2 | C(\eta) \cap \pi_2^{-1}((0,s))$ is one to one. Since s_η is defined by a first order formula, it is a semi-algebraic function of η . For $\eta \notin E$ and $0 < t < s_\eta$, we write $z(\eta,t)$ for the unique point in $\pi_2 | C(\eta) \cap \pi_2^{-1}((0,s_\eta))$.

We have shown that $C(\eta)$ depends semi-algebraically (not algebraically!) on η ; it follows that $z(\eta,t)$ depends semi-algebraically on η and t .

LEMMA 4: (1) $z(\cdot,\cdot)$ is a semi-algebraic function of η and t .

(2) The limit

$$z(\eta) = \lim_{t\to 0} z(\eta,t)$$

exists and is a semi-algebraic function of η .

(3) The limit $\lim_{\eta\to 0} z(\eta)$ exists.

PROOF: To see that $z(\cdot,\cdot)$ is a semi-algebraic function of η and t , note that its domain is $D = \{(\eta,t) \in (0,1)\times(0,1) : t < s_\eta\}$ (which is a semi-algebraic set because $s(\cdot)$ is a semi-algebraic function) and that its graph is $\{w \in C : (\pi_1(w),\pi_2(w)) \in D\}$ (which is a semi-algebraic set because π_1 and π_2 are semi-algebraic - indeed, linear - functions). The existence of the limit $z(\eta)$ follows immediately from piecewise monotonicity of the semi-algebraic functions which are the components of $z(\eta,\cdot)$. To see that $z(\cdot)$ is itself a semi-algebraic function, write its graph as:

$$\text{graph}(z(\cdot)) = \{(\eta,w) : \forall\varepsilon > 0 \text{ , } \exists t \text{ such that } 0 < t < \varepsilon \text{ and } |w - z(\eta,t)|^2 < \varepsilon^2 \}$$

and apply the Tarski-Seidenberg theorem. Finally, the existence of the limit $\lim z(\eta)$ follows immediately from piecewise monotonicity. ∎

We define the *logarithmic solution for the game* Γ to be the strategy part of the limit $\lim z(\eta)$. That is, the logarithmic solution for the game Γ is $\lambda = \sigma(\lim z(\eta)) = \lim \sigma(z(\eta))$.

This procedure depends of course on the prior probability distribution p and on the game Γ . If we fix the game form and consider the game as parametrized by the payoffs u , we may view the logarithmic solution λ as a function of the prior probability distribution p and the payoffs u . Routine modifications of Lemmas 1 - 4 show that $\lambda(p,u)$ depends semi-algebraically on p and u. Summarizing:

THEOREM 2: The logarithmic solution $\lambda(p,u)$ is well-defined, and is a semi-algebraic function of the prior probability distribution p and the payoffs u.

In particular, if we view the prior probability distribution as fixed, and view the logarithmic solution as a function solely of the payoffs u, we conclude that it is continuous at every point of a dense open set whose complement has (Lebesgue) measure zero.

We turn now to the linear tracing procedure. Write

$$G = \{(0,t,q) : q \text{ is an equilibrium of the game } \Gamma(0,t)\}$$

A *linear trace* is by definition a curve in G from $z(0,1)$ (the strategy part of which is the unique equilibrium of the game $\Gamma(0,1)$) to a point of the form $(0,0,q)$ (so that in particular, q is an equilibrium of the game Γ). If a linear trace exists and is unique, we say the *linear tracing procedure is well-defined* and the strategy q is the *linear solution* of the game Γ. Harsanyi and Selten assert that there is always a linear trace from $z(0,1)$ to $(0,0,\lambda)$, where λ is the logarithmic solution. (In particular, if the linear tracing procedure is well-defined then the the linear solution and the logarithmic solution coincide.) To deduce this, Harsanyi and Selten give the following argument. For each η, parametrize the curve $L(\eta)$ by arc length, normalized so that the total length of $L(\eta)$ is 1. For $s \in [0,1]$, let $\omega(\eta,s)$ be the point in $L(\eta)$ whose distance from $z(\eta,1)$ is s. Define $\omega(s) = \lim \omega(\eta,s)$, so that the function ω defines a curve in G, beginning at $z(0,1)$ and ending at $(0,0,\lambda)$.

Unfortunately this argument is not correct. The first difficulty is that, just as above, the set $L(\eta)$ need not be a curve. Even if it is a curve, the function $\omega(\eta,s)$ need not depend on the parameter η in any sort of algebraic or semi-algebraic way, so that $\omega(s) = \lim \omega(\eta,s)$ need not be defined. (Keep in mind that arclength is not an algebraic- or even a semi-algebraic - function!) And even if $\omega(s) = \lim \omega(\eta,s)$ is defined, it will not in general be a continuous function of s, so that we will not obtain the desired curve beginning at $z(0,1)$ and ending at $(0,0,\lambda)$.

However, our construction provides a straighforward route around these difficulties.

THEOREM 3: There is a linear trace from $z(0,1)$ to $(0,0,\lambda)$, where λ is the logarithmic solution. In particular, the logarithmic solution and linear solution coincide whenever the latter is unique.

PROOF: Let Z be the limiting set, as $\eta \to 0$, of $C(\eta)$; i.e.,

$$Z = \{z : \forall \varepsilon > 0, \exists \eta, \exists w \in C(\eta) \text{ such that } 0 < \eta < \varepsilon \text{ and } |w - z|^2 < \varepsilon^2\}$$

It is evident that Z is a semi-algebraic set. As before, we see that every point $z \in Z$ is an equilibrium for the game $\Gamma(\pi_1(z),\pi_2(z))$. On the other hand, $\pi_2(z) = 0$ for every $z \in Z$, so $Z \subset G$. Since $z(\eta,1)$, $z(\eta) \in C(\eta)$ for each η, we see that $z(0,1)$ and $(0,0,\lambda)$ belong to Z. Since each $C(\eta)$ is connected, so is Z. Finally, since Z is a connected semi-algebraic set, it follows from triangulability that every two points of Z lie on a curve in Z. ∎

5. EQUILIBRIUM SELECTION

As we have shown, the tracing procedure yields, for each game, a well-defined logarithmic solution, but the tracing procedure itself constitutes only a part of the entire Harsanyi-Selten equilibrium selection procedure. However, as the sketch below shows, the results of Section 4 yield rather easily that the entire Harsanyi-Selten equilibrium selection procedure is well-defined. (For details on the selection procedure we refer to Harsanyi-Selten [1988], and especially to the Summary of Procedures, Section 5.7.)

To apply the equilibrium selection procedure to a game Γ (in standard form), we begin by constructing, for all sufficiently small $\varepsilon > 0$, the uniformly perturbed game Γ_ε. For each of these games, we apply the appropriate decompositions and reductions, and - if necessary - apply the tracing procedure to find the solution $\zeta(\varepsilon)$ for the game Γ_ε. It is easy to see that the decompositions and reductions are semi-algebraic; thus, in view of Theorem 2, above, the solution $\zeta(\varepsilon)$ depends semi-algebraically on the parameter ε . Piecewise monotonicity assures us that the solutions $\zeta(\varepsilon)$ approach a limit ζ as ε approaches 0 , and this limit is the solution for the original game Γ .

REFERENCES

T. Bewley and E. Kohlberg [1976], "The Asymptotic Solution of a Recursion Equation Occurring in Stochastic Games," *Mathematics of Operations Research* , 1, 321-336.

L. Blume and W. R. Zame [1989], "The Algebraic Geometry of Perfect and Sequential Equilibrium," Working Paper, Center for Analytic Economics, Cornell University.

J. Bochnak, M. Coste, M-F. Roy [1987], *Geometrie Algebrique Reele* , Springer-Verlag, Berlin.

H. Delfs and M. Knebusch [1981a], "Semi-Algebraic Topology over a Real Closed Field I," *Mathematische Zeitschrift* , 177, 107-129.

H. Delfs and M. Knebusch [1981b], "Semi-Algebraic Topology over a Real Closed Field II," *Mathematische Zeitschrift* , 178, 175-213.

R. Hardt [1980], "Semi-Algebraic Local Triviality in Semi-Algebraic Mappings," *American Journal of Mathematics* , **102**, 291-302.

J. C. Harsanyi [1975], "The Tracing Procedure: A Bayesian Approach to Defining a Solution for n-Person Noncooperative Games," *International Journal of Game Theory* , **4**, 61-94.

J. C. Harsanyi and R. Selten [1988], *A General Theory of Equilibrium Selection in Games* , MIT Press, Cambridge.

W. Hildenbrand [1974], *Core and Equilbria of a Large Economy* , Princeton University Press, Princeton, N.J.

H. Hironaka [1975], "Triangulations of Algebraic Sets," in *Algebraic Geometry* , Proceedings of Symposia in Pure Mathematics, American Mathematical Society, Providence.

E. Kohlberg and J. F. Mertens [1986], "On the Strategic Stability of Equilibria," *Econometrica* , **54**, 1003-1037

D. M. Kreps and R. Wilson [1982], "Sequential Equilibria," *Econometrica* , **50**, 863-894.

S. Lojasiewicz [1964], "Triangulations of Semi-Analytic Sets," *Annali Scuola Normale Superiore di Pisa* , **18**, 449-474.

S. Lojasiewicz [1965], "Ensembles Semi-Analytiques," *Publication I.H.E.S.*

J. Milnor [1968], *Singular Points of Complex Hypersurfaces* , Princeton University Press, Princeton.

A. Seidenberg [1954], "A New Decision Method for Elementary Algebra," *Annals of Mathematics* , **60**, 365-374.

R. Selten [1975], "Re-examination of the Perfectness Concept for Equilibrium Points in Extensive Games," *International Journal of Game Theory* , **4**, 25-55.

L. Simon [1987], "Basic Timing Games," Working Paper, University of California at Berkeley.

A. Tarski [1931], "Sur les Ensembles Definissables de Nombres Reels," *Fundamenta Mathematica* , 17, 210-239.

E. van Damme [1987], *Stability and Perfection of Nash Equilibria* , Springer-Verlag, New York.

L. van den Dries [198], "A Generalization of the Tarski-Seidenberg Theorem, and Some Non-definability Results," *Bulletin of the American Mathematical Society* , 15, 189-193.

H. Whitney [1957], "Elementary Structure of Real Algebraic Varieties," *Annals of Mathematics* , **66**, 545-556.

A PERSPECTIVE ON RENEGOTIATION IN REPEATED GAMES

by

Dilip Abreu and David Pearce*

Abstract: In this paper we discuss the conceptual foundations of one approach to modelling renegotiation in repeated games. Renegotiation–proof equilibria are viewed as social conventions that players continue to find beneficial after every history. The theory can be understood in terms of stationary stable sets of credible deviations.

Self–enforcing agreements play an important role in many areas of economics, and in the social sciences more generally. An extensive literature uses the repeated game model to explore formally what can be achieved by self–enforcing agreements among rational individuals. In a serious challenge to this work, a number of authors beginning with Farrell (1984) have argued that most of the agreements previously considered are vulnerable to *renegotiation*: after some histories of play, the participants will scrap the original implicit contract, and forge a new agreement. While the validity of the attack is widely accepted, the appropriate definition of a renegotiation–proof equilibrium is less clear. The most widely explored approach has been pursued by Farrell and Maskin (1987), Bernheim and Ray (1987), van Damme (1986), Blume (1988), Bernheim and Whinston (1986), and Benoit and Krishna (1988). Other possibilities are studied by Cave (1986), Asheim (1988), DeMarzo (1988) and Bergin and MacLeod (1989). Most of these papers focus on definitions and characterizations, with less space devoted to discussing the conceptual foundations for the respective definitions. Here we try to explain the thinking behind the approach we have taken in our own work on renegotiation (Pearce (1987) and Abreu, Pearce and Stacchetti (1989)) and to pinpoint the differences in perspective that distinguish it from the standard formulations in the literature.

Logically prior to the issue of renegotiation–proofness is the question of whether *any* cooperation among players is consistent with their rationality. We touch briefly on this subject before describing the Farrell–Maskin (1987) definition of a weakly renegotiation–proof set. A discussion of the contrasting reasoning that shapes their formulation and ours, respectively, leads us to suggest a framework that unifies several of the concepts introduced in Pearce (1987) and Abreu, Pearce and Stacchetti (1989). This interpretation of our work shows the influence of Greenberg (1988) and Asheim (1988). Finally, we apply our line of reasoning to finitely repeated games. Throughout, discounted repeated games with perfect monitoring are the subject of inquiry, but the same considerations apply to simple imperfect monitoring models.

---

*We would like to thank Reinhard Selten for encouraging us to write this essay and our co–author Ennio Stacchetti with whom some of the ideas presented here were developed. We are grateful for the hospitality of the ZiF at the University of Bielefeld, and for the financial assistance of the Sloan Foundation and the National Science Foundation.

COOPERATION AND SELF–ENFORCING AGREEMENTS

Suppose that some simultaneous game G is repeated indefinitely, and denote by $G^{\infty}(\delta)$ the resulting supergame in which each player's payoff is the discounted sum (according to the discount factor δ) of the infinite stream of his payoffs in the component games. An equilibrium of $G^{\infty}(\delta)$ is said to display *cooperation* if, at some point on the equilibrium path, behavior does not correspond to any Nash equilibrium of the stage game G. In other words, in some period at least one player is not playing a myopic best response to others' choices, presumably because he expects other players' future choices to vary according to his current play. Is such variation consistent with rationality?

Güth, Leininger and Stephan (1988) argue for the imposition of subgame consistency (see Selten (1973) or Harsanyi and Selten (1988)) on solutions of infinitely repeated games. This principle requires that if two subgames are isomorphic, behavior of rational players should be identical in the two games. Notice that all subgames of $G^{\infty}(\delta)$ are isomorphic: each is a perfect copy of $G^{\infty}(\delta)$ itself. Thus, subgame consistency rules out the variation in behavior that might permit cooperation to arise.

Rationality in a strategic context is a notoriously delicate concept, and the term has been used in many ways. Certainly the theorist can impose as an axiom the doctrine that the internal structure of any given game (or subgame) determines a unique profile that is consistent with player rationality; from this starting point, one can build clean and tractable theories. But we do not see the necessity for such an approach, nor does it seem the most attractive way to idealize intelligent human behavior. Consider a social convention, or arrangement, that creates the presumption that certain actions will be followed by particular responses. If this gives incentives that allow players to cooperate to their mutual profit, greedy players might be expected to adopt the convention and perpetuate it. Of course, not all arrangements can inspire confidence; an agreement should be self–enforcing. Exactly what this means is admittedly a difficult question. Subgame perfection (Selten (1965, 1975)) is a necessary feature of a self–enforcing agreement, but it does not consider the possibility that the group (as opposed to an individual) might in concert abandon the equilibrium plan. This is precisely the issue that the literature on renegotiation in supergames addresses.

RENEGOTIATION AND CREDIBLE DEVIATIONS

A theory of renegotiation attempts to identify those equilibria that can survive the threat of renegotiation in every contingency. The definition of a weakly renegotiation–proof set given by Farrell and Maskin (1987) is representative of the most popular approach in the literature. A *weakly renegotiation–proof set* is "... a collection S of perfect equilibria and all their continuation equilibria, such that at no point in the game tree of any equilibrium in S would players all prefer moving to another member of S." (Farrell and Maskin (1987), page 3). The elements of S are interpreted as the agreements the players might credibly make.

Think for a moment of the particularly simple case in which G is a symmetric game and all of the subgame perfect equilibria of $G^{\infty}(\delta)$ are strongly symmetric (all players expect the same payoffs as one another in any given subgame), or in which attention is restricted to such equilibria. Here a weakly renegotiation–proof set cannot contain two equilibria with different values, because they would be Pareto–ranked. Since continuation payoffs are constant, there can be no cooperation in this setting.

This seems to us unnecessarily pessimistic. Suppose G has a unique Nash equilibrium with value 0, and $G^{\infty}(\delta)$ has subgame perfect equilibria with average discounted values ranging from -4 to 5, for example. Suppose further that there is an equilibrium σ all of whose continuation values lie between 2 and 4,[1] inclusive, and that *all* perfect equilibria of $G^{\infty}(\delta)$ include continuation values of 2 or less in *some* contingencies. We would consider σ to be renegotiation–proof, for the following reason. If, in a subgame in which the continuation value of σ is 2, someone proposes renegotiation to an equilibrium γ with a higher value, he is implicitly making the case that the value 2 is needlessly severe, and can be abandoned. If it can be abandoned now, *in violation of a prior agreement*, it can presumably be abandoned in any future contingency. But γ (like all equilibria of $G^{\infty}(\delta)$) specifies the value 2 or worse in some subgame, so γ itself is vulnerable to renegotiation. In other words, the suggestion that the group can credibly renegotiate away from 2 is internally inconsistent. To say that an equilibrium with value 4 may be credible as part of an ongoing plan, is *not* to imply that it is believable as a proposed *deviation*.

Notice that Farrell and Maskin (1987), Bernheim and Ray (1987) et al. do not distinguish between the credibility of an ongoing plan, and its credibility as a *breach* of a social convention. We favor making the distinction, both on logical grounds and for its intuitive appeal. It is easy to give examples drawn from common experience illustrating the operation of the distinction. Society threatens jail sentences for embezzlement, and routinely carries out the threat (which is costly to all concerned). Not jailing the innocent is also part of the convention. No one concludes that since it is credible (in a society that needs to deter embezzlement) to let a person go free if he has committed no crime, it must also be credible not to jail embezzlers. The latter act of setting free is a breach of the social code, and undermines the public's belief in the code's relevance. Similarly, it is credible for a parent both to reward a child's good behavior and to punish bad behavior. Punishment is necessary since the child cannot take seriously the idea that although misbehavior has gone unpunished in the past, in the future punishment will be sure and swift.

The principal objection to the distinction we propose is that it requires a lack of stationarity in the set of strategy profiles that are credible: a given profile is believable after some histories, but not after others. Most of the literature on renegotiation assumes that

[1]As Pearce (1987) and Abreu, Pearce and Stacchetti (1989) show, equilibria all of whose continuation values lie strictly above the Nash value, are not exceptional.

because the repeated game is stationary, so too is the set of things that are credible or plausible. Carried to its logical conclusion, ***this argument rules out cooperation entirely.*** Suppose that anything that is plausible or credible at one point in the supergame is also credible in any other contingency. Presumably this applies to strategy profiles, suggestions, threats, promises and so on. Since each player has access to the same negotiation statements after some history h as he had at the beginning of the game, and exactly the same ones are plausible, each player should be able to secure the same continuation payoff as he negotiated at the beginning of $G^{\infty}(\delta)$. But if no one's continuation payoff ever varies, there are no incentives for cooperation.

We suggest that the natural way to impose stationarity is to require that the set of credible ***deviations*** is always the same. A credible deviation is a strategy profile to which the group believes it could defect, in ***violation*** of the previously negotiated plan. Recall that a history of play has no "physical" relevance in the subgame that follows it; it matters only via the social convention. Since a deviation by the group finds no support in the convention it is *breaching*, the credibility of the deviation can hardly be history–dependent.

Even if the group believes it can deviate to a particular profile, it may not want to do so. As usual, it is hard to determine what a group "wants." Various criteria may be used to decide whether or not the deviation will occur; the Pareto criterion is the one usually invoked in the literature. We shall return to a discussion of this and alternative criteria later. For the moment suppose that $\succ$ is a partial ordering of the set of supergame profiles, with the following interpretation. If x is a profile specified by the social convention in force, and y is a credible deviation with $y \succ x$ (read "y dominates x"), then the group will abandon x in favor of y. Thus, a self–enforcing agreement must have the property that none of its continuation equilibria is dominated by any credible deviation. It is also desirable to explain why a profile is ***not*** considered a credible deviation: after some history it is dominated by some credible deviation. The last two remarks suggest a formalization in terms of ***stable sets*** of credible deviations, in the spirit of von Neumann and Morgenstern (1947). Greenberg (1988) has emphasized the applicability of the stable set approach to diverse strategic problems. Asheim (1988) pursues Greenberg's lead in the context of repeated games and renegotiation–proof equilibria. The novelty in our work is the distinction we make between ongoing agreements and credible ***deviations.*** This leads to quite a different theory from Asheim's, the latter being more in the spirit of Farrell and Maskin (1987), Bernheim and Ray (1987) et al..

STABLE SETS OF CREDIBLE DEVIATIONS

Let S be the set of pure strategy profiles of G, and Σ be the set of pure strategy profiles of $G^{\infty}(\delta)$. For any $\sigma \in \Sigma$ and history h, $\sigma|h$ denotes the profile induced by σ on the subgame following h. Fix a (strict) partial ordering Ω of Σ. For each set of supergame strategy profiles $\Omega \subseteq \Sigma$, define

$$D(\Omega) = \{\sigma \in \Sigma \mid \exists h \in \bigcup_t S^t \text{ and } \gamma \in \Omega \text{ such that } \gamma \succ \sigma|h\},$$

the set of profiles dominated after some history by elements of Ω. For singleton sets $\{\sigma\}$ we abuse notation by writing $D(\sigma)$.

An equilibrium strategy profile σ cannot be credible as a deviation if it dominates one of its own continuation equilibria: the continuation equilibrium would be abandoned in favor of σ. Consequently, we confine attention to the set of ***consistent deviations***:

$$\Gamma = \{\sigma \in \Sigma \mid \sigma \text{ is a subgame perfect equilibrium and } \sigma \notin D(\sigma)\}.$$

We are now ready to define a set of credible deviations, and a renegotiation–proof equilibrium associated with the set.

DEFINITION: A set Ω of consistent deviations is a ***set of credible deviations*** if

(i) $\Omega \cap D(\Omega) = \phi$ Internal Stability

(ii) $\Gamma \backslash \Omega \subseteq D(\Omega)$ External Stability

Nothing in Ω is dominated by any element of Ω, whereas all consistent deviations not in Ω are dominated from within Ω.

DEFINITION: The set of equilibria that are ***renegotiation–proof*** relative to a credible set of deviations Ω is $R(\Omega) = \{\sigma \in \Sigma | \sigma \text{ is a subgame perfect equilibrium and } \sigma \notin D(\Omega)\}$.

Notice that the difference between Ω and $R(\Omega)$ is that the elements of $R(\Omega)$ are not required to be consistent. Recall our distinction between credibility as part of an ongoing plan, and credibility as a deviation. An element of $R(\Omega)$ may dominate one of its continuation equilibria, say $\sigma|h$. This does not threaten the credibility of σ, because the group does not believe it can ***deviate*** to σ from an ongoing agreement. If it believed it could, it would also deviate from $\sigma|h$ to σ; this shows the internal inconsistency of believing σ to be a credible deviation.

The specific theory that emerges from this framework depends on the choice of $\succ$, the dominance relation. In the special environment in which G is symmetric and attention is restricted to strongly symmetric equilibria of $G^\infty(\delta)$, $\succ$ should simply order the equilibria according to the players' unanimous ranking. The resulting theory coincides with the definition given by Pearce (1987): an equilibrium σ is renegotiation–proof if the infimum of its set of continuation values is at least as high as the corresponding infimum for any other equilibrium. As long as the threat of permanent reversion to myopic Cournot–Nash behavior would suffice to sustain some cooperation, this theory (unlike the standard ones) also predicts cooperation (Pearce (1987)). Thus, even in this setting where there is no argument about which dominance criterion to use, solutions depend critically on whether or not deviations are distinguished from ongoing plans.

When players are not unanimous about the ranking of a continuation equilibrium and a credible alternative, most authors assume that the deviation will occur if and only if all players strictly prefer it to the status quo. Letting the relation $\succ$ in the framework above be the Pareto partial ordering is not compatible with the existence of solutions. Pearce (1987) proves the existence of a weaker solution that combines the Pareto rule with the idea of credible deviations, but does not impose external stability. Similarly, Farrell and Maskin (1987) et al. use the Pareto partial ordering in the standard formulation without requiring external stability. Asheim (1989) imposes both internal and external stability with a Pareto dominance criterion, but does not prove existence, showing instead that stationarity is incompatible with existence.

Despite its popularity, the Pareto rule seems a dubious criterion for predicting the occurrence of renegotiation. In this respect it is a convincing sufficient condition: if everyone strictly prefers a credible alternative to the status quo, the latter is sure to be abandoned. But to require unanimous consent before a deviation occurs is extreme. In particular, the Pareto criterion is in conflict with one's sense that in situations with multiple equilibria, the outcome should depend on the bargaining positions of the participants. In general this involves highly complex considerations.

Abreu, Pearce and Stacchetti (1989) (hereafter APS) introduce a bargaining approach while sidestepping some of its difficulties by focusing on symmetric games. Although a multitude of factors may contribute to bargaining power, presumably this is the same for each player. In a static setting, equal bargaining power might be expected to lead, under modest assumptions, to a symmetric outcome. Similarly, in the repeated game, a player might reasonably insist on "equal treatment," except to the extent that allowing relative payoffs to vary with the history is in his interest. An arrangement that exploits variations in shares is credible only if a player finds the convention useful even when his fortunes are lowest. We assume that an individual will not object to a continuation payoff of w, say, if in *every* subgame perfect equilibrium, someone must accept w or less in some subgame. This motivates the solution concept studied by APS. A subgame perfect equilibrium σ is a ***consistent bargaining equilibrium*** if the lowest continuation payoff received by any player in any subgame is at least as high under σ as for any other subgame perfect equilibrium. Intuitively, this means that after no history can anyone protest that he is suffering an unnecessarily severe punishment.

In the framework introduced earlier, let $\succ$ be the Rawlsian ordering: for any σ, γ, $\sigma \succ \gamma$ if and only if $\min_i v_i(\sigma) > \min_i v_i(\gamma)$, where for any $\xi \in \Sigma$, $v_i(\xi)$ is player i's associated (average) repeated game payoff. One can check that there is a unique set Ω of credible deviations and that $R(\Omega)$ is exactly the set of consistent bargaining equilibria. Existence is guaranteed by a proposition we establish below. In games with imperfect monitoring, the flexibility gained by exploiting asymmetric continuation rewards can be dramatic, and hence a consistent bargaining equilibrium can easily be asymmetric. But APS show that in games of perfect monitoring satisfying some regularity assumptions, solutions are always strongly symmetric. This result leads to entirely elementary characterizations of maximal credible collusion in terms of the data of the stage game G.

The drawbacks of using the Rawlsian ordering as a dominance criterion are that it is a somewhat extreme way to model bargaining power, and that it is quite unconvincing in naive extensions to asymmetric games. An alternative that comes immediately to mind is the Nash bargaining solution (Nash, 1950). In cases in which some equilibrium of the stage game provides a natural threat point, letting $\succ$ be the ordering induced by the Nash product yields an interesting theory. Since the priorities of the group are summarized by the Nash solution, renegotiation will occur in any contingency in which the Nash product is "unnecessarily" low. More precisely, for each equilibrium of $G^{\infty}(\delta)$, one can compute the Nash product of the differences between players' average discounted payoffs and their respective threat point payoffs. A subgame perfect equilibrium σ withstands potential renegotiation if the minimum value of the Nash product, over all subgames, is at best as high as the corresponding minimum for any other perfect equilibrium $\succ$. Solutions in this sense always exist, as the theorem given below establishes.

A closer look reveals that, despite its attractive features, this theory is less than satisfactory. It is sabotaged by the inadequacy of the Nash bargaining solution as an expression of players' bargaining power in this context. To see this vividly, think of a "dummy player" whose actions have *no* effects on others' payoffs, but whose payoffs are affected by others' choices. Two games differing only by the addition of a dummy player can have quite different Nash bargaining solutions, despite the fact that the dummy player has no bargaining power. This is unacceptable in a positive theory of self–interested strategic behavior.

To our knowledge, no solution concept has yet been formulated that appropriately summarizes players' relative bargaining positions in a repeated game. One could at best hope for a concept that would rank equilibria on the basis of attractive principles; unanimity amongst theorists regarding the most appealing ranking is unlikely to be achieved. Thus, we are not able to propose a particular theory of renegotiation in general repeated games. Rather, we are advocating an *approach* to the subject, one that is formalized in the "stable sets of credible deviations" framework presented earlier. If in some context there is a dominance rule $\succ$ that seems suitable, the framework yields a corresponding set of renegotiation–proof equilibria of the infinite horizon game.

We prove the existence and uniqueness of a (stable) set of credible deviations when the strict partial ordering $\succ$ is derived from a continuous,[2] complete ordering R on $\boldsymbol{R}^n$. (That is, $\sigma \succ \gamma$ if and only if $[v(\sigma)\ R\ v(\gamma)$ and $\sim(v(\gamma)\ R\ v(\sigma))]$ where $v(\sigma) \in \boldsymbol{R}^n$ is the repeated game payoff–vector associated with the strategy profile σ). This assumption seems to us to provide the most coherent setting within which to discuss the issue of renegotiation.

---

[2]That is, for all $w \in \boldsymbol{R}^n$ the sets $\{x \in \boldsymbol{R}^n | x\ R\ w\}$ and $\{x \in \boldsymbol{R}^n | w\ R\ x\}$ are closed.

(A1) The strict partial ordering $\succ$ on Σ is derived from a continuous, complete ordering R on R^n.

The proof also uses the following standard assumption.

(A2) The strategy sets S_i and the payoff functions $\Pi_i: S_1 \times \ldots \times S_n \to R$ of the stage game G are compact and continuous, respectively. Furthermore G has a Nash equilibrium in pure strategies.

We confine attention to pure strategy equilibria of $G^{\infty}(\delta)$. The last part of (A2) serves only to guarantee that the repeated game has a pure strategy equilibrium.

PROPOSITION: Under (A1) and (A2) a unique set of credible deviations exists.

PROOF (Sketch): The proof uses the idea of self-generating sets and associated results from Abreu, Pearce and Stacchetti (1986), and mimics Proposition 1 of Pearce (1987). The reader who wishes to follow the argument closely will need to consult the latter.

First note that under (A2), the ordering R can be represented by a continuous function $u : R^n \to R$ (see Debreu (1954)). Let Θ be the set of all compact self-generating sets. For $X \in \Theta$ let $\underline{X} = min\ \{u(x)|x \in X\}$. Let $M = \{\underline{X}|X \in \Theta\}$. As in Proposition 1 of Pearce (1987) it follows that there exists $W \in \Theta$ such that $\underline{W} = sup\ M$. Let Ω be the set of all equilibria σ such that $u(v(\sigma)) = \underline{W}$ and $u(v(\sigma|h)) \geq \underline{W}$ for all $h \in \bigcup_t S^t$. Let γ be a consistent deviation. Then $u(v(\gamma)) \leq u(v(\gamma|h))$ for all h. Since for any equilibrium γ, the set $C(\gamma)$ of continuation values of γ, is a self-generating set, it follows that $u(v(\gamma)) \leq \underline{W}$. Furthermore $\gamma \in \Omega$ if and only if $u(v(\gamma)) = \underline{W}$. Clearly Ω satisfies internal and external stability and is the unique set of consistent deviations to do so.

Q.E.D.

FINITELY REPEATED GAMES

For any positive integer T, $G^T(\delta)$ denotes a discounted supergame consisting of T successive periods of G. Repeated games of this kind are generally considered the most promising setting for an uncontroversial treatment of renegotiation-proofness. The applicability of backward induction arguments has resulted in virtual unanimity among those writing on renegotiation in finite horizon games. We give the recursive definition by Bernheim and Ray (1987) of a Pareto-perfect equilibrium; the formulations of Benoit and Krishna (1988), van Damme (1986), Farrell and Maskin (1987) and Asheim (1988) are identical or equivalent. An equilibrium of $G^1(\delta)$ is Pareto-perfect if it is not Pareto-dominated by any other equilibrium of

$G^1(\delta)$. An equilibrium of $G^T(\delta)$, $T = 2,3,\ldots$, is Pareto–perfect if it is Pareto efficient within the set of subgame–perfect equilibria all of whose continuation equilibria after the first period are Pareto–perfect in $G^{T-1}(\delta)$. Benoit and Krishna have derived elegant characterizations of the Pareto–perfect set for large T and $\delta = 1$.

We find the consensus unconvincing. Recall that a history–dependent equilibrium is an arrangement unsupported by any structural consequences of play: all subgames of length t are isomorphic. The arrangement is adopted and retained only as a device for sustaining cooperation. But in the final period of a finite horizon game, it is common knowledge among the players that the convention no longer serves this purpose: there are no succeeding periods, so cooperation is impossible. Why, then, should players be guided in the last period by a convention that has outlived its usefulness?

Pareto perfection requires unanimity for renegotiation to occur. Therefore a solution may exhibit behavior in certain final–period subgames of $G^T(\delta)$ that differs dramatically from what one would predict as the probable outcome of the component game G. An example illustrates the point:

2

1		b_1	b_2	b_3	b_4	G
	a_1	8,8	0,9	0,0	0,0	
	a_2	9,0	5,5	6,0	0,–10	
	a_3	0,0	–10,0	7,1	0,0	
	a_4	0,0	0,6	0,0	1,7	

In the game G shown above, there are three pure strategy Nash equilibria, having respective values (5,5), (7,1), and (1,7). We submit that the equilibrium (a_2,b_2) having value (5,5) is far more likely to be played than is either of the others. This is attractive on the basis of symmetry, and considerations of utilitarianism, riskiness and self–signalling claims (Farrell (1988)) only strengthen the case. By playing a_3, player 1 risks receiving -10. Moreover, claiming to play a_3 and exhorting player 2 to respond with b_3 is unconvincing: if player 1 were intending to choose a_2, he would still want 2 to choose b_3. Suggesting the profile (a_2,b_2), on the other hand, is unambiguous.

The reader may check that the following instructions form a Pareto–perfect equilibrium of $G^2(1)$: play (a_1,b_1) in the first period, followed by (a_2,b_2) if no one has deviated or if both have, (a_3,b_3) if player 2 alone deviates, or (a_4,b_4) if player 1 alone deviates. Ex ante, this scheme is attractive, promising high payoffs for both participants. But how credible is the promise? If player 1 were to cheat in the first period, he should subsequently announce his intention to play a_2 in the terminal period. The arguments at his disposal seem likely to prevail over player 2's plaintive reminders about the initial agreement.

Not all games are as easy to analyze as the component game G just discussed. Theorists may find it difficult to predict with any confidence what will occur even in a single static game. But this does not justify the construction of equilibria in which final–period payoffs are manipulated to support cooperation earlier in the repeated game. Behavior in the last period is more likely to be determined by the players' view of their bargaining position in the one–shot game. Immutability of final period play then spreads to the other periods by the usual backward induction arguments.

CONCLUSION

To achieve cooperation in a repeated game, players must adopt a social convention specifying how behavior will vary according to the history of play. Thus, although all subgames of an infinitely repeated game are identical, players' views of what is credible are history–dependent. In particular, a strategy profile may be credible as part of an ongoing arrangement, but not if instead it constitutes a breach of the social convention. Since a deviation is "unauthorized," its credibility is arguably independent of the social convention, and hence of the history. For infinitely repeated games, this suggests that one look for a stationary stable set of credible deviations; these, in conjunction with a rule summarizing how the group will choose between an ongoing equilibrium and a credible alternative, determine which agreements are actually self–enforcing. We find the Pareto rule unsuitable for this role; what is needed is a criterion reflecting players' bargaining power. The notion of a consistent bargaining solution (APS (1989)) in symmetric games is in this spirit; it can be given a stable set interpretation using the framework developed here. A satisfying definition of renegotiation–proof equilibrium for asymmetric repeated games awaits the development of more sophisticated theories of bargaining for these games.

REFERENCES

Abreu, D., D. Pearce, and E. Stacchetti (1986): "Optimal Cartel Equilibria with Imperfect Monitoring," *Journal of Economic Theory*, 39, 251–269.

Abreu, D., D. Pearce, and E. Stacchetti (1989): "Renegotiation and Symmetry in Repeated Games," Cowles Foundation Discussion Paper.

Asheim, G. (1988): "Extending Renegotiation–Proofness to Infinite Horizon Games," mimeo, Norwegian School of Economics and Business Administration.

Benoit, J.–P., and V. Krishna (1988): "Renegotiation in Finitely Repeated Games," mimeo, Harvard Business School.

Bergin,J., and W. B. MacLeod (1989): "Efficiency and Renegotiation in Repeated Games," mimeo, Queen's University.

Bernheim, B.D., and D. Ray (1987): "Collective Dynamic Consistency in Repeated Games," mimeo, Stanford University.

Bernheim, B.D., and M. Whinston (1987): "Coalition–Proof Nash Equilibria II: Applications," *Journal of Economic Theory*, 42, pp. 13–29.

Blume, A. (1987): "Renegotiation–Proof Theories in Finite and Infinite Games," mimeo, University of California, San Diego.

Cave, J. (1987): "Long Term Competition in a Dynamic Game: The Cold Fish War," *Rand Journal of Economics*, 18, pp. 596–610.

van Damme, E. (1989): "Renegotiation–Proof Equilibria in Repeated Prisoner's Dilemma," *Journal of Economic Theory*, 47, pp. 206–207.

Debreu, G. (1954): "Representation of a Preference Ordering by a Numerical Function," in Decision Processes, R.M. Thrall, C.H. Coombs, R.L. Davis, eds.. New York: Wiley, pp. 159–165.

DeMarzo, P.M. (1988): "Coalitions and Sustainable Social Norms in Repeated Games," mimeo, Stanford University.

Farrell, J. (1984): "Credible Repeated Game Equilibria," unpublished manuscript.

Farrell, J., and E. Maskin (1987): "Renegotiation in Repeated Games," Working Paper 8759, Department of Economics, University of California, Berkeley.

Greenberg, J. (1988): "Social Situations: A Game–Theoretic Approach," forthcoming, Cambridge University Press, Cambridge.

Güth, W., W. Leininger, and G. Stephan (1988): "On Supergames and Folk Theorems: A Conceptual Discussion," ZiF–Working Paper: Game Theory in the Behavioral Sciences 1987–88, 19, University of Bielefeld.

Harsanyi, J.C., and R. Selten (1988): "A General Theory of Equilibrium Selection in Games," forthcoming.

Nash, J. (1950): "The Bargaining Problem." *Econometrica*, 18, pp. 155–162.

von Neumann, J., and O. Morgenstern (1947): Theory of Games and Economic Behavior, Princeton: Princeton University Press.

Pearce, D.G. (1987): "Renegotiation–Proof Equilibria: Collective Rationality and Intertemporal Cooperation," Cowles Foundation Discussion Paper No. 855.

Selten, R. (1965): "Spieltheoretische Behandlung eines Oligopolmodells mit Nachfrägetragheit," *Zeitschrift für die gesamte Staatswissenschaft*, 12, pp. 301–324.

Selten, R. (1973): "A Simple Model of Imperfect Competition where 4 are Few and 6 are Many," *International Journal of Game Theory*, 3, pp. 141–201.

Selten, R. (1975): "Reexamination of the Perfectness Concept for Equilibrium Points in Extensive Games," *International Journal of Game Theory*, 4, pp. 22–55.

ON SUPERGAMES AND FOLK THEOREMS: A CONCEPTUAL DISCUSSION

by

Werner Güth, Wolfgang Leininger and Gunter Stephan*

1. Introduction

That strategic rivalry in a multiperiod decision situation might be quite different from one day affairs is well accepted in game theory. Repeating the game over and over allows players to repond to others' actions and forces each player to visualize the reactions of his opponents. Hence, if a game is repeated an infinite number of times, the threat of retaliation becomes an enforcement mechanism which assures the cooperative behavior of the agents. This is, in a very crude and simple fashion, the message of the Folk Theorem.

Folk Theorems are the starting point for several developments in game theory (see for example Aumann, 1981 or Fudenberg and Maskin, 1986) but, to the authors' knowledge, the conceptional basis of the Folk Theorem has not received very much attention until recently. It is, therefore, the aim of this paper to discuss the question: What is the conceptional background of the Folk Theorem and its related results?

Game theory could be viewed as a theory of rational decision behavior in strategic conflicts. A first principle of rationality is that players should play an equilibrium strategy, since every non-equilibrium solution is, by definition, a self-destroying prophecy. A second principle of rationality is that solutions should not depend upon incredible threats. This requirement leads to the concept of subgame-perfectness as introduced by Selten (1965). Finally, a rational player should judge options by their consequences and should make his decisions dependent upon strategically relevant past events only. Since the consequences cannot precede the decision, and since only part of any previous action really affects future options, rationality must be based upon future events and those past actions which directly influence the present and future game situation. However, it is to some extent debatable what constitutes a "strategically relevant" event, and what not. Hence, it is the basic aim of this paper to elicit and comment on the view of "strategic relevance" underlying Folk Theorems and to point to the consequences of other concepts of strategic relevance.

Before going into a detailed discussion, we argue in Section 2 why a conceptional discussion of Folk Theorems is important. We characterize the conceptional difficulties of the Folk Theorem and

* We are grateful to W. Böge, E. v. Damme, J. Friedman, B. Kalkofen and R. Selten for critical comments. The usual disclaimer applies.

the implications of opposite viewpoints in a non-technical way. The more technical part of the paper starts in Section 3 which contains such fundamental concepts as subgame-perfectness and an explanation of the Folk Theorem in a formal model. Since we are concerned with the conceptional basis of Folk Theorems, considerations are restricted to the most basic version of the so-called Folk Theorem instead of the most general one. (For a review of the present state of the art, see the recent publications of Fudenberg and Maskin, 1986 and van Damme, 1987). Section 4 discusses two consistency concepts and their relationship to the Folk Theorem in some detail and Section 5 covers some concluding remarks.

2. Folk Theorems: a non-technical discussion

Folk Theorems are a last resort for social scientists who believe that the outcome of economic processes has to be efficient. The main argument is that if all agents would envision an inefficient result, they could, and would, improve their prospects by aiming at an efficient outcome, simply by cooperating or by changing the property rights (see for instance Coase, 1960). One could speak of an "efficiency school" whose most important base is in the main theorems of welfare economics which claim the efficiency of competitive allocations.

The extreme opposite position is to view social interactions as being typical prisoner's dilemma type situations where individual selfishness implies inefficient results. According to such mistrust in the "efficiency school", competitive markets are extreme idealisations and do not adequately reflect real markets. Consequently, there is no reason to expect that the actual markets yield efficient results.

Folk Theorems provide new arguments for the "efficiency school". Even on a market which has structural similarities to a prisoner's dilemma game, agents will achieve efficient results provided they interact repeatedly. Thus inefficiency is restricted to the rare case of non-competitive spot markets where agents are not expected to meet again over the course of time. As such, accepting Folk Theorems means to concede a strong argument in favor of the "efficiency school".

However, Folk Theorems do not only claim efficient outcomes, but rather that almost every outcome is possible in equilibrium. In particular, those outcomes that result from (subgame) consistent behavior.

As mentioned above and as will be explained in more detail later, the axiom of subgame-consistency implies a specific assumption, whether past decisions are considered as strategically relevant or not. If (subgame) consistency is assumed, then only those past actions which influence the rules of the present situation matter for present and future decision behavior. In the extreme case, which will be considered in this paper mainly to exemplify our argument, no past decision matters in this way; i.e., the past decisions do not matter at all and all time dependencies are strategically induced. As a consequence subgame-consistency will often justify the ad-hoc assumption of stationary decision behavior and can - in this restricted context - also be used to justify the concept of subgame perfect equilibrium points.

Folk Theorems implicitly claim the opposite position: It is required that all previous actions matter, regardless of whether they have a direct impact on the present situation or not. Note that such a requirement makes the modelling of social decision problems an extremely demanding task. At any point in time one has to provide the whole history, including the history of the market, to describe a situation adequately.

In addition, if one relies on Folk Theorems in claiming efficiency, one refutes outcomes which satisfy the requirement of subgame-consistency. On the other hand, accepting the axiom of subgame consistency means to deny Folk Theorems (and as a consequence very often efficiency). Since subgame-consistency in many situations is a rather convincing requirement, this represents a sacrifice for the efficiency result of which one should be aware.

3. Folk Theorems in game theory

3.1 Preliminaries

Consider a I-person game $G = (S_1,...,S_I,u_1,...,u_I)$ in the normal form. As usual, $S_i \neq \phi$, $i = 1,...,I$, denotes the strategy space of player i. u_i: $S \rightarrow \mathbb{R}$, is player i's pay-off function, where $S = X^I_{i=1}S_i$ indicates the cross product over all players.

It is supposed that game G is repeatedly played a finite or an infinite number of times. G^T denotes a finite (T-fold) repetition of game G starting from period $t = 1$ and ending at period $t = T$. Analogously, G^∞ is the infinite repetition of game G which begins in period 1. Following the convention, G^∞ is called the supergame of G and, for simplicity, we assume:

Assumption 1. All moves are observable ex post.

Hence, before a player makes a move, he not only knows his own past moves, but has complete knowledge about all other previous moves.

A sequence of strategy vectors $H^t = (s^1,...,s^{t-1})$, $t \leq T$, with $s^\tau = (s^\tau{}_1,...,s^\tau{}_I) \in S$, is called a history of the game G^T if the strategies s^τ have actually been played in periods $\tau = 1,...,t\text{-}1$ (for convenience set $H^1 = \{0\}$). Since in view of Assumption 1 all moves are observable ex post, H^t is common knowledge when a player has to move in period t, and for any period t the set of options open to player i can be identified with his strategy space S_i.

A strategy for player i in the game G^T is a function which selects, for any history H^t of the play, an element of S_i. Formally, the strategy ${}^Ts_i$ of player i is a sequence of mappings $\{{}^Ts_i{}^t, t=1,...,T\}$

$$(3.1) \qquad {}^Ts_i{}^t: S^{t-1} \rightarrow S_i \\ H^t \mapsto s_i{}^t,$$

by which for any period t, a unique strategy $s_i{}^t \in S_i$ is assigned to any history H^t. S^{t-1} is the (t-1)-fold Cartesian product of S.

Any I-tuple of strategies ${}^{T}s = ({}^{T}s_1,...,{}^{T}s_I)$ inductively defines an outcome path $a = (a^1({}^{T}s),...,a^T({}^{T}s))$ of the game G^T as follows:

$$(3.2) \qquad a^1({}^{T}s) = s^1,$$

and

$$(3.2a) \qquad a^t({}^{T}s) = {}^{T}s(a^1({}^{T}s),...,a^{t-1}({}^{T}s)) = {}^{T}s^t(s^1,...,{}^{T}s^{t-1}).$$

With this notion, the pay-off of playing the game G^T can be directly given on the set of strategy vectors of G^T. To keep the numbers of parameters as low as possible, we neglect impatience to win and define player i's pay-off by functions of the von Neumann- Morgenstern-type: in the game G^T by:

$$(3.3) \qquad U_i^T({}^{T}s) = \Sigma^T_{t=1} u_i(s^t)/T,$$

and in the supergame G^∞ by:

$$(3.3a) \qquad U_i^\infty({}^{\infty}s) = \liminf(\Sigma^T_{t=1} u_i(s^t)/T)$$

Given these conventions, a Nash-equilibrium of G^T is an I-tuple of strategies ${}^{T}s$ such that for all i and any strategy ${}^{T}s'_i$:

$$U_i^T({}^{T}s) \geq U_i^T({}^{T}s'_i, {}^{T}s_{-i}),$$

where $({}^{T}s'_i, {}^{T}s_{-i})$ denotes the original strategy I-tuple but with the i's component replaced by ${}^{T}s'_i$.

Obviously, (3.3) and (3.3a) are very specific pay-off functions. Other functions might be used if Folk Theorems and supergames G^∞, consisting of infinitely many repetitions of a base game, are considered. Rubinstein (1979) for example, uses the Ramsey-Weizsäcker overtaking criterion. Fudenberg and Maskin (1986) discount future periodic pay-offs in order to a observe well-defined pay-off for G^∞. In addition, they investigate games G^T, $T < \infty$, where cooperative results can be stabilized by introducing some very specific form of incomplete information in the spirit of Kreps and Wilson (1982). The incomplete information approach is based upon the assumption that players might be irrational, i.e., one possible type of player will choose non-optimally. However, since we want to keep the formal framework as simple as possible, considerations are restricted to the most simple cases.

3.2 Subgame-perfectness

Any history H^t, $t = 1,...,T$, might be the start of a subgame $G^T(H^t)$. By ${}^{T}s(H^t)$ we denote the strategy vector of the subgame $G^T(H^t)$ which is induced by the strategy vector ${}^{T}s$ of G^T as follows:

$$(3.4) \qquad {}^{T}s^1(H^t) = {}^{T}s^{t+1}(H^t)$$

and

$$(3.4a) \qquad {}^{T}s^{\tau+1}(H^t)(s^1,...,s^\tau) = {}^{T}s^{t+\tau}(H^t,s^1,...,s^\tau)$$

for all $\tau < T-t-1$.

This intuitively means that ${}^{T}s(H^t)$ stipulates, for the subgame, the same decisions as ${}^{T}s$ for the game G^T in the part forming $G^T(H^t)$.

A strategy vector ${}^{T}s(H^t)$ is a Nash-equilibrium point of the subgame $G^T(H^t)$, if for each player i the strategy ${}^{T}s_i(H^t)$ is the best reply to the strategy constellation $({}^{T}s_j(H^t))_{j \neq i}$ of all other players j.

If one accepts that rational players will always choose equilibrium points, then a natural consequence is to assume that rational players will also behave in the same way in any subgame as well. This is especially important in games where commitments are impossible. Since, in those cases, behavior in a subgame depends only upon the subgame itself, for an equilibrium to be sensible it is necessary that this equilibrium induces an equilibrium in every subgame.

Definition 1. A strategy vector ${}^{T}s$ of game G^T is called a subgame-perfect equilibrium, if (1) ${}^{T}s$ is a Nash-equilibrium of G^T, and (2) if ${}^{T}s$ induces Nash-equilibria ${}^{T}s(H^t)$ for all subgames $G^T(H^t)$ of G^T.

For later use we also define: A subgame-perfect equilibrium of G^T or G^∞, is respectively called stationary if it prescribes the same strategy choices for the stage game G.

To explain Definition 1, in a subgame-perfect equilibrium no player can improve his pay-off in any subgame by an isolated change of strategy. Selten (1965) has introduced the notion of subgame-perfectness as a rationality argument to exclude that in sequential games Nash-equilibria are sustained through incredible threats. Subgame-perfectness implies that threats implicit in the description of the players' strategies are credible since all players make use of the information on the previous play and the structure of the remaining subgame to determine credible future moves (see Hellwig and Leininger, 1987).

To illustrate the concept of subgame-perfect equilibria let the game G^T be the T-fold repetition of the well-known prisoner's dilemma game. For $I = 2$ and strategy sets $S_i = \{s_i^1, s_i^2\}$ the pay-off functions u_i are illustrated by a bi-matrix (see Table 3.1).

	s^1_2	s^2_2
s^1_1	$(1,1)$	(β,α)
s^2_1	(α,β)	$(0,0)$

Table 3.1: The two person prisoner's dilemma game[1]

The strategy vector ${}^{T}s = ({}^{T}s_1, {}^{T}s_2)$ which selects for all histories H^t

1) Note that 1 is the row and 2 is the column player. For any strategy vector s = (s1,s2) the pay-off vector is listed in the form u(s) = (u1(s),u2(s)); > 1, 0 > ß.

$$(3.5) \qquad {}^{T}s_i{}^{t}: H^t \mapsto s_i{}^2$$

is a subgame-perfect equilibrium point as can be shown by backward induction. Consider first the case $T < \infty$. Obviously, s_i^2 is the best reply regardless of the other player's move in a one stage game. Therefore, the strategy vector ${}^{T}s(H^T) = (s_1{}^2, s_2{}^2)$ defines an equilibrium point of $G^T(H^T)$. Let $t < T$ and suppose that, in all periods $\tau > t$, both players choose $s^2 = (s_1{}^2, s_2{}^2)$ regardless of the history H^τ. Obviously, this implies that players 1 and 2 should also use their s^2 strategies in period t, and the conjecture from above is proven for $t < T$. For $T = \infty$, the strategy vector ${}^{T}s$ as defined by (3.1) is also a subgame-perfect equilibrium point of G^∞, since a player i cannot do better than choosing $s_i{}^2$, if his opponent j ($\neq$ i) plays $s_j{}^2$ regardless of what has happened before.

3.3 Folk Theorem

Next we formalize a Folk Theorem for the model above. Given the game $G = (S_1,...,S_I,u_1,...,u_I)$, the minmax value

$$(3.6) \qquad v_i = \min_{s_{-i}} \max_{s_i} u_i(s_1,...,s_n)$$

of player i is the lowest pay-off level to which he can be brought down by his opponents. Alternatively, it is the minimal pay-off he can guarantee himself by a unilateral strategy choice. Furthermore,

$$(3.7) \qquad H(G) = \{x = (x_1,...,x_I) \in \mathbb{R}^I / \text{ there is } s \in S,\ x_i = u_i(s) \text{ for all } i = 1,...,I\}$$

denotes the set of feasible pay-off vectors in the base game G and CH(G), its convex hull. The set

$$(3.8) \qquad H^*(G) = \{x \in CH(G)/x_i > v_i \text{ for } i = 1,...,I\}$$

is the set of (strictly) individual rational pay-off vectors in G with the following properties: (1) the pay-off exceeds the minimax values v_i for all players $i = 1,...I$, and (2) can be realized by the players, if they jointly can randomize.

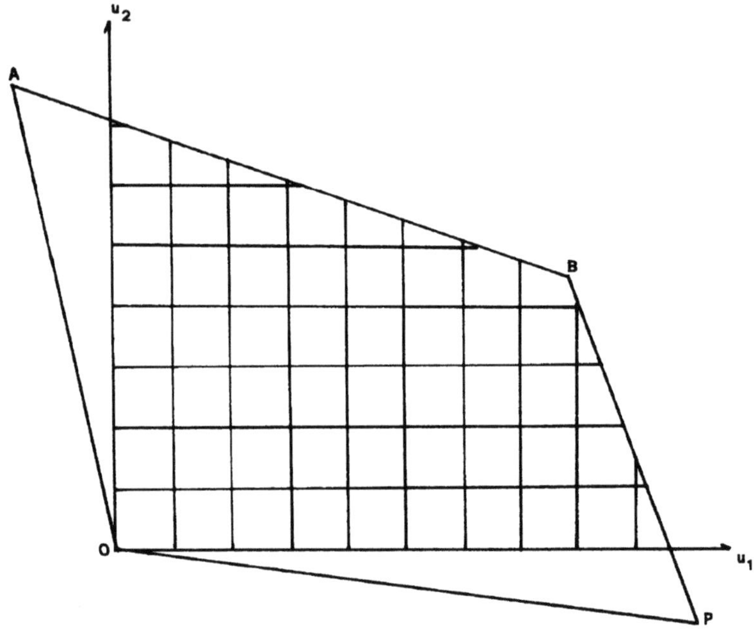

Figure 3.2: The set CH(G) and $H^*(G)$ for the prisoner's dilemma game

To provide a graphical illustration of the set $H^*(G)$ and CH(G), Figure 3.2 shows these sets for the prisoner's dilemma game G as determined by Table 3.1. The minimax value v_i of both players is 0, the set CH(G) is the quadrangle OPBA, and the set $H^*(G)$ is the shaded area in Figure 3.2.

Folk Theorem. Let G^∞ be the supergame of the game G with the pay-off function U^∞ defined in (3.3a). For any pay-off vector x in $H^*(G)$, there exists a subgame-perfect equilibrium point ${}^\infty s$ of G^∞ such that $U^\infty({}^\infty s) = x$.

Instead of an exact proof (see for example, Aumann and Sharley, 1976, Rubinstein, 1979), we only illustrate the essential idea of the Folk Theorem with the help of the prisoner's dilemma game (see again Table 3.1,) where $s^2 = (s_1^2, s_2^2)$ is the unique equilibrium point yielding the minmax values. Now let $x \in H^*(G)$ be a pay-off vector which can be realized by a rational point randomization over S (see (3.8)). Clearly, x can be induced by an appropriate plan $\{s^t/t = 1,2,...\}$ of strategy vectors in periods $t = 1, \ldots$. This plan can be chosen to determine a subgame-perfect equilibrium path ${}^{\infty}s$ of G^{∞} which supposes that all players stick to this plan as long as no deviation has occurred previously. However, if a deviation has occurred, they would react by playing s^2. Hence, if a player deviates in period t from the plan, his periodic pay-off function will be v_i forever and his pay-off $U^{\infty}{}_i$ will be $v_i = 0$. Since playing s^2 forever is a subgame-perfect equilibrium, the punishment behavior after deviating together with the plan $(s^1, \ldots)$ defines a subgame-perfect equilibrium point of G^{∞} which proves the Folk Theorem for the prisoner's dilemma game.

For other games the situation might be more complicated since deviations from a plan $(s^1, \ldots)$ inducing some pay-off vector $x \in H^*(G)$ may require punishments which are also harmful for the punishers themselves. In order to eliminate incentives to deviate from the punishment behavior, punishers have to be threatened by punishments for deviating from the punishment behavior which in turn have to be stabilized by punishment for deviating from the punishment of punishers etc. Thus in general, an infinite sequence of punishments for deviations might be needed to construct a subgame-perfect equilibrium point inducing a pay-off vector $x \in H^*(G)$.

It is important to observe that, in order to sustain and implement those punishments, strategies have to be conditioned on the past behavior of all players. Such strategies are called history-dependent. This means that, in the above example, those strategies which permanently threaten the play of the base game equilibrium after a deviation are conditioned on the entire history of play at each stage. If, for example, strategies could only memorize the previous 20 periods, no punishments for a deviation would be possible from the 21th period after deviation. Hence, the philosophy underlying the Folk Theorem takes all past actions that were possibile in the previous rounds of play of G to be "strategically relevant" for any future play of G.

4. Consistency and Folk Theorems

Game theory might be viewed as the theory of individually rational decision behavior in strategic conflicts. As a first requirement of rationality, the subgame-perfectness condition was introduced into our considerations. Subgame-perfectness as given in Definition 1 means that rational players who rely on the equilibrium concept will behave in the same way in all decision situations unless it is possible to bind oneself for future decisions. By this, we mean that solutions are excluded which are based upon incredible and unreasonable threats (see Selten 1975). A second rationality requirement will be imposed in the following section. It will be supposed that rational players only determine those equilibrium solutions which are, in a certain sense, independent of strategically irrelevant features of the game (see Selten, 1973, Harsanyi and Selten, 1988). In particular, this means that players behave similarly in similar game situations.

4.1 Isomorphic subgames and subgame-consistency

Roughly speaking, the notion of subgame-consistency reflects the rationality requirement from above. It imposes upon a non-cooperative game the following: (1) Once a subgame is reached, the former parts of the game should be treated as strategically irrelevant and the solution of the subgame should depend upon its specific rules only; and, (2) two subgames which are strategically equivalent induce the same decision behavior. The first condition effectively rules out the use of history-dependent strategies in the subgame; i.e. decisions in the subgame cannot depend on the particular path that leads to this subgame. In particular, in a repeated game the decision in the t-th repetition cannot depend upon whether a deviation has occurred before t or not.

Usually, two subgames are called strategically equivalent if an isomorphism between these two subgames can be defined. A isomorphism itself consists of a set of one-to-one mappings between the set of players and the their strategy sets in the two subgames and a linear transformation by which the pay-off functions of the two subgames are related (see Harsanyi and Selten, 1988). In the context of repeated games, isomorphy between subgames can be defined as follows

Definition 2. Two subgames $G^T(H^t)$ and $G^T(H^\tau)$ of the game G^T are isomorphic, if

(4.1) $$T - t = T - \tau.$$

Note that (4.1) does not imply $H^t = H^\tau$. It is nevertheless sufficient to allow the following isomorphism between two subgames $G^T(H^t)$ and $G^T(H^\tau)$: Both subgames have the same set of players who have the same strategy sets, namely the (T-t)-fold product of the strategy space S_i of the base game G. We now simply map H^t onto H^τ and map the vector of pay-off functions in $G^T(H^t)$, that means

$$1/T[\Sigma^{t-1}_{m=1} u(s^m) + \Sigma^T_{m=t} u(s^m)],$$

with $H^t = (s^1,...,s^{t-1})$, is mapped onto

$$1/T[\Sigma^{t-1}_{m=1} u(s'^m) + \Sigma^T_{m=t} u(s^m)],$$

the vector of pay-off functions in $G^T(H^\tau)$ with $H^\tau = (s'^1,...,s'^{t-1})$. These pay-offs differ by the constant term $1/T[\Sigma^{t-1}_{m=1} u(s^m) - \Sigma^{t-1}_{m=1} u(s'^m)]$ determined by the two histories. Thus these two subgames are not different from any strategic viewpoint.

Intuitively, Definition 2 says that two subgames are isomorphic if they have the same set of players and the same length, and thus open the same strategic possibilities and cover the same information conditions, independent of their history. This means, what really matters is the number of repetitions of the base game G and the rules of the base game. Once the subgame is reached, its strategical features are independent of its history.

Note an important consequence of Definition 2 for repeated games: For the supergame G^∞, condition (4.1) implies that all subgames are isomorphic and any subgame is isomorphic to the supergame G^∞ itself. Regardless of the history H^t, the game situation $G^\infty(H^t)$ is always identical from a strategical viewpoint since, after finitely many repetitions, the players still have to play infinitely many times. However, for a finitely repeated game G^T $(T < \infty)$, this result does not hold true. Obviously, no true subgame can be isomorphic to the game itself and there exists non-isomorphic subgames. This property has severe impact if the results of finite and infinite repetitions of the base game G are considered (see below).

To formalize the concept of subgame-consistency, let L be a solution function which assigns the set of solution candidates ${}^{T}s(H^t)$ to any subgame $G^T(H^t)$ of game G^T:

Definition 3. A solution function L is called subgame-consistent if it prescribes the same behavior (solution candidates) for isomorphic subgames.

What are the implications of the subgame-consistency condition? Assume, for example, that the solution function requires that any solution candidate of a game be a Nash-equilibrium point. If the solution function is subgame-consistent and if all subgames are isomorphic, then it will necessarily select only subgame-perfect equilibria, since the equilibrium property is transmitted from the solution set of the game itself to each subgame solution set.

Lemma 1. If all subgames of a game are isomorphic to the game itself and if the solution function is subgame-consistent and selects Nash-equilibria only, then any solution is subgame-perfect.

Since any subgame of the supergame G^∞ is isomorphic to the supergame G^∞ itself, this result implies:

Corollary 1. Let a solution function for the supergame G^∞ be subgame-consistent and select Nash-equilibria only. Then any solution is subgame-perfect.

In addition, any subgame-consistent solution function prescribes the same solution set to isomorphic subgames. In view of Corollary 1, any equilibrium solution of the supergame G^∞ must be a subgame-perfect equilibrium, and the solutions of the subgames $G^\infty(H^t)$ must be identical for each history H^t, t = 1,2,... . Thus, given a history H^t, the initial move vector ${}^{\infty}s^1(H^t)$ must be independent of H^t and, consequently, every subgame-consistent equilibrium play of G^∞ must be stationary.

Corollary 2. If the solution function of a supergame G^∞ is subgame-consistent, then any equilibrium play is stationary.

We have seen that, for the supergame G^∞, subgame-consistency implies: Any equilibrium solution is subgame-perfect and stationary. We now let ${}^{\infty}s = ({}^{\infty}s_1,...,{}^{\infty}s_I)$ be an arbitrary stationary strategy vector in G^∞; i.e., we observe the same move vector $s^t = (s_1{}^t,...,s_I{}^t)$ in any period t = 1,2,... . Obviously, ${}^{\infty}s$ is a subgame-perfect equilibrium point of G^∞ only if the stationary move vector s^t is an equilibrium point of the base game G. Otherwise, if s^t would not be an equilibrium point of G, at least one player could gain repeatedly by deviating unilaterally. Hence, by combining the results from above

Lemma 2. If the solution function is subgame-consistent, then the set of subgame-perfect equilibria of G^∞ coincides with the set of stationary equilibria.

The results from above cause a problem for the conceptional background of Folk Theorems. For the supergame G^∞, starting from the prisoner's dilemma game G as stated in Table 3.1, Lemma 2 implies that players will constantly choose $s^2 = (s_1^2, s_2^2)$. Therefore, the alternative pay-off vectors in the shaded area of Figure 3.2 cannot be induced by subgame-consistent solution functions which satisfy the equilibrium property. This demonstrates that the Folk Theorem becomes void if a subgame-consistent solution concept based on the equilibrium property is demanded. Moreover, at the present stage of our discussion, the Folk Theorem seems to be conceptually inconsistent. The Folk Theorem requires subgame-perfect equilibria but, in view of Corollaries 1 and 2 for the supergame G^∞, subgame-perfection is implied by the subgame-consistent application of the equilibrium property. This puzzle is resolved by the following observation: In the context of the supergame G^∞, the imposition of subgame-consistency is equivalent to the requirement of subgame-perfectness only if (in both cases) history-dependent strategies are considered. For example, with respect to those (and only those) strategies, subgame perfectness is nothing more than the subgame-consistent application of the equilibrium property in G^∞. The Folk Theorem, however, explicitly allows for history-dependent strategies and those strategies account for the additional subgame-perfect equilibria in Figure 3.2. In contrast, the notion of subgame-consistency, as defined above, by definition rules out history dependent strategies.

As is well-known from recent research (Friedman, 1985, Benoit and Krishna, 1985), there are certain analogues to the Folk Theorem for finitely repeated games G^T, $T < \infty$, if the base game G has at least two equilibria that differ sufficiently in pay-offs. These analogues also require history-dependent strategies. Here, however, we are interested in the implications of the use of history-independent strategies only. Again we observe that, for those strategies in G^T, all subgame-perfect equilibria are subgame-consistent; but now, somewhat surprisingly, perhaps more than just stationary equilibria survive.

Since history-independent strategies in G^T depend on t but not on H^t, the strategies ${}^Ts^t_i(\,)$ can be written, according to (3.1), as ${}^Ts^t_i$ and we obtain (by backward inductive reasoning):

Lemma 3: The set of subgame-perfect equilibria of G^T consists of all strategy vectors $(({}^Ts_1^t,\dots,{}^Ts_I^t))^T_{t=1}$ such that for any t, $({}^Ts_1^t,\dots,{}^Ts_I^t)$ is an equilibrium of the base game G. Moreover, any subgame-perfect equilibrium is subgame-consistent.

Note that now subgame-perfectness and subgame-consistency do not imply stationary equilibria. How did this difference come about? Recall (4.1) and the definition of the isomorphism giving strategically equivalent subgames: It implies that two subgames of G^T of different length, hence a different number of repetitions, cannot be isomorphic. Thus for those subgames, subgame-consistency does not impose any further restriction at all. In fact, one can say that there are no isomorphic subgames of G, since, as far as history-independent strategies are concerned, all subgames starting in the same period can be identified. This contrasts the fact that in G^∞ all subgames are isomorphic.

An unfortunate consequence of this outcome is that G^∞ is a bad approximation to G^T for large T if subgame-consistency is imposed. However, for T approaching infinity, the set of subgame-perfect (and subgame-consistent) equilibrium pay-offs converges to the convex hull of the equilibrium pay-offs of the base game G, whereas only extrenal points of this set are equilibrium pay-offs in G^{∞}. This follows directly from Lemmas 2 and 3. The next section will contain a closer look at this discontinuity and suggest a consistency concept that will remove it.

4.2 Substructures and truncation consistency

Until now, we have only considered subgames of the supergames G^{∞} and G^{T} which result after the (finitely) repeated play of the base game G. For the supergame G^{∞}, we have seen that any of these games is isomorphic to G^{∞} itself. In the case of a finite repetition of the base game, we could observe that different non-isomorphic subgames result. This explains the differing strength of the subgame-consistency criterion in these two situations.

However, complimentary to any game's subgame is the remaining substructure of the original game, after the subgame has been removed (and been replaced by a pay-off vector according to a feasible solution of the subgame). These truncated versions of the original game represent substructures of the game which are not subgames in the proper sense, but nevertheless are formally well-defined games. Moreover, such truncated games play a role in the decision calculus of any player who employs backward induction. Not surprisingly then, a subgame-perfect equilibrium not only induces an equilibrium in any subgame, but also an equilibrium in any truncation of the game that results from replacing a subgame by the pay-off vector of the equilibrium induced in this subgame.

This additional consistency property is referred to as truncation consistency (see Harsanyi and Selten, 1988) and implies that, as far as the subgame-perfect solution concept is concerned, the analysis of the full game can be equivalently replaced by the analysis of its subgames and the corresponding truncations. It is easily conceivable then that isomorphisms occur between subgames and truncations which should be taken into account when defining the criterion of subgame-consistency.

Consider the following example: Suppose G is a base game with two equilibria, e_1 and e_2. Now consider the game G^2 that consists of playing G twice. G^2 then has four subgame-perfect and subgame-consistent equilibria: (e_1,e_1), (e_1,e_2), (e_2,e_1) and (e_2,e_2). This follows from Lemma 1 which only considers subgames when defining the isomorphism relation underlying subgame-consistency. However, observe that in G^2 the subgame, which results after the first round of playing G, is isomorphic to the truncation that results from replacing the second round by an equilibrium pay-off of G!

If we were to require that in those two isomorphic games the same solution is played, we would rule out the equilibria (e_1,e_2) and (e_2,e_1) from further considerations. This indicates that, by considering isomorphisms, relations between subgames and truncations the condition of subgame-consistency would be strengthened. Inclusion of such an isomorphism is indeed justified by the above observation that the analysis of the full game is completely and equivalently replaceable by an analysis of subgames and corresponding truncations. If one rationally believes that in a subgame a certain solution applies, one should also accept this solution to be played if the substitution leads to a game isomorphic to the substituted one.

We will now show that a strict application of this consistency principle leads to a characterization of stationary equilibria in finitely and in infinitely repeated games.

Let us first refine our consistency notion (see Definition 3) to include isomorphisms between subgames and truncations. Let us call any subgame or truncation a game-substructure and define

Definition 4[2]**.** Two game-substructures of G^T ($T \leq \infty$) are called isomorphic if they represent the same number of repetitions of the base game G.

and

Definition 5. A solution function is consistent if it prescribes the same behavior in isomorphic substructures.

Note that the definition above not only includes considerations of isomorphisms between subgames and the corresponding truncations (as in the simple example above) but also isomorphisms between any subgame and any truncation that can arise. The implied isomorphism relation with respect to the pay-off structures again simply takes into account that truncating a repeated game and substituting its continuation by an equilibrium pay-off of the continued subgame results in a finitely repeated game with the following pay-off structure: All pay-off vectors that are determined by a play in this finite game are translated in the same way by adding the constant solution pay-off vector of the continuation subgame to all of them. Such an operation should not have any effect on the solution of a game. As before, the number of players and strategy set coincide.

Lemma 4. The set of consistent equilibria of G^∞ coincides with the set of subgame-consistent equilibria of G^∞ which is given by all stationary equilibria of G^∞.

The reason for this result is rather obvious: Truncations of G^∞ at t and τ respectively are not isomorphic whenever $t \neq \tau$. Hence no further restrictions are imposed if the notion of subgame-consistency is strengthened.

Lemma 5. The set of consistent equilibria of G^T ($T < \infty$) is strictly smaller than the set of subgame-consistent equilibria of G^T and consists of precisely of all stationary equilibria of G^T.

Thus, consistency characterizes stationary equilibria in repeated games irrespectively of whether a game is finitely or infinitely often repeated. The intuition behind Lemma 5 is that to any subgame of length t there exists an isomorphic truncation of length t, both requiring the same play. This is,

2) Note, that this means: There exists an isomorphism which maps one substructure into the other (see also the remarks in front of Definition 2).

the subgame that represents the T-th play of G requires the same behavior as the isomorphic truncation representing the first play of G. Next the subgame consisting of the last two repetitions of G is isomorphic to the truncation consisting of the first two plays of G etc..

Our findings so far can be summarized as follows: First, as our strong notion of consistency characterizes stationarity of an equilibrium point in a repeated game, we have shown that the frequently observed ad hoc imposition of stationarity implies profound restrictions on individual behavior in terms of consistency. In particular, in the case of a finitely repeated game these restrictions go beyond subgame-consistency. Secondly, we have shed light on some apparent discontinuities in the behavior of the solution sets if one considers an infinitely repeated game as the limit game of finitely repeated ones. In particular, we have shown that they result from different underlying behavioral assumptions. This disappears if one postulates consistent behavior.

Usually the main philosophical justification for analyzing games with infinite plays (which cannot occur in real life) is that infinite games can serve as good approximations of large but finite games which avoid arbitrary assumptions about a specific upper bound for the possible length of plays. Discontinuities in the solution behavior when changing from large but finite games to infinite games therefore question the basic justification for considering infinite games. In this sense, consistency offers itself as an intuitively convincing but strong rationality condition which can guarantee a sound and justified use of infinite games as approximations of real life situations with possibly long but finite plays.

5. Conclusions

The preceding sections have shown that the crucial conceptual basis of Folk Theorems is the notion of a history-dependent strategy for a subgame of a supergame. In a sense, this is the most general form which a strategy for a subgame can take as it embodies a global view of the full game that is brought to bear locally in any subgame. Those strategies allow players to implement delicate punishment modes which account for the multitude of equilibrium behavior expressed by 'Folk Theorems'. One might argue that the enormous complexity of such strategies casts some doubt on their behavioral relevance. This motivated our look at the implications of a conceptually simpler notion of 'strategy' for a subgame, namely history-independent strategies. Those strategies are, in a sense, the least general for a subgame since they completely abstract away the context (i.e. full game) of the subgame. We have supported this view by arguments that led to the concept of subgame-consistent behavior (Selten, 1973, Selten and Harsanyi, 1988). We showed that the subgame-consistent equilibria of supergames G^{∞} are stationary; yet for finitely repeated games, G^T, $T < \infty$, subgame-consistency does not imply stationarity. However, a unified treatment of finitely and infinitely repeated games is obtained if subgame-consistency is strengthened to consistency. The behavioral assumptions underlying consistency are very strong, yet they lead to a characterization of stationarity of equilibria which is thereby shown to also be a very strong requirement.

In summary, it is of great importance for theoretical analysis whether behavior in a repeated game depends upon post decisions or not. Furthermore, one always has to clarify whether one is trying to investigate how rational players with unrestricted analytical capabilities strategically interact in time or whether one is more interested in explaining how common people behave in such situations. Whether the postulate of history-dependent or history-independent strategies is a good behavioral assumption certainly depends upon the applied context from which a game is abstracted. We have commented here on the implications of the most and least general strategy-specifications. Alas, the "truth" may frequently lie in between: behavior might best be described by strategies that condition themselves on some aspects of past behavior only, e.g. the most recent past, but not the entire course of past events. Unfortunately, the truth may also imply that people do not satisfy even the most basic requirements of individual rationality as in, for instance, the subgame-perfect equilibrium property.

6. References

Aumann, R.J. (1981): "Survey on repeated games". Essay in Game Theory and Mathematical Economics in Honor of Oskar Morgenstern, Bibliografisches Institut Mannheim: 11-22.

Aumann, R.J and L.S. Shapley (1976): Long term competition- a game theoretical analysis, Mimeo, Hebrew University.

Benoit, J.P. and V. Krishna (1985): "Finitely repeated games". Econometrica 53:1007-1029.

Coase, R.H. (1960): "The problem of social cost". Journal of Law and Economics 3:1-44.

Damme v., E. (1987): Stability and perfection of Nash-equilibria. Springer Verlag, Berlin etc..

Friedman, J.W. (1985): "Trigger strategy equilibria in finite horizon supergames". Journal of Economic Theory 35:390-398.

Fudenberg, D. and E. Maskin (1986): "The Folk Theorem in repeated games with discounting and incomplete information". Econometrica 54:533-554.

Harsanyi, J.C. and R. Selten (1988): A general theory of equilibrium selection in Games.To appear.

Hellwig, M. and W. Leininger (1987): "On the existence of subgame-perfect equilibrium in infiniteaction games of perfect information". Journal of Economic Theory 43:55-75.

Kreps, D.M. and R. Wilson (1982): "Sequential equilibria". Econometrica 50:863-894.

Rubinstein, A. (1978): "Equilibrium in supergames with the overtaking criterion". Journal of Economic Theory 21:1-9.

Selten, R. (1965): "Spieltheoretische Behandlung eines Oligopols mit Nachfrageträgheit". Zeitschrift für die gesamte Staatswissenschaft 12:301-324.

Selten, R. (1973): "A simple model of imperfect competition, where 4 are few and 6 are many". International Journal of Game Theory 2:141-201.

Selten, R. (1975): "Reexamination of the perfectness concept for equilibrium points in extensive games". International Journal of Game Theory 4:25-55.

IT'S NOT WHAT YOU KNOW, IT'S WHO YOU PLAY[1]

Joel Sobel

I. Introduction

Game Theory has little to say about what makes a player successful. A game specifies a set of players, a strategy set for each player, and payoffs. The analysis of a game does not explain why one individual assuming the role of a particular player should do better than another. But we often say that some people are better managers than others, that some people are better negotiators, etc. At a shallow level, there is an easy explanation. Certain tasks, including performance in strategic settings, require specific skills. People who have those skills perform the tasks better than those who do not.

This paper investigates the idea that differential ability to play a game may be determined by the number of strategies in the game that an individual can effectively use. I have little to say about how the differential ability arises. Rather I assume that some individuals have access to a different subset of the (theoretically) available full strategy space. One special case of this assumption arises when individuals may differ in their access to information.

Within this framework it is tempting to say that a better manager is someone who has a large set of available strategies. Better players are those who can do more. It is not surprising that this is not true in general. The inability to make a commitment not to use a potentially dangerous strategy can lower payoffs.

Section II discusses zero-sum games. For these games it is possible to say unambiguously that more strategies are better than less. However, even for zero-sum games it is not possible to completely order the subsets of all possible strategies. One subset of the strategy space may be better than another against one opponent, but worse against another opponent. In order to know who the better player is, you must specify who the opponent is.

[1]I thank Vince Crawford, Ron Harstad, Mark Machina, Max Stinchcombe, and Glenn Sueyoshi for helpful conversations. I am grateful to NSF and the Sloan Foundation for financial support.

Section III presents examples that demonstrate how losing the ability to use a strategy may have any effect on the payoff of a player or its opponent. These examples are simple.

In Section IV I study a particular example that is designed to show the importance of the context in which a game is played. Two players must coordinate their actions with each other and with a target. A player may be made worse off if it gains information about the target that its opponent does not have. In this setting there are incentives for becoming a better player (in the sense of knowing more) may exist only if the other player's information improves at the same time. There are implications for this analysis in settings where individuals from different cultural backgrounds interact. Different cultures lead to different ways of solving problems. Different ways of solving problems may lead to different solutions or, if the problem solvers are fallible, different kinds of mistakes. Identifying situations in which different ways of solving problems systematically lead to poor outcomes may provide a better understanding of why disagreements arise.

II. Zero-Sum Games

Game theory provides a complete and relatively noncontroversial theory of zero-sum games. Nevertheless, many zero-sum games remain interesting as diversions to both participants and spectators. Although game theory tells us the value of particular games, there still seems to be uncertainty about who will win a soccer match or chess championship. On the other hand, there is often a general consensus about whether one side or the other has a better chance of winning a contest that theoretically has a zero value. How can this be? Several answers are fully consistent with the theory. Players may be using mixed strategies in the equilibrium. For example, the value of the game may be strictly between one and negative one, but payoffs either 1 (victory) or -1 (defeat). Any play of the game will have a definite winner but there will be uncertainty ex ante about who that will be. Second, the game may depend on something that is not known by observers (which horse in the race has a bad leg) or even the participants. The outcome of the game will therefore be a random variable. Third, what we think of as a game may in fact be the combination of many related, smaller games. We may be quite sure that Steffi Graf will win a tennis match against a lesser opponent without believing that she will win every point she serves. A complicated allocation of effort problem makes the outcome of one point in a tennis match depend on the current score.

These explanations do not provide a satisfactory answer to why I have little chance to win a chess game against Gary Kasparov. There is no need, because the game is played under perfect information, to use random strategies. Uncertainty plays no role in the game. Kasparov will win the chess game because he is a better chess player than I. But what does this mean? The practical problem with chess is that the strategy space is so large that no one can find an optimal strategy. Chess champions are better able to recognize winning strategies than the rest of us. Yet all chess players use dominated strategies because they do not have the mental capacity and the time to find better ones. Game theory does not have anything to say about how human players would or should play this game. Indeed, without enriching the standard theory, there is no way to explain why imposing a limit on the total time a player takes to move should affect outcomes. In this section I propose one way to model differences in ability and investigate some implications for zero-sum games. My aim is to be able distinguish winners from losers.

In this section I consider only two-player zero-sum games in which each player has a finite set of pure strategies. A game is described by a finite matrix M that represents the payoffs of the row player. When I present examples, it is convenient to assume that M is skew symmetric. That is, that M is a square matrix and if M' is the transpose of M, then $M + M' = 0$. These games describe fair contests. In principal, players have access to the same set of strategies and payoffs do not depend on which player chooses rows and which chooses columns. The value of skew-symmetric games is zero. When examining skew-symmetric games one can talk about a better player, rather than a better row player or a better column player.

I assume that the players differ in their ability to play the game. There are many ways to interpret ability differences. Perhaps the simplest is to assume that players are able to use only a subset of the available strategies. A player may not know that other strategies exist or may not be skilled or knowledgeable enough to use all of the strategies. The theoretical description of the game as given by the matrix M, which I call the original game, will be different from the game with strategy limitations that is actually played, which I call the real game. If two players have different sets of available strategies, then even if the original game is skew-symmetric, there is no reason to expect the real game to have zero value.

A special case of this assumption is that players differ in the amount of information that they have. Assume that the state of the world determines the payoffs of the game. A player who knows the state of the world can select a different action in every state. A player who knows nothing must play the same action in every state. In general, requiring that a strategy be measurable with respect to a given partition on the set of available states limits the set of strategies. The finer the partition on information, the more strategies a player has available.

In this section differential information is the only reason I give to explain why some players have more strategies than others. I do not explain why one player may have access to more information than another. One other explanation for differences in available strategies appears in the literature. Several authors (for example, Ben-Porath [1986], Gilboa and Samet [1989], and Kalai [1987], see Kalai's paper for extensive references) have analyzed the effect of limiting players to strategies that can be implemented by finite automata. These papers study infinitely repeated games, where the entire set of strategies in the original game cannot be described using finite automata. In finite matrix games a different notion of what determines a strategy's complexity is needed. In chess, players do not construct full contingent plans. Rather they have only a general idea of what they will do three or four moves in the future. In order to deal with the difficulty of specifying any strategy, it may be necessary to deal explicitly with extensive-form representations of the game and construct models of what it means to use partial plans. My goal and accomplishment is more modest.

This section studies how the size of a player's strategy set influences the player's success. For zero-sum games it is a simple consequence of the maxmin theorem that increasing the set of strategies available to a player increases the player's value (and hence lowers the value of the game to the opponent).[2] The reason that increasing the set of strategies is good for a player in a zero-sum game is that each player can compute its best strategy by solving an optimization problem that does not depend on its opponent's strategy choice: It maximizes its security level. Since increasing the number of available strategies cannot reduce the value of the optimization problem, a player can only gain when it has more strategies available. In a zero-sum game, one player's gain is the other player's loss, so enlarging one player's strategy set cannot help the other player. The same argument guarantees that more information (in the Blackwell-Girshick [1954] sense) is better than less: If a player's information improves, then it is still able to do what it did before. Consequently, its security level cannot go down.

It follows that the strength of players in zero-sum skew-symmetric games can be partially ordered. One player is better than another if that player has more strategies. The question remains whether a complete order on player's strategy spaces could exist.

It is easy to give an example of a game in which the abilities of players are not completely ordered. Consider Example 1, which is the child's game "stone, paper, scissors." If one player can only use Up (or Left), another player Middle (or Center), and a third only Down (or Right), then it is apparent that the first player always beats the third player, the second player always beats the first player, and the third player always beats the second player. The skills of the players are not ordered. You cannot identify the better player unless you know who the player's opponent is.

---

[2]Selten [1960, 1964] introduced the postulate (on solutions to games) of strategic monotonicity, which requires that a player's payoff does not increase when it is forbidden to one of its moves. He shows that the value of two-player, zero-sum games satisfies the property of strategic monotoncity.

EXAMPLE 1

	Left	Center	Right
Up	0	-1	1
Middle	1	0	-1
Down	-1	1	0

It is also possible to construct a variation of Example 1 in which the players differ only in their access to information. The game is equally likely to take on one of the three forms below.

GAME A

	Left	Center	Right
Up	0	6	3
Middle	-6	0	3
Down	-3	-3	0

GAME B

	Left	Center	Right
Up	0	-3	-3
Middle	3	0	6
Down	3	-6	0

GAME C

	Left	Center	Right
Up	0	3	-6
Middle	-3	0	-3
Down	6	3	0

information, the game obtained when players of different types are paired is no longer skew-symmetric. There is no reason for the value of the game between a player of type a and a player of type b, for example, to have zero value. Indeed, this game has a positive value to the type b player. When the original game is equally likely to be A, B, or C, each player has twenty seven pure strategies: A choice of which of three actions to take to in each of three contingencies. Each type of player has access to only nine strategies in the real game. To analyze a game between players of different types, recall that a player has a dominant strategy when it knows which game it is playing. I can restrict attention to a three by three game in which the strategies of a player is the action taken when uninformed about the payoff matrix. When a type a player choose rows while a type b player selects columns, the expected payoff matrix is:

a versus b

	Left	Center	Right
Up	-1	0	1
Middle	-2	-1	2
Down	0	-1	2

The three strategies of player a represent how it plays when it does not receive information. If the player knows that A describes payoffs, then it plays the dominant strategy of Up. I obtained the payoffs in the a versus b matrix by averaging payoffs in games A, B, and C under the assumption that a would always play Up given A and that b would always play Center given B. Similarly I can construct payoff matrices for the games that arise when a type b player chooses rows and a type c player picks columns, and when a type c player chooses rows while a type a player picks columns.

b versus c

	Left	Center	Right
Up	-1	0	1
Middle	-2	-1	2
Down	0	-1	2

c versus a

	Left	Center	Right
Up	-2	0	1
Middle	-1	1	0
Down	-2	2	1

It is a simple exercise to compute the optimal strategies in these games. A type a player selects its dominant strategy (Up) when it is informed. Otherwise it randomizes equally over its other two strategies whether it is playing a type b or a type c opponent. Similar reasoning determines the optimal strategies for the other types of player. b expects to win on average when it plays a; the value of the a versus b game is -1/2 (for type a). However, type a expects to win against type c, and type c expects to beat type b. The value of information in the example depends on who you can use it against.

Here is one way to understand the example. When type a is not informed, it can infer that it is not a good idea to play Up. So type a avoids a large loss in the event that it must play Game C. As a result, type c's information is less valuable against type a than against type b.

The reason for treating zero-sum games separately is that there is little hope for general results in non-zero sum settings. In the next section I briefly describe simple examples that show that restricting one player's set of strategies in a two-player game may be valuable to both, neither, or precisely one of the players. I should point out that the observation that a larger strategy set makes for a better player does not even hold in zero-sum games if players make mistakes. For example, one way in which to interpret skill is the ability to implement a strategy successfully. Players could differ both in the number of strategies from the original game that they can use, and the probability that they succeed in using a particular strategy when they attempt to do so. One possible assumption is when a player fails to implement a strategy then it plays a randomly selected alternative strategy. A player who is unable to play a strictly dominated strategy in a zero-sum game even by mistake may be better off than a player who can use the strategy, is smart enough never to do so consciously, but occasionally makes mistakes. Perhaps this explains why in some games people talk about a category of mistakes that could only be made by experienced players.

III. Examples

In this section I describe some simple examples which demonstrate that restricting the number of strategies available to a player can have any effect on the payoffs of that player or on the other players.

The results of the previous section show that in zero-sum games losing a strategy hurts the player and helps its opponent. Losing a strategy can hurt both players if it makes it impossible to attain a high payoff. Example 2 is the simplest such example. If Player 2 cannot use Right, then both players receive the payoff one. Otherwise they both receive two. Examples 3 and 4 are situations in which losing the ability to play a dominant strategy is advantageous to player one. In

Example 3 when both strategies are available the equilibrium outcome is (D,L); both players obtain the payoff two. If player one loses the ability to use D, then the equilibrium is (U,R) and both players' payoff increases. Example 4 is similar in that when player one loses D, the equilibrium moves from (D,L) to (U,R). In this example player two is made worse off when player one's strategy set becomes smaller.

EXAMPLE 2

Left	Right
1,1	2,2

EXAMPLE 3

	Left	Right
U	1,3	3,4
D	2,2	4,1

EXAMPLE 4

	Left	Right
U	1,1	1,2
D	2,4	4,3

The examples show that a player cannot expect adding more strategies to lead to better results. What makes a player better in these games depends upon the game. Similar examples can be constructed when the players differ in their access to information: Better information doed not guarantee that a player will receive a higher payoff in a non-zero sum game.

Gilboa and Samet [1989] obtain a striking version of this result in a repeated game setting. They show that if only one player in a two-player repeated game has access to just those strategies that can be implemented by a finite autonoma, then it can approximate the highest payoff consistent with the other player attaining its pure strategy minmax value. Hence they identified a class of games where limits on strategies is strategically advantageous. This type of weakness is not an advantage in zero-sum games as Ben-Porath [1986] has shown.

Neyman [1989] provides a framework where increasing the size of the strategy set is in a player's advantage. Neyman embeds a game into a larger game in which player one's strategy set depends on the state of nature. Suppose that player one's strategy space is strictly larger in one state than in another, but that opponents cannot distinguish these states. Neyman shows that player one does at least as well when it has more available strategies. If opponents take into account the possibility that you have additional strategies, then it is better to have them.

Neyman's result suggests that it it cannot hurt to acquire another strategy provided your opponent does not know about it.

IV. A Coordination Game

I study a two-player coordination game. Player i selects an action x_i and has preferences given by

$$U_i(x_i, x_j, T) = -a_i(x_i - x_j)^2 - b_i(x_i - T)^2,$$

where $i = 1$ or 2, player $j \neq i$ takes the action x_j, and T is a random variable. I assume that $a_i + b_i = 1$ and that a_i and b_i are nonnegative.

A player wants to minimize the sum of two losses. First it wishes to choose a number that is as close as possible to the number selected by its opponent. Second, it wants to be as close as possible to a target T. The game is motivated by a game described in Rubinstein [1988] where two prisoners wish to plan an escape. They cannot communicate and the optimal date for the escape is months in the future. Rubinstein does not specify preferences.

If $b_i = 0$ for $i = 1$ and 2, then players would only wish to coordinate with each other. In a sense, this game is more difficult to solve than the one in which coordinating on T is also important. If preferences are independent of T, then it the game has multiple Nash Equilibria. Since a player's optimal response will always be a pure strategy (equal to the expected value of the other player's strategy), all Nash Equilibria involve perfect coordination. However, any strategy of the form $(x_1, x_2) = (x, x)$ will be an equilibrium. Without any outside knowledge, it may be difficult to focus on a particular equilibrium. If either b_1 or b_2 is strictly positive and if T is common knowledge, then the game has a unique Nash Equilibrium in which $x_i = T$ for both i. To see this, note that a player i's optimal response (x_i) to player j's strategy must be a pure strategy that satisfies

$$x_i = a_i\bar{x}_j + b_iT, \tag{1}$$

where $\bar{x}_j$ is the expected value of x_j. Using (1) for $i = 1$ and $j = 2$ and for $i = 2$ and $j = 1$ demonstrates that $x_i = x_j = 0$. One effect of including the target T in preferences is to make a particular equilibrium focal.

I am interested in the game that arises when the value of T is not common knowledge. In Rubinstein's game the failure of common knowledge arises because the prisoners must count the number of days until the optimal time for an escape and they might make a mistake (or worry that the other prisoner might make a mistake). I also wish to discuss the case in which players just figure out the value of T, but because of limited reasoning abilities they may not arrive at a correct answer. The purpose of the example is to demonstrate that it is not necessarily advantageous to become better at estimating T; the ability to count the number of days until it is optimal to escape is less valuable when the other player lacks this talent.

I postulate a simple error structure. I assume that players have a common prior on T. For notational convenience I assume that the expected value of T is zero. Players either learn the value of T exactly or they learn nothing. Four events are possible: both players learn the value of T; only player one learns the value; only player two learns the value; and neither player learns the value. I want to allow the possibility that there is correlation over when the players learn; I denote the probabilities of the four possible events by, respectively, B (both), P_1 (only player one), P_2 (only player two), and N (neither).

Here is a concrete way to think about the error structure. Assume that T is the realization of an exponentially distributed random variable. Players must compute T, but they have a limited amount of time and they make mistakes. If they succeed in computing T, then they are certain of its value. Otherwise, they learn nothing. Sobel [1989] discusses ways in which to perform this type of task. The form of the error structure keeps computations relatively straightforward without destroying the message of this section. It has been used recently in other applications (for example Hausch [1988]).

It is not difficult to compute the unique Nash Equilibrium of this game. Player i's strategy consists of a function $x_i(t)$ that represents what it does when it learns that the realized value of T is t, and a number z_i, which is what it does when it learns nothing about T. As in the case where T is common knowledge, the form of the players' utility functions guarantees that mixed strategies are not used in equilibrium. In order to be an optimal response the strategies must satisfy, for $i = 1$ or 2, $i \neq j$:

$$x_i(t) = a_i\{[B/(B+P_i)]x_j(t) + [P_i/(B+P_i)]z_j\} + b_i t \tag{2}$$

and

$$z_i = a_i\{[P_j/(P_j+N)]\bar{x}_j + [N/(P_j+N)]z_j\}, \tag{3}$$

where $\bar{x}_i$ is the expected value of $x_i(t)$. The ratios of probabilities that appear in these formulas are conditional probabilities. For example, if player one learns that the realized value of T is equal to t, then it believes that the probability that player two learned the same thing is $B/(B+P_1)$. When player one has this information, it expects player two to pick $x_2(t)$ if informed and z_2 otherwise. b_i does not appear in the expression for z_i since the expected value of T is assumed to be equal to zero. Solving equations (2) and (3) is not difficult. Taking the expected value in the first equation (for both $i = 1$ and 2) yields a system of four independent homogeneous linear equations for the four variables z_1, z_2, $\bar{x}_1$, and $\bar{x}_2$. Hence these quantities must all equal zero. Substituting $z_1 = z_2 = 0$ back into (2) and solving yields

$$x_i\ (t) = c_i\ t \tag{4}$$

where

$$c_i = \{a_i\ [B/(B+P_i\ )]b_j\ +\ b_i\}/\{1\ -\ a_1[(B/(B+P_1)]a_2[(B/(B+P_2)]\} \tag{5}$$

An easy computation shows that c_i is between zero and one, and strictly between zero and one if a_i, b_i, and P_i are strictly positive. Hence an informed player shades its strategy. Rather than setting x_i equal to the known value of the target, it reduces its guess to take account of the possibility that the other player is not informed.

Equations (4) and (5) can be used to find the expected utility of a player of this game. Player i's expected payoff is the sum of four terms, depending upon who has information. If both players have information, then player i's payoff is

$$-\mathrm{var}(T)[a_i\ (c_i\ -\ c_j\ )^2 + b_i\ (1 - c_i\ )^2], \tag{6}$$

where var(T) is the variance of T. The expression reflects that a player loses utility from two sources: the failure to coordinate with the other player and the failure to match the target. Similar expressions can be found when exactly only player i is informed:

$$-\mathrm{var}(T)[a_i\ c_i^2\ +\ b_i\ (1-c_i\ )^2], \tag{7}$$

when only player j is informed:

$$-\mathrm{var}(T)[a_i\ c_j^2\ +\ b_i\ ], \tag{8}$$

or when neither player is informed:

$$-\mathrm{var}(T)b_i\ . \tag{9}$$

The expected utility of player i is the weighted sum of (6) through (8) with the terms being weighted by, respectively, B, P_i, P_j, and N.

In the example there is one improvement of information that unambiguously leads to higher expected payoffs to both players. It is apparent from expressions (6)-(9)

that reducing the variance of T is beneficial to both players. Indeed, reducing var(T) increases a player's welfare no matter which subset of players acquires information. There are other ways in which the information structure may change. Two general kinds of change are of interest to me. In one change, the information of the players improves in a way that increases correlation. In the other, one player acquires more information independently from the other.

There are different ways to think about the comparative statics exercise. I concentrate on specific changes; other approaches yield the same qualitative results. To simplify the formulas I will assume that the players have identical preferences and symmetric information. That is, I will consider the case in which $a_1 = a_2$, $b_1 = b_2$, and $P_1 = P_2$. First, consider a change that increases the correlated information players have. This type of change occurs if B increases by ε and N decreases by ε. Such a change holds constant the probability that either player is the only one to acquire information, while increasing the probability that both players acquire information at the same time. Simple computations, using (6), (7), (8), (9), and some algebraic manipulation, show that the change in the utility of player i from this change is

$$\mathrm{var}(T)[bc^2 (3-2c)]. \tag{10}$$

(10) is strictly positive. That is, a correlated improvement in information increases the players' equilibrium expected utility.

Another kind of improvement in information is a change that benefits precisely one of the players. One way to imagine this kind of change is to increase P_1 by ε while reducing N by ε and holding the other probabilities fixed. Manipulation reveals that the derivative of player i's equilibrium expected utility with respect to this type of change is

$$\mathrm{var}(T)[1 - 2ap + a(2a^2 - 1)p^2]b^2 (1-ap)^{-3} (1+ap)^{-1}, \tag{11}$$

where $p = B/(B+P_1)$ is the conditional probability that both players are informed given that player one is. Since $1 - 2ap + a(2a^2 - 1)p^2$ is positive when a and p are close to zero but negative when these numbers are close to one, an independent increase in one player's information has an ambiguous effect on utility.

One reason that an independent increase in information can be detrimental to player one is that it raises the probability that only player one is informed while lowering the probability that no one is informed. Neglecting the effect that this change has on equilibrium strategies, it may have a negative effect on player one's utility: Player one gains the ability to better approximate the target, but becomes less coordinated with player two in the process. The direct effect of a correlated increase in information (given by (6) minus (9)) is always positive in the symmetric case. This fact turns out to be enough to show that the total effect of the change also leads to an increase in utility.

There are other ways in which to describe changes in the information structure that increase the correlation between the agents' information. Epstein and Tanney [1980] introduce the idea of correlation-increasing transformations. In my context this type of transformation would increase B while holding fixed the probabilities of obtaining information $(B + P_i)$. This change always increases a player's expected utility (an increase in B compensated by equal decreases in P_1 and P_2 and an equal increase in N changes a player's expected utility by $var(T)(2ac^3)$). Considering this type of change is important if the players could somehow choose among alternative information structures that leave unchanged the total probability that they acquire information. It is not the way to ask whether more information is preferred to less.

The example has a simple message. If coordination with your opponent is important, correlated information is more valuable than independent information. A good player in this situation is likely to be one who is known to have information that is similar to its opponent's information. One cannot identify a good player without knowing the context in which the game is played.

The failure to know T may be the result of some limitations on the abilities of the players. The nature of the limitations could depend upon the training or other characteristic of the player. There many be many different ways in which to perform the task of collecting information about T. Different players may collect information in different ways. The success of a player in the game will depend not only on how it obtains information, but on the technology used by its opponent. The example points out circumstances where a shared technology for solving a problem may be advantageous to both players. In situations where players may make mistakes and have access to different technologies for solving problems, it is therefore important to analyze the nature of the game and the technology of the other players.

Reference

Elchanan Ben-Porath, "Repeated Games with Bounded Complexity," manuscript, 1986.

David Blackwell and M. A. Girshick, Theory of Games and Statistical Decisions, New York: Dover, 1954.

Larry G. Epstein and Stephen M. Tanny, "Increasing generalized correlation: A Definition and Some Economic Consequences," *Canadian Journal of Economics*, vol. 13, February 1980, 16-34.

Itzak Gilboa and Dov Samet, "Bounded Versus Unbounded Rationality: The Strength of Weakness," *Games and Economic Behavior*, vol. 1, June 1989, 170-90.

Donald B. Hausch, "A Common Value Auction Model with Endogenous Entry and Information Acquisition," manuscript, 1988.

Ehud Kalai, "Artificial Intelligence and Strategic Complexity in Repeated Games," Northwestern University Discussion Paper, October 1987.

Abraham Neyman, "Games without Common Knowledge," manuscript 1989.

Ariel Rubinstein, "Comments on the Interpretation of Game Theory," ICERD Discussion Paper TE/88/181, August 1988.

Reinhard Selten, "Bewertung Strategischer Spiele," Zeitschrift für die gesamte Staatswissenschaft, 116. Band, 2. Heft, 1960, 221-82.

Reinhard Selten, "Valuation of n-Person Games," chapter 27 in Advances in Game Theory edited by M. Dresher, L. S. Shapley, and A. W. Tucker, Princeton: Princeton University Press, 1964, 577-626.

Joel Sobel, "How to Count to One Thousand," manuscript, 1989.

GAME THEORY AND THE SOCIAL CONTRACT*

by

Ken Binmore

The skill in making and maintaining Commonwealths, consisteth in certain rules, as doth Arithmetique and Geometry, not (as Tennis-play) on Practise only.

Hobbes, Leviathan

Abstract: The ST/ICERD blue series contains an earlier discussion paper with the same title called "Part I". This sought to defend the Rawlsian position on social contract issues by an appeal to Rubinstein-type bargaining arguments employed "in the original position". The current paper is *not* "Part II" of this earlier paper, but a re-appraisal of the whole problem. The re-appraisal continues to defend a version of the Rawlsian position against Harsanyi's utilitarian alternative, but no longer makes the error of emphasizing technical matters from the theory of non-cooperative bargaining models. It is necessary that the right ideas be borrowed from this theory, but the crucial issue is that "commitment" be properly modeled. The bulk of the paper is concerned with justifying the manner in which the term "properly modeled" is to be interpreted. The treatment forces close attention to be given to questions of the inter-personal comparison of utilities. This treatment may be of some independent interest, since it leads to some unifying considerations for cooperative bargaining solutions. No background in game theory or knowledge of mathematics beyond elementary algebra is required to read the paper. On the other hand, the arguments do often hinge on points which need to be thought about with some precision.

1. Introduction

A *game* is being played by a group of individuals whenever the fate of each individual in the group depends not only on his own actions but also on the actions of the other members of the group. *Game Theory* is concerned with identifying rational behavior in such situations. As such, it seems ideally suited as a "tool for the moral philosopher" (Braithwaite 1955). However, with some notable exceptions[1], it has seen little effective use in this role. Probably this is because moral philosophers seek

*The original version of this paper (Binmore 1984) was prepared for the *Public Choice Institute* held in Halifax, Nova Scotia, in the summer of 1984. The current much revised version arises from a presentation at the *Game Theory Conference* held in Bergamo, Italy, in the summer of 1987. I am grateful to B. Barry, M. Perlman and L. Sacconi for helpful comments.

more from game theorists than game theory is yet able to offer. On the other hand, game theorists are not entirely empty-handed and applications in moral philosophy are particularly attractive because it is an area which allows a ready justification of various simplifying assumptions to which game theorists have been forced in order to make progress.

Game theories are normally classified as being *cooperative* or *non-cooperative* in character. In cooperative theories, it is supposed that *binding* contracts can be agreed before the play of the game and attention concentrates on identifying an optimal contract (or contracts) given certain axioms thought to embody appropriate principles of rational *collective* choice. This approach abstracts away two questions. What makes the contracts binding? Why do individuals adhere to the principles said to be collectively rational? Answers to these questions are increasingly seen as crucial to a game-theoretic analysis. For this reason non-cooperative theories nowadays command much more attention among game theorists. Such theories attempt to justify behavior at a very much more primitive level than cooperative theories. Assumptions are made only about rational choice by *individuals*. The behavior of a coalition then has to be explained entirely in terms of the individual choices of its members. The ultimate goal is to specify all relevant actions, which will include what players say and do in achieving cooperative outcomes, as possible moves in a game governed by a set of *formal rules*. Players are then assumed to make *individually* optimal decisions given the constraints imposed by these rules and their knowledge or expectations about the decisions of other players. The aim of this reductionist approach is, not to displace cooperative theory, but to provide its conclusions with strategically sound underpinnings in those cases for which this is possible. (See the introduction to Binmore/Dasgupta (1987) for a lengthier exposition of this point of view.)

The purpose of this paper is to apply simple ideas from non-cooperative game theory to *social contract* issues. Barker (1947) observes that the social contract theories of Locke, Hume and Rousseau may be criticized as too mechanical, juristic and *a priori* to serve as an adequate explanation of political society. What is attempted here is unashamedly mechanical, juristic and *a priori*, but it is not intended as an *explanation* of political society. It is a parable about how things *would be* under certain ideal circumstances. The exercise is therefore rhetorical rather than scientific. The fact that the paper contains discussion of how certain matters *are* should not be allowed to obscure this point. The purpose of such discussion is simply to make it clear why certain *idealizing* assumptions are thought to be interesting. Nor is much borrowed from the work of Locke and his successors. In fact, the paper is often quite alien to their lines of thought. Chiefly this is because the paper seeks to restrict its reliance on assumptions about "natural law" to a minimum. In speaking of the number systems, Kronecker remarked that "God created the natural numbers: the rest is the work of man." Here the word "natural" is being used to indicate something which is outside the power of man to alter. Whatever thoughts a man might have, Kronecker believed that the natural numbers will remain as they must be. But what is there underlying the conduct of human affairs that could not conceivably have been otherwise? Presumably only what is programmed in our genes. Perhaps, for example, a respect for the property of others is indeed built into our hardware, but there certainly seems room for doubt on such questions. In any case, the interest, in this context, of employing non-cooperative game theory rather than cooperative theory, lies in the fact that heroic assumptions about "natural law" are unnecessary. But this, of course, is not to say that the assumptions it does require will be found any more palatable.

Rousseau (1763) described his aspirations as follows:

> "In this enquiry, I shall always try to bring together what right permits with what interest prescribes, so that justice and utility are in no way divided."

In the approach adopted here, "what right permits" (or duty compels) is built into the rules of a *hypothetical game* played by members of society. "What interest prescribes" appears as a consequence of their strategic choices within this game. "Justice" and "utility" thereby jointly determine an equilibrium outcome of the game. The "social contract" is to be understood as the system of commonly held conventions[2] necessary to sustain this equilibrium.

Notice that ethical questions are firmly separated in this treatment from strategic questions. The former are taken account of *only* in the structure of the hypothetical game to be played. In discussing the solution of the game, it is therefore inadmissible to call upon such ethical notions as "duty" or "obligation" to explain why a social contract is honored. Such duties and obligations as are built into the social contract are honored, not because they *must* be honored, but because it is in the interest of the players to honor them. The social contract is therefore seen as operating *by consent*. Nothing *enforces* a social contract beyond the enlightened self-interest of those who regard themselves as a party to it. In particular, the players should not be supposed to be able to make commitments (i.e. binding promises) about their future conduct, unless some explicit, exogenous mechanism is available to *enforce* the commitment. Without such a mechanism, the social contract will necessarily have to be *self-policing*.

There may be doubts about the viability of such an approach. To quote Gauthier[3] (1986), "Were duty no more than interest, morals would be superfluous". The proper reply is that a social contract can serve to *coordinate* the behavior of those acting in their own enlightened self-interest, and that far more can be achieved in this simple way than is commonly supposed to be the case. Schelling (1960) and Lewis (1969) give examples which make this point very sharply, but the general contention is simply that of Hume (1789). His example of two men rowing a boat which neither can row alone is usually aptly quoted in this context.

A set of commonly understood coordinating conventions, unbuttressed by outside enforcement, might at first seem a flimsy basis on which to found a society, but it turns out that a very wide range of socially desirable behavior admits a robust explanation in terms of equilibrium play in a suitable game (see Section 2). Of course, it is also true that a wide range of socially undesirable behavior is also typically sustainable as an equilibrium in the same game. Society's problem is therefore to coordinate on a *good* equilibrium rather than a *bad* one.

An alternative objection is that, in excising ethical considerations from the analysis of the game, the baby has been thrown out with the bathwater. After all, are not ethical considerations what the debate is all about? But this would be to miss the central point of the approach. The purpose of a social contract argument is presumably to explain *why* we should see ourselves subject to ethical imperatives beyond those built into our biological hardware. Such ethical imperatives should therefore emerge as *consequences* of a social contract argument rather than being assimilated into its *premises*.

The device of a hypothetical game is found in the work of both Harsanyi (1953, 1955, 1958, 1977) and Rawls (1958, 1968, 1972). The hypothetical game[4] they consider involves the members of society agreeing on a social contract in the "original position" behind a "veil of ignorance". The veil of ignorance conceals from the parties to the agreement the actual roles, in the society being agreed to, that each will personally occupy. Justice therefore appears in the rules of the game simply as "symmetry in the original position". Harsanyi deduces a form of *utilitarianism* from this set-up. Rawls deduces an alternative theory based on his "difference principle", which is here referred to as the *maximin criterion* since it requires maximizing the welfare of the least well-off, after certain basic rights and liberties have been assured. The consensus view (e.g. Sen 1987) seems to be a recognition that Harsanyi's case is soundly argued (given his premises) but disquiet with some of the implications of his conclusion. Rawls' argument, however, is seen as less persuasive in consequence of his refusal to accept the validity of orthodox Bayesian decision theory in this context.

It is not original to draw attention to difficulties about "timing" in Rawls' treatment (e.g. Hare 1975, Calvo 1978, Howe and Roemer 1981, Rodriguez 1981). Similar criticisms can also be directed at Harsanyi. Both take an essentially *static* view of the negotiation in the "original position", no provision being made for renegotiation if discontent emerges over time. Rawls (1971, p. 162) is quite explicit in arguing that members of society should regard themselves as *committed* to the hypothetical agreement reached in the original position. Harsanyi is less forthcoming on this point, but his argument is actually equally dependent on such a commitment assumption.

In this paper, it will be argued that, without such commitment assumptions, the situation looks very different. Utilitarianism ceases to be sustainable, at least in the form proposed by Harsanyi. Instead something much closer to Rawls' position emerges from an entirely orthodox treatment. Indeed, within the utility framework used by Harsanyi (his version of the theory of extended sympathy), it will be argued that orthodox reasoning leads precisely to Rawls' maximin criterion in the two-person case.[5]

The case against commitment assumptions is made in the body of the paper. But it is worth noting at this point that the line taken on this question is very much in the spirit of what Rawls considers under the heading of "strains of commitment" (1972, p. 145, p. 176). His remarks on the difficulty in sustaining commitments, which echo those of Schelling (1960), are simply pushed to their logical conclusion. This leaves his attachment to certain commitment requirements "in the original position" looking decidedly anomalous.

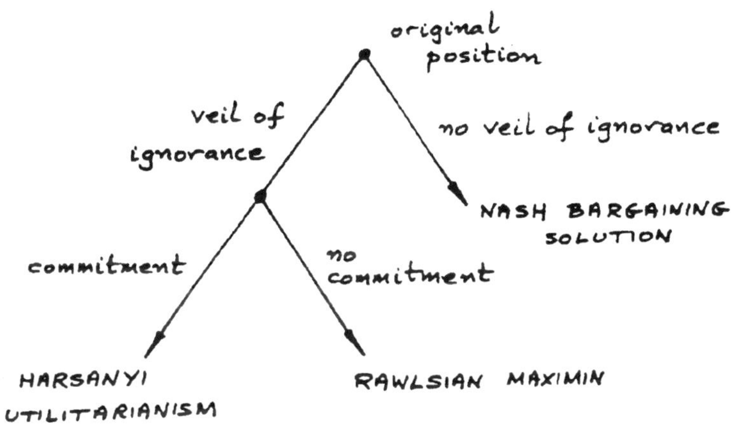

Figure 1

Fig. 1 illustrates how the various ideas mentioned in this introduction are seen as interacting. The reference, in Fig. 1, to the Nash bargaining solution (Subsection 3.1) signals that the bargaining hypothesized in the original position is to be modeled explicitly (albeit crudely). This is a contradistinction to what one might call the "ideal observer" approach of Harsanyi and Rawls.[6] The difference is not substantive as long as all of the symmetries that Harsanyi deduces from his theory of extended sympathy are available. But, if one either follows Rawls in regarding the theory of extended sympathy as providing too abstract a setting for a profitable discussion or one declines to follow Harsanyi as far as he would like to be followed, then one has to contemplate the possibility of asymmetries in the original position. It is then not always immediately obvious how an ideal observer would arbitrate the appropriate bargaining problem: nor how to characterize the consequences of a failure to agree in the original position.[7] Explicitly formulated bargaining models, on the other hand, continue to yield results. Not only this, they yield results which are free of value-judgements about "rational" collective action.[8]

It is, of course, no accident that all of the arrows in Fig. 1 point to concepts which have been characterized axiomatically by social choice theorists. Particular reference should be made to the work of D'Aspremont/Gevers (1977), Arrow (1978), Deschamp/Gevers (1978), Gevers (1979), Hammond (1976a, 1976b), Roberts (1980a, 1980b) and Sen (1970, 1976). Although such work is not usually classified as game-theoretic, there is a sense in which it can be regarded as the *cooperative* analog of the *non-cooperative* approach of this paper. Such a viewpoint allows the non-cooperative model studied in this paper to be seen as a possible aid in evaluating the significance of some of the abstract, idealizing axioms used in this branch of social choice theory.

There is also relevant work which is overtly classified as belonging to cooperative game theory. This is the literature on cooperative "solution concepts" for bargaining problems. Particular mention should be made of Raiffa (1953), Isbell (1960), Kalai and Smorodinsky (1975), Kalai (1977), Myerson (1977), and Roth (1979). The last of these references summarizes much of the earlier work. The non-cooperative approach attempted here also sheds some light on the significance of the axioms which appear in this literature and links it with the social choice work mentioned in the previous paragraph.

To be more precise, it is necessary, in carrying through the paper's program of justifying the Rawlsian maximin criterion within an orthodox game-theoretic framework, to make a very close examination of certain questions concerning the inter-personal comparison of utilities on which Harsanyi is largely silent and which Rawls mostly evades. I hope that the manner in which the paper seeks to resolve these questions will be found to be of independent interest. In any case, the methodology employed allows a fixed bargaining problem to be formulated in superficially different ways, depending on arbitrary mathematical decisions about how preferences are described. The paper's defense of the Rawlsian maximin criterion for two persons extends only to the case when utilities have a specific "extended sympathy" interpretation. But the same outcome can also be described simply in terms of the raw personal preferences of the two individuals involved. Within such a description, the outcome can be characterized in a fashion which *unifies* some of the ideas which are central to "cooparative bargaining theory". In brief, there is a sense in which the Rawlsian maximin criterion is equivalent to a hybrid "solution concept" incorporating features of both the Nash bargaining solution and of the parallel idea of a "proportional bargaining solution" (Roth, 1979, p. 78). The first of these notions excludes inter-personal comparisons of utility altogether and the second requires full inter-personal comparison. The hybrid notion allows for intermediate positions.

(In the extended sympathy framework within which the Rawlsian maximin applies, it is not necessary to commit oneself as to what intermediate position has been adopted.)

There is yet a third relevant literature: that in which social welfare questions are expressed directly in terms of the ownership of physical and economic resources. The basic idea in this work is that an allocation of resources should be deemed "equitable" or "envy-free" if no individual would wish to exchange the bundle of resources allocated to him or her with the bundle allocated to anyone else. This criterion has received a lot of attention, not least because its implementation calls for the use of the market mechanism. In criticizing Dworkin (1981) on this subject, Varian (1985) has provided a comprehensive bibliography. (See also Thomson and Varian 1984). Some very brief discussion of how this literature fits within the framework of this paper is provided in Section 4. It is hoped that devotees of the market mechanism (as other than a device for achieving efficient outcomes cheaply) will found this discouraging.

These references to the literature may arouse suspicions that this is a technical paper. However, the paper contains no theorems as such and much effort has been expended in reducing the technical expertise required to some elementary algebra. Wherever possible the discussion proceeds via examples illustrated with diagrams: and it is nearly always confined to the two-person case. Mathematicians will be impatient with the lack of rigor, but this is a cross which has to be borne. In so far as there is mathematical argument, its purpose is simply to demonstrate that the ideas presented are coherent. The aim is never to demonstrate the precise conditions under which some proposition holds. Such questions do not seem to me to be very interesting at the level of abstraction attempted in the paper. Nor is it assumed that the reader is well-informed on game-theoretic issues. All ideas from game theory are therefore treated *de novo*. However, it will certainly not be an easy paper for those with no background in the area at all.

The paper is organized as follows. Section 2 reviews the manner in which socially desirable outcomes can be sustained as equilibria in non-cooperative games and suggests that commentaries on right, liberties and the like belong in a discussion of how such equilibria are *sustained* rather than in a discussion of how a suitable equilibrium should be *selected*. Section 3 is the main part of the paper. It is divided into three subsections. The first provides a non-technical summary of the relevant theory of non-cooperative bargaining models. The second expands and develops Harsanyi's utilitarian theory but goes on to criticize, not only Rawls rejection of Harsanyi's use of orthodox Bayesian decision theory, but also Harsanyi's dependence on over-strong assumptions concerning symmetry and the comparison of utilities. The third subsection presents my own approach. Section 4 looks more closely at the inter-personal comparison issues and the connection with "cooperative bargaining solutions". Section 5 briefly explores some wider issues and Section 6 summarizes the conclusions.

2. Equilibria

In motivating a chapter on games which are played repeatedly by the same agents, Friedman (1986) quotes Hume (1739)

> "... I learn to do a service to another, without bearing him any real kindness; because I foresee, that he will return my service, in expectation of another of the same kind, and in order to maintain the same correspondence of good offices with me or with others."

The theory of repeated games seeks to give formal expression to such insights by demonstrating that a wide variety of apparently selfless or altruistic behavior can be reconciled with a world-view which sees individuals as players in a complex non-cooperative game, each relentlessly pursuing his or her own personal goals in the knowledge that others are doing the same. Following Aumann (1981), the fundamental theorem in this area is called the "folk theorem".

The fact that repeated games are mentioned signals the important fact that a study which confined its attention to *static* models would be inadequate. *Time* has to be taken explicitly into account, because the mechanism which keeps players on the straight-and-narrow path of righteousness is their anticipation that, if they deviate, their deviation will precipitate *future* behavior on the part of all concerned which will render the deviant worse off than if he or she had not deviated in the first place.

As a relief from the repeated Prisoners'[9] Dilemma as usually served up as an illustration in this context, an alternative dish based on an "overlapping generations" model of Samuelson will be offered. Samuelson's concern was with a question that a social contract theory ought to be able to answer: namely, why is money valuable? But the example is used here for more general purposes. Imagine a world in which only two people are alive in any single time period, a mother and a daughter. Each individual lives precisely two periods and gives birth (presumably parthogenetically) at the midpoint of her life. At birth, each daughter is endowed with two units of a good. If the good were not perishable and only her original endowment were available for her consumption, each individual would choose to consume one unit in her youth and one unit in her maturity. But the good perishes after precisely one period if not consumed before that date.

One equilibrium calls upon each individual to adopt the strategy of consuming her entire endowment in her youth. Each individual then endures a miserable old age but each is optimizing given the strategy choices of the others. A more "socially desirable" outcome would be for each daughter to give her mother one of her two units of the endowment good. Each individual would then be able to enjoy one unit of consumption in each period of her life. But is such behavior sustainable as an equilibrium outcome?

Following Shubik (1981), imagine that each individual conditions her choice of action on what has happened in the past. Suppose, in particular, that each individual uses the strategy of giving one unit of her endowment to her mother if and only if her mother behaved in the same way in the previous period. Otherwise, the individual consumes her entire endowment herself. Observe that no player has an incentive to deviate from this choice of strategy unless someone else also deviates. A collection of strategies, one for each player, with this property is called a *Nash equilibrium.* If players were able to make commitments (unbreakable promises or threats) to the use of particular strategies, no further discussion would be necessary. It would certainly be optimal for each player to make a commitment to play the Nash equilibrium strategy described above provided she believed that all other players were certain to make the same commitment. But suppose she were *unable* to make such a commitment. (See Schelling (1960) for an extended discussion of the practical difficulties in making attempted commitments stick.) Then, if a mother were to deviate, her daughter would not *want* to punish her mother by withholding the gift of one unit of the consumption good, because she would then be punished in *her* turn by *her* daughter. Clearly, it is not enough that there be threatened punishments[10] for deviant behavior; the threats must be *credible*. Players must believe that, if they *were* to deviate, the punishments *would* be inflicted. And for this to be the case, the punishing behavior required from each player must be optimal for that player, given the contingency that triggers it. This means that it is necessary to look for equilibria sustained by strategies from which nobody would wish to deviate now or in the future (provided nobody else deviates now or in the future) *even though there may have been deviations in the past.* A minimal requirement on such a collection of strategies is that they constitute a *subgame-perfect equilibrium* in the sense of Selten (1975). This means that the strategy collection must not only be a Nash equilibrium for the game as a whole, it must also induce a Nash equilibrium on every subgame (whether or not that subgame is reached in equilibrium). Operationally, this usually means that players who fail to carry out their "duty" to punish a deviant have to anticipate being labeled as deviant themselves and hence becoming liable to punishment in their turn (Abreu 1986). In a nutshell, the answer to *Quis ipsos custodes custodiet?* is simply: more guardians.

In the overlapping generations game, very simple subgame-perfect strategies exist. It is only necessary to require that each individual gives one unit to her mother if and only if *no-one* has *ever* deviated from this behavior in the past. However, such a "grim" punishment schedule, in which punishment is visited, not only on the original sinner, but on all her descendants, is not very appealing. Once a punishment phase has been entered, it can never be escaped. In the following alternative scheme only "the guilty" are punished and, after a deviation, matters can return to the ideal situation. A conformist is defined to be an individual who gives her mother one unit of her endowment unless her mother is a non-conformist, in which case she gives her mother nothing. If it is known whether the mother alive at some particular time t is to be deemed a conformist or a non-conformist, then the definition allows the status of all later mothers to be determined inductively. A subgame-perfect equilibrium in the game subsequent to time t is then obtained if the players all adopt the strategy of always conforming (see Costa 1987).

The preceding discussion of the "overlapping generations" game was intended to illustrate the manner in which apparently altruistic behavior can be sustained without any need to attribute anything but selfish motives to the individuals concerned. It is tempting to reject such "explanations" out of hand, on the grounds that they are inconsistent with what introspection tells us about our own inner motivations, but some pause for thought is necessary here. The *hymenoptera* (ants, bees, wasps, etc.) provide spectacular examples of seeming altruism. Worker bees offer the ultimate sacrifice of their own lives in stinging a wasp which inades the hive. Nobody, however, attributes such behavior to ethical sensibilities on the part of the workers. The workers are programmed by their genes and their program is seen as being tailored by evolution to optimize the survival opportunities of the gene package which determines it. (Hence Dawkins (1976) "Selfish Gene"; see also Axelrod (1984)). And once one has learned from the biologists that the *hymenoptera* have the unusual property that each individual shares more genes with a sister than with a daughter, their propensity to behave socially becomes readily comprehensible.

With human societies, of course, matters are more complex, at least, in some respects. We are the victims, not only of our biological programming, but also of our social programming. The ideas (or, more generally, what Dawkins (1976) calls "memes") which infest our heads and govern our behavior, can be seen as competing for survival rather like genes. Ideas which generate "successful" behavior will be replicated into other heads, while ideas which are less "successful" will become unpopular and go out of fashion. But nothing says that an idea needs to embody a correct explanation of *why* it is successful in order that it *be* successful. Nor is there any reason why an idea that is successful in one environment should necessarily be successful if transferred to another.

Consider, for example, the mother/daughter relationship in the "overlapping generations" model. Each mother may teach her daughter that children have a "moral duty" to sustain their elderly parents and to impress this same moral duty on the grandchildren. In such a society, each individual would truthfully be able to reply, if called upon to explain her altruistic behavior, that they were acting from a sense of moral duty. But to accept this answer at face value would be to miss the essential point: namely, what is it that allows this sense of moral duty to survive?

In brief, there is no *necessary* contradiction in recognising that human conduct is regulated, to a large extent, by social norms or biological imperatives, and *simultaneously* making the claim that the *same* conduct is sometimes "as-if" it were motivated by selfish, but enlightened, goals. Old-fashioned, Chicago-style economics presumably would not work at all if this were not the case. Of course, none of this is to deny that many stimuli exist, in the real world as well as in psychology laboratories, to which people respond in a manner which is in no way explicable as optimizing behavior (unless one follows Hobbes in defining this so that such an explanation becomes tautologous). To claim that the Chicago school is sometimes right is not to claim that they are *always* right.

In so far as game theory is relevant to the real world, it can only be to situations in which there is at least some chance that the behavior to be explained has had some opportunity to adapt to its environment. There is therefore necessarily a gap between the ideal world in which game theory operates and the world-as-it-is, and this gap is important to much of the discussion which takes place in a social contract context. For example, in the ideal world of the game theorist, where everybody can be counted on to act in their own enlightened self-interest, the concern of so many writers with the preservation of individual rights and liberties -- often at the expense of other valuable features of the "good society" -- becomes much less pressing. This is not to say that appropriate rights and liberties will not be preserved in the equilibrium on which a game-theoretically ideal society coordinates (although the actual rights and liberties preserved, and the manner in which their preservation is achieved, are not issues which will be central to this paper): only that, in a rational society, it would not be necessary to surrender welfare opportunities in order to guard against *irrational* infringements of the equilibrium arrangements. One does not, for example, need to sacrifice speed to stability in the design of boats if the passengers can be relied upon to sit down when the boat shows signs of rocking.

Returning now to the idealized situation exemplified in the overlapping generations model, one may imagine a mother justifying the ***status quo***[11], in which each daughter gives her mother half her endowment, by drawing attention, à la Hobbes[12], to its advantages compared with the hypothetical ***state-of-nature***[13], in which each daughter consumes her entire endowment herself. She would argue that a social contract should do more than coordinate behavior on an equilibrium: it should coordinate behavior on an equilibrium which is desirable compared with the alternative of operating whatever equilibrium will ensue from a failure to agree on something better.

Fig. 2 provides a schematic example. Players I and II have three Nash equilibria in pure strategies[14] to consider: (row 1, column 1), (row 2, column 2), (row 3, column 3). Suppose they were to bargain over which equilibrium to use. Then the result will depend on what is taken to be the state-of-nature equilibrium. If the state-of-nature equilibrium were (row 3, column 3), then the natural[15] equilibrium on which to coordinate is (row 2, column 2). On the other hand, if the state-of-nature equilibrium were (row 1, column 1), then nothing will persuade player II to shift to any other equilibrium. What should be emphasized is that there is no point in the players agreeing to implement, for example, the strategy pair (row 3, column 2). Player I will not honor such an agreement since he has an incentive to deviate to row 2. Nor does it make sense to locate the state-of-nature at a strategy pair like (row 2, column 3). This pair is not an equilibrium and so cannot be what is anticipated to happen if there is no agreement.

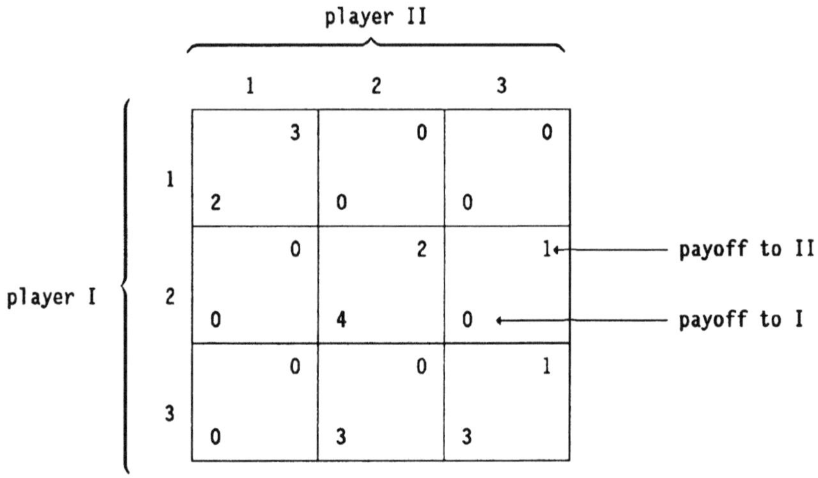

Figure 2

In what follows, a social contract will be called upon to do more than just select one equilibrium given one particular state-of-nature, as in the preceding static example. Instead, a social contract will be required to make the rules for coordinating on an equilibrium in the future *contingent* on the rules as they exist at present. Each possible equilibrium set-up is then to be regarded as a possible state-of-nature from which the social contract should be able to guide society to something more enlightened. Note that this means that attention has to be limited to utopias which are reachable *from the state as currently constituted*. Alternative utopias, which would have been available if things were not as they are, have to be put aside as unattainable. The reason that some utopias may not be available is the insistence that all behavior specified by the social contract should be in equilibrium. In particular, the behavior required to accomplish *the transition* from the current state-of-nature to its utopian replacement must be in equilibrium. But this will not be the case if the transition requires some group of individuals to surrender power or privileges voluntarily without guarantees that such a surrender will bring compensating rewards.

Axelrod's (1984) "Evolution of Cooperation" has popularized some of the mechanisms which biologists see as instrumental in shifting animal populations from one equilibrium to another. In brief, a population programmed to operate an efficient equilibrium can sometimes be invaded by mutants who operate a more efficient equilibrium amongst themselves but behave towards the original population as these behave to each other. In a biological context, the better payoffs obtained at the more efficient equilibrium are identified with a relatively greater number of offspring on average. The mutant population therefore gradually increases at the expense of the original population until the latter is displaced altogether. In games played by conscious optimizers, the corresponding problem has hardly been studied at all. If the game of Fig. 2, for example, is played repeatedly, with the strategy pair (row 3, column 3) having been used in the past, a change of beliefs on the part of players I and II is necessary if they are to shift to (row 2, column 2). How is such a change of beliefs to be engineered? Some progress on such questions has been made by Aumann and Sorin (1985), but many difficulties remain.

However, this paper will proceed as though the considerations relevant to the following archetypal example were valid in general. The example concerns trading without trust. Two criminals agree to exchange a quantity of heroin for a sum of money. If it is assumed that each is free at any time to walk away with whatever is in his possession at that time without penalty, this can be seen as an agreement to move from one equilibrium to another. How is this transition to be accomplished?

Obviously, there is no point in the buyer handing over the agreed price and waiting for the goods. Somehow, they have to arrange for a flow between them, so that the money and the drug change hands *gradually*.[18] Moreover, the rate at which heroin is exchanged for money at any time must be regulated by the requirement that it never be in the interests of either criminal ever to call a halt to the exchange in order to renegotiate the terms of trade. At each time, therefore, the deal which would be negotiated, given the *current* distribution of goods, must lead to the same *final* outcome as the original deal.

Of course, the social contract, as envisaged by Rawls or Harsanyi, is not *actually* negotiated. It is seen as an agreement which would *necessarily* be reached by rational agents negotiating in certain *hypothetical* circumstances. The conditions under which these hypothetical negotiations are seen as taking place, in the "original position", are discussed in the next section. What needs to be dealt with here is the manner in which such a convention for coordinating behavior is to be slotted into the framework outlined earlier in this section.

The first point concerns *timing*. Hare (1975) has perhaps overstated the case in describing Rawls' (1972) use of tenses in discussing the "original position" as baffling, but it is certainly true that a treatment like that attempted here does require careful attention to *when* negotiations are hypothesized to take place. What is assumed here is that the social contract is potentially renegotiated continuously *at all times*, rather than being determined, once and for all, at some indefinite time in the far past. Such an assumption is, of course, necessitated by the use of a conceptual framework within which commitment is disbarred, but it seems to me, in any case, attractive to have a view of social justice which permits appeals at any time against judgements made in the past.

However, what happened in the far past does not become altogether irrelevant. One must imagine some primeval state-of-nature from which society developed. In principle, a social contract should guide society from this primitive equilibrium to a final utopian equilibrium and sustain society in this final state. Because the transition must be achieved without trust, the passage from the primeval state to the utopian state will be modeled as a *continuous* process. (And here it should be reiterated that to proceed "without trust" in this context is *not* to argue that trust is an unimportant constituent of human relationships, nor is it to argue that trust will be absent from an ideal society: it is only to argue that, because we need to *explain* why enlightened self-interest will generate trust in certain circumstances, we are not entitled to make *a priori* assumptions about its existence.) At any time, members of society will be guided, in selecting their individual contributions towards the building of the new Jerusalem, by a consideration of what would necessarily be agreed *if* a negotiation were to take place in the "original position". (Here the word "original" should be understood to refer, not to a particular *time*, but to a particular *informational state* in which each agent is ignorant of his or her historical role in society.)

To come up with a determinate outcome of a bargaining problem, hypothetical, or otherwise, it is necessary to know what the consequences of a failure to agree would be. Hobbes conceived the consequences of such a failure to agree as being a total degeneration to what might be called a *primeval* state-of-nature. For him, this was a state of anarchy, a war of all against all in which the life of man is "solitary, poor, nasty, brutish, and short". A reversion to such a worst possible state would perhaps make some sort of sense if the hypothetical bargaining in the original position were assumed to be accompanied by hypothetical threats that members of society felt themselves *committed* to carry out if the hypothetical agreement which they reason would have been reached in the original position fails to be honored in the real world. The most effective threats would then be those which are most damaging. Without commitment, however, the Hobbesian notion of a total reversion to a brutish state does not seem sustainable. Instead, it will be assumed that society will remain in its *current* state-of-nature, in which advances made in the past are retained, but further advances available for immediate exploitation are disregarded. The game-theoretic implications of this assumption require further discussion, but this is postponed until Subsection 3.3.

In speaking of what *would* happen *if* there *were* a disagreement, the subjunctive mood is important, it is true that rational individuals *will* always agree under the hypotheses of this paper, but *what* they agree on will depend on what would happen *were* they not to agree. (The situation for a subgame-perfect equilibrium is similar. It is necessary to make good assumptions about what would happen if there were deviations from equilibrium behavior in order to determine what equilibrium behavior is.) Thus a social contract will need to make the current coordinating rules contingent on the current state-of-nature. The transition from the primeval state to the utopian state will be complete when the equilibrium specified by the social contract is the same as the current state-of-nature.

It remains to make the point that the behavior specified by the social contract must be *consistent over time*, or, to be more accurate, if the social contract specifies that primeval state-of-nature S should lead to utopia U via an intermediate state T, then the social contract should also specify that primeval state-of-nature T should also lead to the same utopia U. Otherwise a new appeal to social justice, once T has been reached, will overturn the judgement on social justice made earlier in state S. This again mimics the requirement of a subgame-perfect equilibrium that players should not make incredible assumptions about future contingencies. For a more careful account of what is involved in "renegotiation-proof" equilibria, see Farrel and Maskin (1987).

3. The Original Position

As seen here, the basic questions for a social contract argument are the following:

(1) What outcomes can be sustained as dynamic equilibria over time?

(2) How can society be shifted from an old "state-of-nature" equilibrium to a new "utopian" equilibrium?

(3) Where there are many choices for a new equilibrium, how should this choice be made?

Very little beyond the generalities of the preceding section will be offered on the first two questions. In bargaining terms, these questions are concerned with what deals are feasible, whereas the third question, on which the remainder of the paper concentrates, is about how an optimal choice is to be made from the feasible set. Examples of abstract principles that might be used for selecting an optimal equilibrium from the feasible set of equilibria are Harsanyi's (1977) brand of utilitarianism and Rawls' (1972) maximin criterion.

On feasibility questions, only some passing points will be made. Firstly, although our education invites us to focus on a Humean style bourgeois-liberal society when equilibrium issues are emphasized, there is no reason why a Hobbesian authoritarian society should not be equally viable (and perhaps more stable for the reasons that Hobbes describes). Even the Marxian ideal of the "dictatorship of the proletariat" might conceivably be feasible if suitably interpreted. Secondly, it may be useful to note that the paper's separation of feasibility and optimality questions may serve to defuse some of the criticisms directed against Rawls (and hence, implicitly, against Harsanyi). For example, Nozick's (1974) pre-occupations are almost exclusively with issues that I would wish to classify under questions one and two. These issues are certainly very important but seem to me of doubtful relevance to the Rawlsian maximin criterion which I see as an attempt to tackle question three.

The titles of the following Subsections (3.1, 3.2 and 3.3) are chosen to match the labeling in Fig. 1.

3.1 Nash Bargaining Solution

This will be a brief account of the salient features of the two-person case. Nash (1950) abstracted the problem faced by two individuals in a bargaining situation as that of locating a point n in a set X containing a distinguished point d. The point d is interpreted as the *disagreement outcome*[17] the set X is interpreted as the set of feasible deals and the point n as the "bargaining solution", - i.e., the deal on which rational bargainers will agree.

Nash simplified the problem of rational choice very considerably by appealing to Von Neumann and Morgenstern's (1944) theory of "expected utility maximization". Given a collection of plausible axioms[18] about the nature of rational behavior in risky situations, they demonstrate that, to each possible outcome ω, a number $\phi_i(\omega)$, can be attached in such a way that the actions of an individual i are "as though" he or she were seeking to maximize the *expected* value of $\phi_i(\omega)$. A Von Neumann and Morgenstern utility function ϕ_i is not unique, but if ψ_i is another utility function for the same individual, then $\psi_i = B\phi_i + C$, where B and C are constants with B positive. One is therefore free to make an arbitrary choice of the zero and the unit of an individual's "utility scale", but, beyond this, no room for maneuver exists.

Each possible outcome of the bargaining situation can therefore be identified with a pair of numbers consisting of the Von Neumann and Morgenstern utilities the two bargainers attach to that outcome. The bargaining problem can then be represented geometrically as in Fig. 3A.

The set X in Fig. 3A is drawn to be convex as assumed by Nash (1950). The point n, which is the midpoint of the line segment ab, is the ***Nash bargaining solution***[19] for the problem described by the pair (X,d). Nash obtained n as the unique point consistent with a set of rationality axioms for collective choice in this context (in addition to the Von Neumann and Morgenstern axioms for individual choice). It is popular, following Raiffa (1953), to re-interpret Nash's axioms as maxims for an "ideal observer" called upon to arbitrate the bargaining problem.

However, the Nash bargaining solution is not attractive as an ethical concept. Rawls (1972, p. 135) quotes Sen (1970) to this effect and nothing said here should lead Rawls or Sen to a different view. Nash (1950) did *not* propose an ethical interpretation and this paper will not impute one. Although the Nash bargaining solution is used a great deal in this paper, it is used in the same spirit as one might use the proposition 2 + 2 = 4. Both may have displeasing consequences when used in an ethical argument, but neither should be supposed to be carrying any of the ethical burden of the argument.

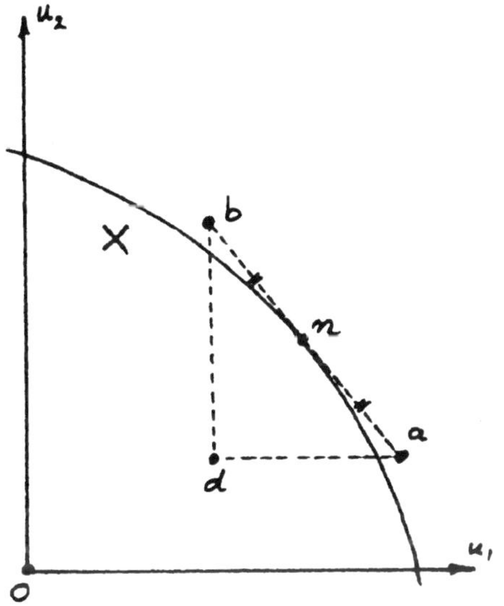

Figure 3A

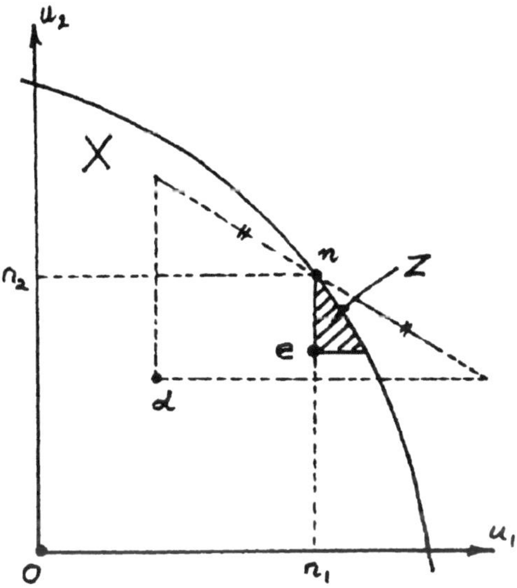

Figure 3B

Indeed, far from seeking to examine ethical questions, Nash was concerned entirely with rational agents, employing whatever bargaining power they have to hand in furthering their own individual interests. His approach does *not* suppose the existence of outside supervision of the bargaining process, nor the services of a referee or arbiter. But it does assume that the bargaining procedure satisfies certain conditions - of which the most important, in the current context, is that it treat the bargainers symmetrically.

To defend this interpretation of the Nash bargaining solution adequately, it is necessary to provide some strategic underpinning of the principles of cooperative behavior which he advances. This requires the analysis of explicit models of relevant bargaining procedures, using *non-cooperative* game theory. If Nash's cooperative principles are correct as descriptions of what rational agents will do in appropriate circumstances, then the predictions they generate should be realized as equilibrium outcomes in suitable formal bargaining games analyzed *non-cooperatively*.

These matters are discussed at length in Binmore/Dasgupta (1987), but perhaps some examples of relevant bargaining procedures will suffice here. Nash himself proposed the very simple procedure in which each bargainer simultaneously makes a once-and-for-all, take-it-or-leave-it demand. If these demands are compatible, they are implemented. Otherwise the disagreement outcome results. Nash showed that, provided there is some vestigial uncertainty about the precise location of the frontier of X, any non-trivial Nash equilibrium of the demand game requires each player to demand approximately the payoffs, n_1 and n_2, assigned to them by the Nash bargaining solution.[20]

Take-it-or-leave-it demands imply the power to make *commitments* in a particularly forceful manner. As noted in Section 2, such an assumption is unattractive, especially if the result is to be applied to bargaining in the "original position" where all the action will only be hypothetical. But, if the bargainers cannot make forward commitments, then *time* enters the picture and account must be taken of how the bargainers feel about time passing without an agreement being reached.[21]

One can model a bargaining procedure in which time plays an active role in various ways. One might, for example, suppose that Nash's demand game is played repeatedly until compatible demands *are* made (paper 8 of Binmore/Dasgupta 1987). However, a model of Rubinstein (1982) (paper 3 of Binmore/Dasgupta 1987) is more appealing. In this model, the bargainers alternate in making proposals which the other may accept or refuse. Subgame-perfect equilibrium outcomes (Section 2) in the Rubinstein model, and a variety of related models, lead to a deal being reached which is an approximation to the Nash bargaining solution *provided that* the players discount time at the same rate and that the interval between successive proposals is sufficiently small (papers 4 and 5 of Binmore/Dasgupta 1987). The latter assumption is reasonable in the context of rational bargaining behavior because unproductive delays between successive proposals are in the interests of neither player.[22] The assumption that players discount time at the same rate simplifies technical issues. However, it is not an assumption of any consequence when bargaining in the original position is considered. (The reason is that bargainers in the original position will not know at what rate they discount time. Instead, each bargainer will be supplied with the *same* probability distribution over the set of possible rates. A model with this feature generates approximately the same outcome as one in which all bargainers have a fixed, identical discount rate, provided that the interval between successive bargaining proposals is allowed to be sufficiently small.)

The theory described above in defense of the use of the Nash bargaining solution seems to me to be very satisfactory as such things go.[23] However, some gilding of the lily is necessary to adapt the theory for the purpose required in this paper. The immediate question concerns the location of the disagreement point d in Nash's formulation. A consideration of this question will identify a substantive difference between the simple Nash demand game and its more sophisticated cousins. The former captures the essence of bargaining with unlimited capacity to make commitments. Such a situation reduces to a race to leave the other player with a take-it-or-leave-it problem. Hence all the action will be telescoped into the first instant after which the players will have left themselves no room for maneuver. Moreover, players will wish to make the option of "leaving-it" as unattractive as possible to their opponent. Hence they will bind themselves to the action which is most damaging to the opponent if the opponent refuses to cooperate. Normally, the most damaging action available will be for one of the players to abandon the negotiations permanently to take up whatever outside option may be available to him or her. In using the simple Nash demand game, it is therefore appropriate to take the disagreement point d to be the pair of utilities the players expect to get if the negotiations break down *permanently*.

Without commitment at all, a different interpretation of the disagreement point d is appropriate. (See Binmore 1983, Shaked/Sutton 1984, Binmore/Shaked/Sutton 1985). The disagreement point d should be located, not at the "outside option point", but at what might be called the "impasse point" - i.e. the pair of utilities the players attach to the event in which they bargain for ever without reaching an agreement and without ever abandoning the negotiation table to take up an outside option. The impasse point may be neatly characterized by observing that, if the players were to make it their agreement outcome, they would be indifferent about *when* the agreement was implemented. Operationally, this means that the set X of Fig. 4A should be replaced by the set Z shaded in Fig. 4B. In this diagram, d represents the impasse point in a bargaining situation *without commitment possibilities* and e represents the pair of "outside option" utilities available to the bargainers. In many cases, the fact that X is replaced by Z will not affect the location of the Nash bargaining solution n. However, Fig. 4B shows a situation in which player 2's outside option is "active". The Nash bargaining solution n, with X replaced by Z, then requires player 1 to concede $n_2 = e_2$ to player 2 to prevent her abandoning the negotiation table, and n_1 represents the utility to player 1 of whatever crumbs remain for him. (If full commitment were possible, the disagreement point would be located at e rather than at d, and hence similar considerations would not arise.)

The point made above is important because, in the discussion of bargaining in the "original position", the disagreement point will be placed at an impasse point based on what is known by the bargainers about the current state-of-nature. The justification is that commitment seems even less defensible in the hypothetical circumstances of the "original position" than elsewhere. With full commitment, the most damaging threat would be to discontinue all cooperation totally and completely, thereby, in the two-person case, precipitating a Hobbesian primeval state-of-nature. Obviously, the conclusions must be expected to be different in the two situations.[24]

3.2 The Veil of Ignorance

This subsection begins by outlining a version of Harsanyi's (1977) formulation of the "original position", together with a derivation of his utilitarian conclusion. Rawls (1972) is hostile to such a conclusion. Nevertheless, it seems to me that the general thrust of his argument would force him to Harsanyi's conclusion if he were to abandon his opposition to orthodox Bayesian decision theory, in which Von Neumann and Morgenstern's utility theory is married to Savage's (1954) theory of subjective probabilities. What carries the argument to its utilitarian conclusion, is chiefly that both authors assume that *commitments* can be made to stick. This is an assumption which has already been attacked and will be attacked again later. In this subsection, however, it will be taken for granted that the players have an unlimited capacity for making any commitments that they choose.

With commitment, it is possible to take a static view of what goes on in the "original position". The players need not confine their agreements to equilibria: nor need they worry about the possibility of renegotiation. Once something is agreed, the players are commited and deviation or reconsideration is ruled out-of-court.

To quote Rawls (1972, p. 17): "... the original position is the initial status quo which insures that the fundamental agreements reached in it are fair." The original position is called upon to give expression to what Rawls terms "the principle of redress". To quote again from Rawls (1972, p. 100): "This is the principle that undeserved inequalities call for redress; and since inequalities of birth and natual endowment are undeserved, these inequalities are to be somehow compensated for."

As a very simple example, which nevertheless gives the flavor of the general situation, consider the problem faced by Adam and Eve in the Garden of Eden. Fig. 4A illustrates a set X whose members consist of the utility pairs that Adam and Eve attach to the various ways of life they can jointly adopt. The asymmetric shape[25] of the set X represents Rawls' "inequalities of birth and natural endowment". The numbers e_A and e_E are the outside option utilities for Adam and Eve respectively - i.e. the utilities they would derive if the attempt to come to joint arrangements were to fail finally and completely. Their problem is to agree on a point x in X on which to coordinate.

If they were to bargain without further constraints, then there is no particular reason why the point x on which they finally settle should compensate the party disadvantaged by the shape of the set X and/or the location of the point $e = (e_A, e_E)$. Even if one postulates that the choice of x is to be delegated to an "ideal observer" who will arbitrate in a "fair" manner, one is still left to provide a definition of what "fair" should mean. Of course, according to Subsection 3.1, if we assume that they bargain *without* regard to fairness questions but instead use whatever bargaining power they have, then the bargaining outcome x should be located at the Nash bargaining solution of X relative to the disagreement point d = e (given that full commitment is being assumed).

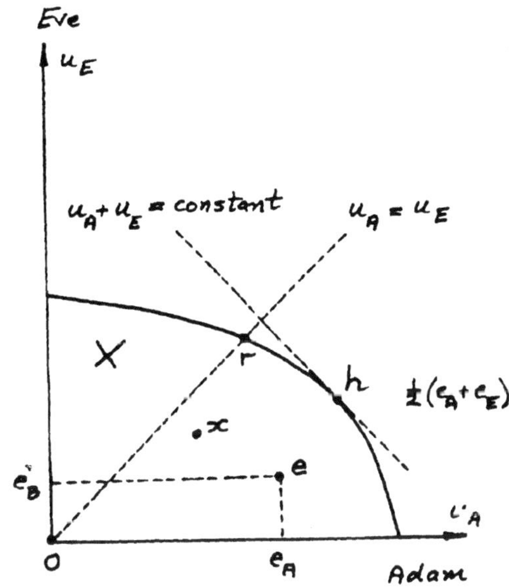

Figure 4A

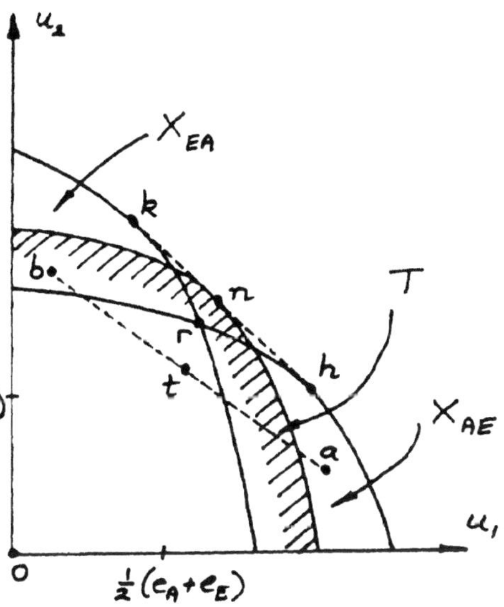

Figure 4B

Harsanyi deals with this difficulty by symmetrizing the situation. The symmetrization is achieved by requiring that negotiation is carried on behind a "veil of ignorance". Beyond the veil of ignorance, the roles that the players occupy in society become unknown. The players therefore have to negotiate a deal without knowing who is Adam and who is Eve. This ought to benefit the naturally disadvantaged party. The question is: by how much?

Fig. 4B is a representation of the problem faced by two persons, labeled 1 and 2, in such an "original position". For definiteness, it will be assumed that player 1 is actually Adam and player 2 is Eve but that this information is withheld behind the veil of ignorance. Their ignorance of their actual roles is reflected in the fact that both players 1 and 2 see their outside option utilities as $(e_A+e_E)/2$. This incorporates the assumption that both attach a probability of 1/2 to the event that player 1 is Adam and player 2 is Eve and a probability of 1/2 to the alternative event that player 1 is Eve and player 2 is Adam. They will be assumed to be able to make agreements which are *contingent* on who turns out to be who. That is to say, they may agree, for example, to implement the utility pair a in set X_{AE} of Fig. 4B *if* player 1 turns out to be Adam and player 2 turns out to be Eve, but to implement the utility pair b in set X_{EA} *if* the roles turn out to be reversed. Since they attach equal probabilities to the two eventualities, they will assign the utility pair t of Fig. 4B to this "contract". The utility pair t lies at the midpoint of the line segment ab. The set[26] T of all such points t, with a in X_{AE} and b in X_{EA}, is shaded in Fig. 4B. It is the set of all utility pairs from which it is feasible for players 1 and 2 to make a joint choice.

Since T is convex and everything is symmetric, it does not matter greatly what theory of bargaining or arbitration is now employed. Any sensible theory will select, as the bargaining outcome, the point n in Fig. 4B, which is the unique, Pareto-efficient, symmetric point in T. To obtain this outcome, the players have to agree on implementing the utility pair h in X_{AE} if player 1 turns out to be Adam and player 2 to be Eve, and to implement k in X_{EA} if the roles are reversed.

An interpretive comment may be helpful. The *actual* situation is as in Fig. 4A and it is the "fairness" of h (for Harsanyi) in *this* situation which is the *substantive* issue. The bargaining in the original position is only *hypothetical*. In defending the point h, Adam tells Eve that, if they *had* been placed in the original position under the specified conditions, then they surely *would have* agreed to make *commitments* to abide by the "contract" agreed to, *whoever turned out to occupy whichever role*. Eve may not like the consequent outcome (especially if $h_E < e_E$), but her commitment would have compeled her to honor the deal. If Eve is agreed that the device of the original position is appropriate and that Adam is right in analyzing what her behavior in the original position would have been, then she therefore ought to act as *though* she were committed to maintaining the outcome h.

In essence, this is Rawls' (1972, p. 167) "slave-holder's argument". A slave complains about her condition of servitude and the slave-holder justifies his position to the slave with the following version of a social contract argument. "If you had been asked to agree to a structure for society without knowing what your role in that society was to be, then you would have agreed to a slave-holding society because the prospective benefits of being a slave-holder would have outweighed in your mind the prospective costs of being a slave. Finding yourself a slave, you therefore have no just cause for complaint." Rawls, of course, denies that a slave-holding society would be agreed to in the original position. But he takes pains to argue that the slave-holder's argument *would be* correct *if it were* true that a slave-holding society *would have been* agreed to in the original position, and explicitly rejects the slave's objection that she sees no grounds for honoring a *hypothetical* contract to which she never *actually* assented and which postulates a lottery for the random distribution of advantage which never *actually* took place. To some extent, my formulation of the original position removes some of the grounds for the slave's indignation. Unlike Rawls and Harsanyi, I allow the structure of society chosen in the original position to be a function of who turns out to be who. Their formulation requires the *same* structure to be chosen regardless of the role assignment. However, to my mind, the slave-holder's argument remains unacceptable.

The slave should point out that, although she and the slave-holder might have *wished* to make commitments in the original position, no mechanism exists which would have allowed them to do so and hence they would not have been *able* to make any commitments. The claim that the slave has an "implicit commitment" to honor a "slave-holding contract" is therefore devoid of any content.

Ellsberg's (1975) kidnapping paradigm is usually quoted in this context. The victim would dearly love to make a commitment to the kidnapper not to reveal his identity so that the kidnapper can release the victim without fear of the consequences. But no commitment mechanism exists. Rawls might respond that no mechanism is necessary, since ethical considerations can be relied upon to secure the honoring of agreements. But do questions of personal integrity take precedence over questions of social justice? Is it not precisely to elucidate such questions that the device of the original position is introduced? In any case, if such ethical considerations exist, they should be "factored out" and built into the structure over which those in the original position negotiate. This is a mathematician's response, but no less valid for that.

In what sense can Harsanyi's proposed resolution of the problem with full commitment be said to be utilitarian? The answer is simple, although it is not perhaps an answer that a dyed-in-the-wool utilitarian would find entirely satisfying. The point h in Fig. 4A is utilitarian in that it is the point in X at which the arithmetic sum

$$u_A + u_E$$

of Adam and Eve's utilities is maximized.

It is immediately apparent that some sleight-of-hand has been visited upon the unwary. The sum of the utilities can only be meaningful if the utilities are comparable. But the Von Neumann and Morgenstern theory makes no provision for such inter-personal comparisons: nor does the Nash bargaining theory.[27] What has been swept under the carpet is Harsanyi's use of his version of the theory of extended sympathy as developed by Arrow (1963), Suppes (1966), Sen (1970) and others.

The theory requires individuals to be able to say whether they would prefer to be Adam in state s or Eve in state t, for all relevant "states of the world". With appropriate rationality requirements, player 1 will then assign a Von Neumann and Morgenstern utility

$$\Phi_1(i,s)$$

to each pair (i,s) representing an individual i in state s. In the case being considered here, the only relevant individuals are Adam and Eve and so either $i = A$ or $i = E$.

A utility $\Phi_1(E,s)$ is to be distinguished from a Von Neumann and Morgenstern utility

$$\phi_E(s)$$

representing Eve's own *personal* preference. However, if player 1 is to take seriously what it would mean to be Eve, then he must admit that, if he were Eve, then his personal preferences over states of the world would then be the same as Eve's personal preferences. This means that $\Phi_1(E,s)$ must represent the *same* preferences over states s as $\phi_E(s)$. But, as noted in Subsection 3.1, Von Neumann and Morgenstern utility functions which represent the same preferences are related. Constants B_{1E} and C_{1E}, with B_{1E} positive, must exist such that, for all s,

$$\Phi_1(E,s) = B_{1E}\phi_E(s) + C_{1E}.$$

Similarly,

$$\Phi_1(A,s) = B_{1A}\phi_A(s) + C_{1A}.$$

It is the coefficients B_{1A}, B_{1E}, B_{2A}, and B_{2E} which are immediately significant. They provide a basis for comparing Adam's and Eve's personal utilities. Observe that, *from player 1's point of view*, one unit on Adam's personal scale is equivalent to B_{1A}/B_{1E} units on Eve's personal scale. Similarly, *from player 2's point of view*, one unit on Adam's personal scale is equivalent to B_{2A}/B_{2E} units on Eve's scale. Since player 1's viewpoint is not necessarily compatible with player 2's, what has been obtained is only an *intra*-personal comparison of utilities rather than an *inter*-personal comparison.[28]

Nothing more than this can be squeezed from Von Neumann and Morgenstern's theory. But Harsanyi (1977) needs to go further and appeals to what game theorists, following Aumann (1987), nowadays call the "Harsanyi doctrine".[29] To caricature the doctrine: it is that human beings share a common psychological inheritance and hence, *if* they shared a precisely identical history of experience, *then* they would arrive at precisely the same judgements. This is more plausible in the circumstances of the original position than it is as a general proposition. One need only suppose that *some* common ground exists and it is this which is retained in the original position, other matters having been "factored out" - i.e. absorbed into the structure about which decisions in the original position are made.

On applying the Harsanyi doctrine to the extended sympathy argument given above, the conclusion is that Φ_1 and Φ_2 must represent the *same* preferences and therefore $\Phi_2 = B\Phi_1 + C$. It follows immediately that $B_{1A}/B_{1E} = B_{2A}/B_{2E}$ and hence that player 1's and player 2's intra-personal comparisons agree.

But if everyone is agreed on what their intra-personal comparisons are, then their common view constitutes an inter-personal comparison of the personal utility scales. Furthermore, since it is not proposed that the extended sympathy utility scales (as opposed to the personal utility scales) be compared, these can be calibrated in whatever manner is convenient. In particular, by scaling up player 3's utility units with a suitable factor, it can be ensured that

$$B_{1A}/B_{1E} = B_{2A}/B_{2E} = 1.$$

This symmetrizes matters entirely. In particular, the symmetry of Fig. 4B is justified and therefore Harsanyi's conclusion.

However, on the way, some flakiness in Harsanyi's description of his conclusion as "utilitarian" has been exposed. The interpretation of the utilities in the arithmetic sum $u_A + u_E$ is not what would immediately occur to an orthodox utilitarian. Moreover, the equal weighting of the two utilities appears only as a mathematical convenience (derived from the rescaling of Φ_2) and has no real-world significance. If Φ_2 had been left alone, then an asymmetric version of Fig. 4B would have resulted and a weighted sum would have replaced the arithmetic sum $u_A + u_E$. Nevertheless, the final outcome, in real terms, would have been exactly the same.

These last remarks prompt what is perhaps a naïve question: how would one actually implement Harsanyi's conclusions in the real world? To answer this question requires a considerably more careful account of where his utility functions come from. But this will be necessary in any case before my own alternative approach can be presented.

First it would be necessary to discover, by observation or experimentation, what Adam's and Eve's personal preferences are. Assuming rationality, these are represented in terms of Von Neumann and Morgenstern utility functions. What utility scales should be chosen? To simplify matters, suppose that both Adam and Eve agree that there is a state H (heaven) which is better for both than all achievable states and a state h (hell) which is worse for both than all achievable states. Common practice is then to calibrate the scales so that

$$\phi_A(h) = \phi_E(h) = 0; \quad \phi_A(H) = \phi_E(H) = 1.$$

This calibration will be taken for granted from now on: but *not* the common practice of proceeding as though such an *arbitrary* decision had established an *a priori* basis for inter-personal comparison.

Next, a similar exercise is necessary for the extended sympathy preferences of players 1 and 2. Assuming, for the moment, that the difficulties posed by the fact that these preferences are unobservable judgements made by hypothetical individuals can be resolved in some way, calibration of the appropriate utility functions will be necessary. To simplify matters again, it will be assumed that all are agreed that Adam would suffer more in hell than Eve, while Eve would enjoy greater bliss in heaven than Adam. The extended sympathy utility scales can then be anchored so that

$$\Phi_i(A,h) = 0; \quad \Phi_i(E,H) = 1$$

$$\Phi_i(A,H) = U_i; \quad \Phi_i(E,h) = 1 - V_i,$$

where $0 \leq U_i \leq 1$ and $0 \leq V_i \leq 1$ $(i = 1,2)$. The reason is that eight equations have been provided to determine the eight constants B_{iA}, C_{iA}, B_{iE}, and C_{iE} $(i = 1,2)$ introduced earlier. In fact, a trivial calculation shows that, for all states s,

$$\left.\begin{aligned} \Phi_i(A,s) &= U_i\phi_A(s) \\ \Phi_i(E,s) &= V_i\phi_E(s) + 1 - V_i. \end{aligned}\right\} \quad (i = 1,2)$$

With the Harsanyi doctrine, it will be true that $U_1 = U_2 = U$ and $V_1 = V_2 = V$. But this account will proceed *without* the Harsanyi doctrine since my own approach dispenses with it. The next step is therefore to obtain an *asymmetric* version of Fig. 4B. This appears as Fig. 5A. The new X_{AE} is the set of all utility pairs

$$(\Phi_1(A,s),\Phi_2(E,s)) = (U_1\phi_A(S),V_2\phi_E(S) + 1 - V_2),$$

where s is any relevant state. The new X_{EA} is the set of all utility pairs

$$(\Phi_1(E,s),\Phi_2(A,s)) = (V_1\phi_E(s) + 1 - V_1,U_2\phi_A(s)),$$

where s is any relevant state. The state-of-nature is denoted by s_0. This determines two points, $e_{AE} = (\Phi_1(A,s_0),\Phi_2(E,s_0))$ and $e_{EA} = (\Phi_1(E,s_0),\Phi_2(A,s_0))$. (Note that, with these renormalizations, the set X of Fig. 4A ceases to be the same as the set X_{AE} of Fig. 4B.[30])

The point w in Fig. 5A is what the players expect to get if the negotiations break down *permanently*, given that each attaches probability 1/2 to ending up as either Adam or Eve. Thus $w = (e_{AE}+e_{EA})/2$. According to Subsection 3.1, it is appropriate to place the disagreement point d at w when it is intended to use the Nash bargaining solution to predict the bargaining outcome in this situation. Concern about the location of the disagreement point was unnecessary earlier because of the symmetry supplied by the Harsanyi doctrine, but here it matters. The Nash bargaining

That is to say, X_{EA} is the set of all utility pairs

$$(\beta u_2 + 1-\beta, \alpha u_1),$$

where (u_1,u_2) is any utility pair in X_{AE}.

The algebra is tiresome, but the corresponding geometry is easy to appreciate. Fig. 6A shows how X_{EA} may be constructed from X_{AE}. The point $P_{AE} = (u,1-v)$ is the utility pair resulting from a state s when player 1 is Adam and player 2 is Eve. The point $P_{EA} = (1-\beta v,\alpha u)$ is the utility pair resulting from the *same* state s when player 2 is Adam and player 1 is Eve. (The numbers λ and μ are defined by $\lambda = \phi_A(s_A)$ and $1-\mu = \phi_E(s_E)$, where s_A is the best *achievable* state for Adam and s_E is the best *achievable* state for Eve.)

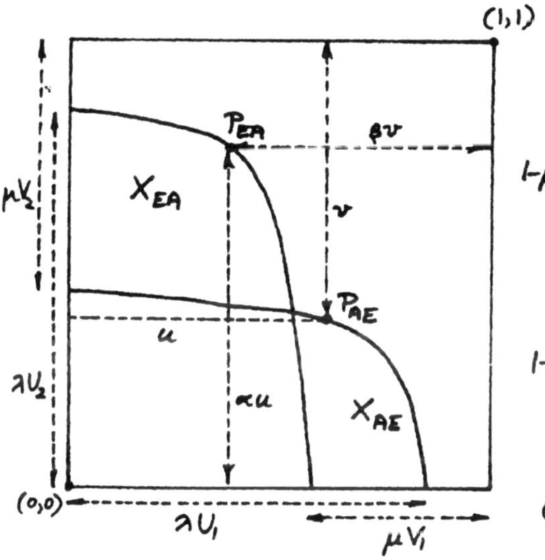

Figure 6A

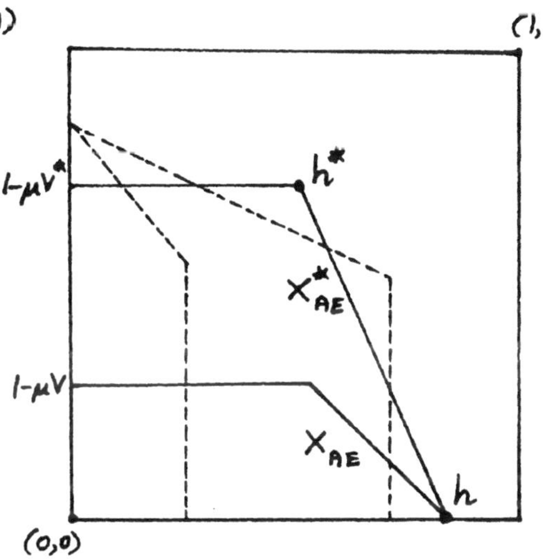

Figure 6B

The above examination of the asymmetric case allows the role of the Harsanyi doctrine to be evaluated more carefully than is possible in the symmetric case alone. It turns out not to be vital at all in providing a resolution of the bargaining problem faced by those in the original position. The problem it addresses is actually an *interpretive* problem: namely, how are the extended symapthy preferences attributed to players 1 and 2 to be elicited? The Harsanyi doctrine seemingly answers this question by making it unnecessary that it be asked. But such an impression is only partly correct.

Given symmetry, it follows that $U_1 = U_2 = U$ and $V_1 = V_2 = V$ (i.e., $\alpha = \beta = 1$). But it is not true that the values of U and V are irrelevant. They

solution at n in Fig. 5B can be achieved by implementing the utility pair h if player 1 turns out to be Adam and player 2 turns out to be Eve, and implementing k if the roles are reversed. Without the Harsanyi doctrine, Harsanyi's argument

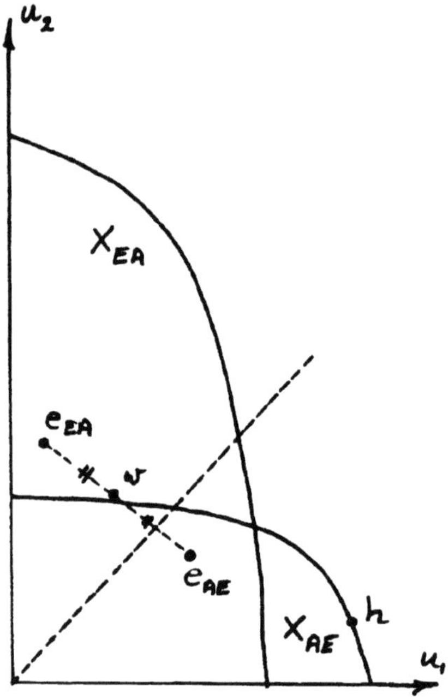

Figure 5A

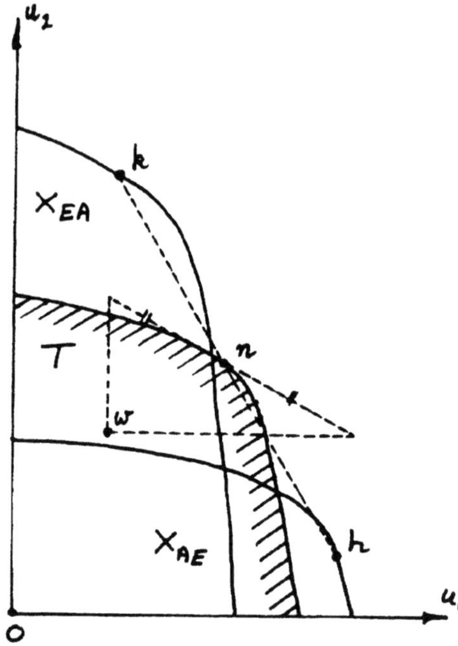

Figure 5B

therefore requires that Adam defend the point h in Fig. 5A against Eve's protestations of unfairness. But this point does not seem to be utilitarian in any obvious sense.

Before leaving Fig. 5A, it will be useful, for later purposes, to make a number of algebraic points. The first is that, in terms of X_{AE}, the set X_{EA} is simply the set of all utility pairs

$$\left[\frac{V_1u_2+(V_2-V_1)}{V_2},\ \frac{U_2u_1}{V_1}\right]$$

where (u_1,u_2) is any utility pair in X. This reveals that, once X is determined, the set X_{EA} depends only on the ratios

$$\alpha = \frac{U_2}{U_1}\ ;\ \ \beta = \frac{V_1}{V_2}\ .$$

affect the actual state implemented by Harsanyi's theory. In Fig. 6B, the parameter U has been kept constant and V has been decreased to V^*. The utility pair chosen by Harsanyi's theory then changes from h to h^*. But the utility pair in the new situation which corresponds to the *same state* remains h. The example also indicates why unease is felt about some of the practical consequences of Harsanyi's version of utilitarianism.

How are we to know what values to assign to U and V? Recall that, with my somewhat fanciful choice of benchmark states, this requires an *intra*-personal judgement to be made about the extent that Eve will suffer in hell as compared with Adam, and a similar judgement concerning their relative felicity in heaven. If I understand Harsanyi, these judgements are to be regarded as *primitives*. They are somehow commonly held and prior to the theory. Otherwise the theory would not be implementable in real terms. But such a viewpoint seems to leave much of the story untold.

Harsanyi (1977, p. 55) asserts that there is an "unavoidable need for inter-personal comparison of utility in ethics". (See also, for example, Hammond 1976b). There is certainly a sense in which analyses of the type attempted in this paper are impossible if there is no common ground at all between the players about the extended sympathy ground-rules. I enlarge on this point in Section 4. But I do not agree that accepting this point means that one must work with *a priori* weightings of utility functions which are derived exclusively from individual or collective judgements that lie completely outside the scope of the formal analysis. This makes a second major point of difference with Harsanyi, in addition to that concerning commitment opportunities.

But, in taking up this position, I seem to be multiplying the difficulties to be faced. There are four constants, U_1, U_2, V_1 and V_2, which cannot be left floating in the air. Harsanyi reduces the number to two with his doctrine and leaves the others to be determined exogenously. But neither of these stratagems is now to be admitted. On top of this, the problem of where the extended sympathy preferences come from has still to be confronted. However, I propose to attempt to make a silk purse out of this sow's ear.

A long digression is necessary to explain how it is proposed to achieve this transformation. It is necessary at this stage for two reasons. The first is that Rawls' position on these questions is to be taken up at the end of this subsection, and it will be helpful to be able to comment on how his criticisms of Harsanyi relate to my own heavily reformulated framework. The second reason is that the next subsection on the consequences of making a no-commitment assumption is already sufficiently closely argued without being further complicated by raising questions of utility calibration.

It seems obvious to me that player 1's extended sympathy preferences should be Adam's and player 2's should be Eve's. After all, player 1 is *actually* Adam and player 2 is *actually* Eve, although they are unaware of these facts. In principle, such an identification requires a thickening of the veil of ignorance so that neither player 1 nor player 2 can deduce their actual identity from an examination of their own extended sympathy preferences. As it happens, however, the argument which follows leads to Adam and Eve being credited with *symmetric* extended sympathy preferences and so this cause of concern is only apparent. But, although this conclusion is the same as that obtained from the Harsanyi doctrine, it will soon be apparent that the route to the conclusion is very different.

Unlike players 1 and 2, who are only hypothetical entities, Adam and Eve can be asked what their extended sympathy preferences are: or, more precisely, what their extended sympathy preferences would be if they were in the informational position of players 1 and 2 respectively. But how is any answer they might choose to give to be verified? And what led them to their answers in the first place? Such elicitation difficulties raise immediate questions of *strategy* which will form the basis for the assumptions to be made in what follows about extended sympathy preferences. The requirement will be that the extended sympathy preferences assigned to players 1 and 2 (and hence to Adam and Eve) must have the property that, if players 1 and 2 had these preferences, then it would be in the interests of neither to report different preferences given that the other reports his or her "true" preferences. This assumption makes a virtue out of the necessity of admitting that extended sympathy preferences are *unelicitable* in a context like that envisaged here. The virtue lies in the fact that it allows values to be assigned to the constants U_1, U_2, V_1 and V_2. But the assumption needs a better defense than the observation that it facilitates the analysis.

The proper defense is that the logic of the whole approach *requires* the assumption. Recall the quote from Rousseau in the introduction, and the comment that "what right permits" is to be built into the *rules* of a hypothetical game (i.e. that played in the original position). All *choices*, on the other hand, are to be made *strategically* - according to "what interest prescribes". It may be objected that preferences are not the subject of choice. Certainly, economists treat an individual's *personal* preferences as a primitive which needs no explanation. Presumably, the understanding is that these personal preferences are historical artefacts generated by evolutionary processes of a socio-biological character. It is not clear to me that this is an adequate defense for treating personal preferences as fixed inputs for the type of analysis attempted here. However, doubts on this question will be put aside until Section 4. But even if the problem of what shapes *personal* preferences is put aside, there still remains the question of what shapes *extended sympathy preferences*. My answer to this question depends on an appeal to the socio-biological "explanation" of self-consciousness, although I hope it will not be thought that the paper collapses if this is rejected as far-fetched.

Very briefly, it is thought that social interaction among the primates requires that individuals incorporate a "simulator" which allows them to put themselves "in the shoes" of their fellows.[31] Thus, some form of extended sympathy is seen as a basic biological building block of the human mind. Within kin-groups sharing genes, there will therefore be a tendency to act "as one" - although not, of course, to the extent practiced by the social insects of the order *hymenoptera* with their usual genetic arrangements. But what of extended sympathy between unrelated individuals? It exists to facilitate coordination[32] and, from the socio-biological perspective, *for no other reason*. A particular level of success for a group, as compared with other groups, can be achieved in various ways. Some of these will favor individuals with certain extended sympathy viewpoints at the expense of others. The favored individuals will then be selected by evolution. Thus the extended sympathy preferences of an individual will be determined *strategically* in the long-run. Individuals may well truthfully report that they are not misrepresenting what lies deep within their hearts. But, if an individual did *not* have strategically optimal, extended sympathy views, then he or she would not be around to do any reporting.

This has been a very impressionistic sketch of what I see as the socio-biological viewpoint, but perhaps it is adequate to make it clear that there is an intellectually viable defense of the *principle* of making extended sympathy preferences the object of strategic choice. Of course, there will remain an unbridged chasm between these generalities and the particularities required later on.

After the preceeding flights of fancy, some algebra may be welcome. The no-commitment considerations of the next subsection make it *necessary* that the *strategic* choices of the extended sympathy preferences attributed to players 1 and 2 render the state-of-nature s_0 *role-neutral*. The term "role-neutral" indicates a state which, if implemented, would leave players 1 and 2 indifferent as to which role they were to occupy in that state. Thus, to say that s_0 is role-neutral means that $e_{AE} = e_{EA}$. This requirement reduces the number of floating constants from four (U_1, U_2, V_1 and V_2) down to two. That is to say, in regard to reducing the number of constants, it achieves the same effect as the Harsanyi doctrine as a result of assigning a leading role to the state-of-nature, which his *a priori* symmetry assumption relegates to the rank of a spear-carrier.

Put $\phi_A(s_0) = a$ and $\phi_E(s_0) = 1-b$. (Thus, $e = (e_A, e_E) = (a, 1-b)$ in Fig. 4A.) In this subsection, attention will be confined to the case when $a+b > 1$. Then

$$e_{AE} = (\Phi_1(A,s_0), \Phi_2(E,s_0)) = (U_1 a, 1-V_2 b)$$

$$e_{EA} = (\Phi_1(E,s_0), \Phi_2(A,s_0)) = (1-V_1 b, U_2 a).$$

The role neutrality requirement $e_{AE} = e_{EA}$ then yields the two equations

$$U_1 a + V_1 b = 1$$

$$U_2 a + V_2 b = 1.$$

(Since $0 \leq U_i \leq 1$ and $0 \leq V_i \leq 1$, e cannot be role-neutral when $a+b < 1$.) The two equations eliminate two floating constants by making V_1 and V_2 functions of U_1 and U_2. The geometry is again easy to appreciate. The restriction is that the two points, q_1 and q_2, in Fig. 7A must lie on a fixed "constraint line".[33]

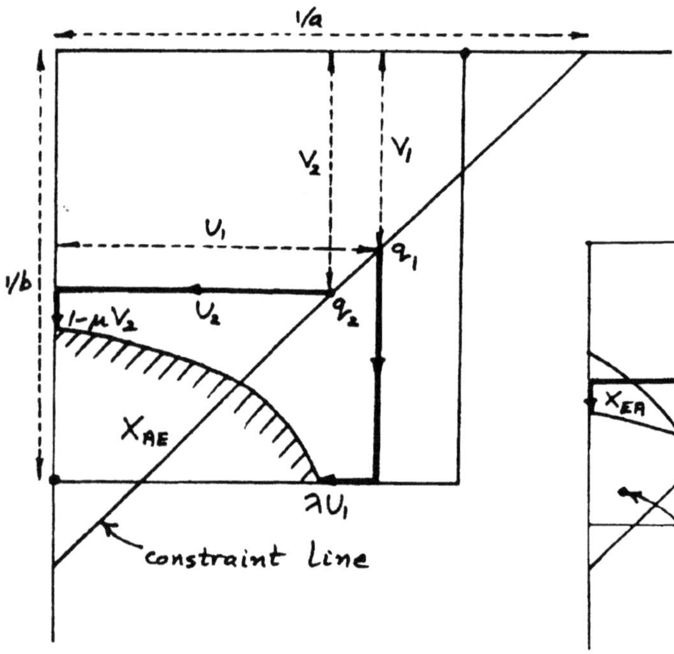

Figure 7A

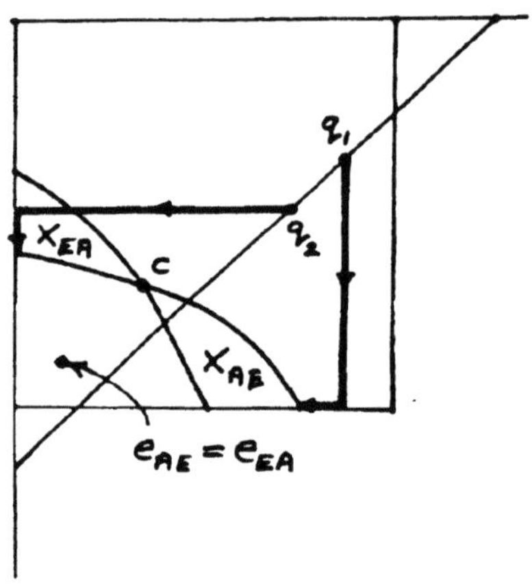

Figure 7B

What of the remaining constants? That is to say: how are q_1 and q_2 located on the constraint line? Again the answer is that these locations are determined by strategic considerations. And, as their precise locations are immediately relevant to what follows on Rawls, it will be worth pursuing the point further, although not with the rigor that a mathematician would think adequate.

The no-commitment assumptions of the next subsection make it a strategic necessity that the extended sympathy preferences attributed to players 1 and 2 render the state s_0 role-neutral. The manner in which this requirement constrains the extended sympathy preferences of players 1 and 2 is what has just been discussed. To proceed further, another conclusion must be mortgaged from the next subsection.

The argument given there identifies the outcome of rational bargaining in the configuration of Fig. 7B with the "cross-over" point c. Note that, since c will not typically be role-neutral with respect to values of U_1, U_2, V_1 and V_2 which make s_0 role-neutral, the implementation of c will involve two different states in the general case: a state s_{AE} to be implemented when player 1 is Adam and player 2 is Eve; and a state s_{EA} to be implemented otherwise. Since both generate the same utility pair c, players 1 and 2 will be indifferent between these two states. However, all that is immediately relevant about the analysis of the next subsection is that the rational bargaining outcome is a *symmetric* function of the bargaining configuration. This is made inevitable by the fact that the analysis of the next subsection is based on the use of the Nash bargaining solution (see Subsection 3.1), which is symmetric in the same sense. This does *not* mean that the bargaining *outcome* is necessarily symmetric. (For example, the cross-over point c in Fig. 7B assigns *different* utilities to players 1 and 2). It simply means that, if the bargaining positions of players 1 and 2 were reversed, then so would be the utilities assigned to them by the bargaining analysis. Sometimes a collective choice rule with this property is called *anonymous* since it takes no account of how the players are named or labeled. The reason for making this point, about the rational bargaining outcome being a symmetric function of the situation, is that it has consequences for the strategic choice of extended sympathy preferences attributed to players 1 and 2.

The value of the bargaining outcome, and hence the final payoffs to players 1 and 2, depends on the locations of q_1 and q_2 on the constraint line in Fig. 7A. Players 1 and 2 are free to choose q_1 and q_2 strategically. They are therefore involved in a game. Recall from Section 2 that, for q_1^* and q_2^* to be *Nash equilibrium* strategies in this game (Section 2), the requirement is that q_1^* be a best possible choice for player 2 given that player 1 is to choose q_2^*, and that q_2^* is a best possible choice for player 2 given that player 1 is to choose q_1^*. No eyebrows will be raised at the observation that the Nash equilibria of *symmetric* games always include a *symmetric* Nash equilibrium. But the choice problem faced by players 1 and 2 *is* symmetric. The reason is that the bargaining outcome (for example, c in Fig. 7B) depends on X_{AE} and player 1 in the same way that it depends on X_{EA} and player 2. But X_{AE} is determined by U_1 and V_2 in the same way that X_{EA} is determined by U_2 and V_1. Finally, U_1 and V_1 are determined by the location of q_1 on the constraint line and U_2 and V_2 by that of q_2. Hence, the choice of where to place q_1 on the constraint line looks exactly the same to player 1 as the choice of where to place q_2 looks to player 2. Thus, if players 1 and 2 have *symmetric* preferences, then the game they are playing is *symmetric* and therefore has a *symmetric equilibrium*.

The argument above demonstrates that there are *symmetric* choices $U_1 = U_2 = U^*$ and $V_1 = V_2 = V^*$ which are strategically coherent, although this clearly will not be true if these parameters are chosen arbitrarily. Fig. 8A is Fig. 8B redrawn in more detail for the symmetric case. Observe that X_{EA} is simply the reflection of X_{AE} across the main diagonal of the square. It is *not* true that any such symmetric configuration represents a possible equilibrium. For example, when the constraints $0 \leq U_i \leq 1$ or $0 \leq V_i \leq 1$ are not active in equilibrium, the configuration of Fig. 8B becomes necessary. (This is explained in an appendix along with other related technical points of a peripheral nature.) Nor has it been claimed that *only* symmetric configurations can be strategically coherent, but the possible existence of strategically coherent asymmetric configurations is not studied in this paper.

The immediate point to be extracted from this long aside on the consequences of making the choice of extended sympathy preferences a matter of strategy, is that it leads all the way back to Fig. 4B (compare this with Fig. 8A) but without an appeal to the Harsanyi doctrine. Even the state-of-nature point e_{AE} gets symmetrized at d^*. This is because $d_1^* = U^*a$ and $d_2^* = 1-V^*b$. But $U^*a + V^*b = 1$ by the role-neutrality requirement on s_0. Thus $d_1^* = d_2^*$. Of course, a symmetric conclusion would not emerge if no *a priori* symmetry assumptions at all were present. But the basic symmetry requirement for the above analysis, namely that U_1 and U_2 are drawn from the same range, and similarly for V_1 and V_2, seems relatively mild compared with the Harsanyi doctrine.

In spite of the fact that a symmetric situation has been achieved, there remain major points of difference with Harsanyi's theory. The most obvious is that it is not Harsanyi's "utilitarian point" h of Fig. 4B which is to be defended as the solution to the problem, but the *cross-over point* c^* (labeled r in Fig. 4B). It will be necessary to bear this in mind during the following discussion of Rawls' attitude to these matters. Equally important is the role played by the state-of-nature in providing a basis for intra-personal comparison of utilities. This issue requires some further comment.

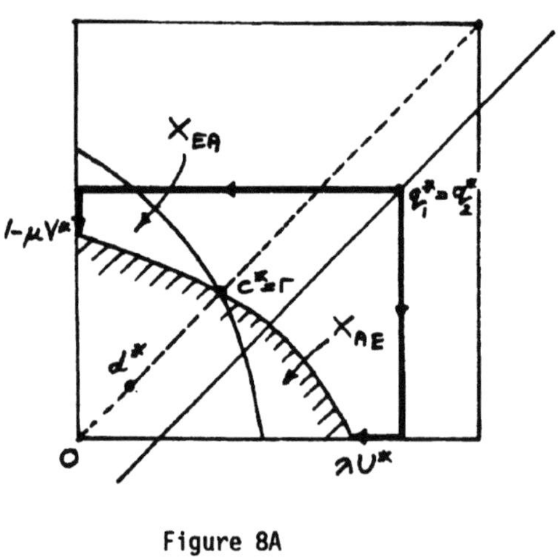

Figure 8A

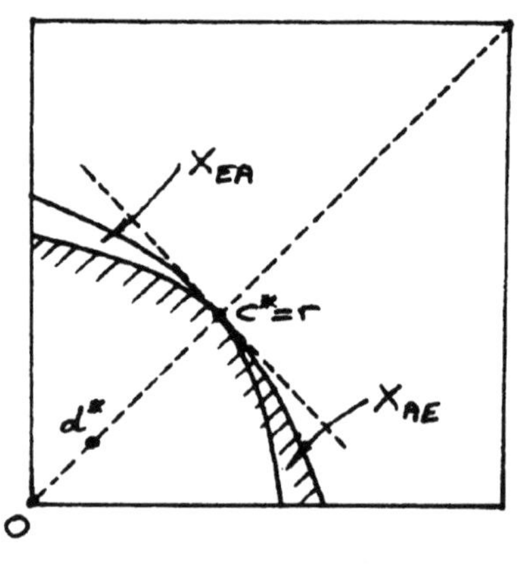

Figure 8B

The symmetric consequences of the strategic choice of q_1 and q_2 resolve most of the outstanding questions concerning intra-personal comparison. The implementation of the cross-over point $c^* = r$ in Figs. 8A and 8B is also greatly simplified because it joins the state-of-nature point d^* in becoming *role-neutral*. In fact, *all* the points on the main diagonal of these figures are role-neutral in the symmetric case. What this means in terms of implementation, is that $s_{AE} = s_{EA}$, and so players 1 and 2 will not actually make a role-contingent choice of what state to implement. They will agree to operate the *same* state no matter who turns out to be who. But there is one intra-personal comparison issue which the symmetrization complicates. The two role-neutrality equations for s_0, namely $U_1a + V_1b = 1$ and $U_2a + V_2b = 1$, are reduced to the single equation $U^*a + V^*b = 1$. Thus, one of the constants is still floating. But the loss of determinancy is only apparent. As observed earlier, the preceding equilibrium argument shows only that symmetric equilibria exist[34]: not that *all* symmetric situations are in equilibrium. The argument therefore needs to be milked again to tie down the last constant. But further consideration of this point can wait until after Rawls' view of these matters has been discussed.

So far, in this subsection a version of Harsanyi's formulation of what goes on "in the original position" has been presented, with close attention to the fine detail of the utility theory foundations. Criticism of three of his building blocks was offered: namely

(1) The assumption of full commitment opportunities;
(2) The "Harsanyi doctrine";
(3) The unexplained parametrization of extended sympathy preferences.

In addition, the ground was prepared for the manner in which it is proposed to deal with these difficulties in the next subsection. It remains to discuss Rawls' (1972) cricitisms of Harsanyi's formulation, in so far as these remain valid in respect of my own reformulation, and to comment on how his *maximin solution* fits within the general picture. It seems to me that Rawls' unhappiness with Harsanyi's theory, apart from his dislike of its consequences, centers on building blocks (2) and (3). Building block (1) is exlicitly defended by Rawls in the slave-holders argument discussed above. It is in taking up this position that I feel that Rawls is on fundamentally unsound ground. This is not to say that I am willing to concede anything whatever to his attacks on orthodox Bayesian decision theory in the current context: only that, if he were willing to push his views on the "strains of commitment" to their logical conclusion, he would not *need* to make such attacks.

If a naïve view is taken and problems of utility comparison are ignored, then the Rawlsian *maximin solution* is located at the point r in Fig. 4A. The letter r is chosen for Rawls but it might equally aptly honor Raiffa since the point r is located at one of the "arbitration points" axiomatized by Raiffa (1953). The point r

also appears in Fig. 4B where it is what I call a cross-over point - i.e., a point common to the Pareto-frontiers of X_{AE} and X_{EA}.

However, Rawls does not take a naïve view on utility comparison questions. Indeed, he is so averse to assumptions like Harsanyi's second and third building blocks, that he will have no truck with utility theory at all. Instead, he works in terms of what he calls "primary goods". But it does not seem to me that he can be allowed to dispense with utility theory and then proceed as though all the problems for which utility theory was created do not exist, however much (extended) sympathy one may feel for his distrust of the type of legerdemain with utilities with which much of this subsection has been concerned.

For example, if the axes in Fig. 4B are re-interpreted in terms of primary goods (however defined), then orthodox decision theory would require the players to be heavily risk-averse if the point r is not to be Pareto-inferior to a lottery attaching equal probabilities to h and k as advocated by Harsanyi. Rawls therefore needs to deny the validity of orthodox Bayesian decision theory in this context. Consequently, he insists that those in the original position use only *objective* probabilities (i.e. probabilities obtained by the observation of long-run frequencies) and eschew the use of *subjective* probabilities altogether. The aim is to rule out Harsanyi's lottery on the grounds that, since no *actual* coin is tossed, any evaluations made in the original position about prospective role assignments will necessarily be subjective in character. He buttresses this position by recalling the well-known difficulties with Laplace's "principle of insufficient reason" and goes on to defend his advocacy of the maximin criterion (rather than the maximization of expected Von Neumann and Morgenstern utility) with a reference to Milnor's axiomatization of "complete ignorance" (Luce and Raiffa 1957, p. 297).

It is not denied that Rawls' views on this subject have considerable force when applied to the arcane practices of naïve Bayesians in general: nor that the unthinking application of the Harsanyi doctrine to all situations is bound to generate precisely the same paradoxes as the Laplace principle. Indeed, I have written at length elsewhere (Binmore 1987a, Section 6) about the difficulties in the foundations of game theory that arise from a naïve use of Bayesian principles. But all these problems arise from a thoughtless application of a theory designed for use with "closed universe" problems to "open universe" problems. In brief, for a "closed universe" problem, it is necessary that there be no unresolved doubts about the nature of the fundamental domain of uncertainty. In Milnor's axiomatization of "complete ignorance" such doubts are embodied in his "column duplication" axiom. His theory, as its title suggests, is therefore very much an "open universe" one.

But should we follow Rawls in classifying the informational situation in the original position as "open"? I think not. After all, if we are not invoking the Harsanyi doctrine to explain the structure of extended sympathy preferences, then the fundamental domain of uncertainty is just a two-element set {AE,EA} whose members represent the two possible role assignments. No doubts can exist about what this set is, because it is given *a priori* as part of the definition of the original position. Indeed, it is *because* such issues can be dealt with so unambiguously, that the subject matter of this paper is so attractive to a game theorist.

One cannot then simply deny that individuals will make decisions *as though* maximizing the expected value of a utility function relative to subjective probabilities for the two role assignments AE and EA. Such a denial requires rejecting one or more of Savage's (1951) axioms for Bayesian decision theory.[35] But if they are to be rejected in this simplest of all possible cases, then they have to be rejected *always*. Perhaps they ought to be. But, if so, their rejection should be based on their failings as rationality principles rather than because a rather distant and doubtful consequence is found displeasing.

The Savage theory only implies that rational individuals will make decisions *as though* AE and EA occur with certain probabilities. It does not require that these subjective probabilities be equal. However, in requiring that those in the original position do attach equal subjective probabilities to the role assignments, one is *not* claiming universal validity for the Laplace principle or the Harsanyi doctrine. Indeed, the whole idea that the informational circumstances of those beyond the veil of ignorance has somehow to be *deduced* from primitive ur-assumptions strikes me as misguided. In particular, it should be a matter of *definition* that the subjective probabilities assigned to the role assignments be equal. Everything is then up-front. Fairness is being defined as "symmetry in the original position": nothing more and nothing less.

These remarks are perhaps over-emphatic. But I agree very much with Rawls (1972, p. 121) that:

> "We should strive for a kind of moral geometry with all the rigor that this name connotes."

I am also much in sympathy with the flavor of his "second principle of justice" - in particular, the difference principle or maximin criterion - as a device for settling questions of social justice. (His "first principle of justice" lies outside the scope of this paper for the reasons mentioned at the beginning of Section 3). Perhaps the "moral geometry" by means of which a version of the maximin criterion is defended here lacks the rigor it should ideally have. But at least nothing unorthodox has been concealed in the woodwork.

Of course, my claim to be in the Rawlsian camp bears up no better under close inspection than Harsanyi's claim of being a Benthamite. Harsanyi's utilitarian conclusion and my derivation of the maximin criterion both depend on suitable utility renormalizations being made (the set X of Fig. 4A being replaced by X_{AE} of Fig. 4B). In both cases, the renormalization is perhaps the most natural available in the circumstances, but critics will not find this an adequate defense. To clarify the point, it may be helpful to indicate what the outcome of the analysis looks like when no renormalization is made after Adam's and Eve's personal utility functions ϕ_A and ϕ_E have been anchored so that $\phi_A(h) = \phi_E(h) = 0$ and $\phi_A(H) = \phi_E(H) = 1$.

The *un*-renormalized set X of feasible payoff pairs appears in Figs. 9A and 9B. (The square in these diagrams is *not* the same square as that of Figs. 6, 7 and 8.) The state s* implemented by my analysis is then that which corresponds to the Nash bargaining solution n in these figures. (See the appendix for a justification.) The disagreement point is taken to be the unrenormalized state-of-nature utility pair e = (a,1-b). The Nash bargaining solution is *not* calculated relative to the whole set X. Instead, it is calculated with the shaded set Z in Figs. 9A and 9B serving as the set of feasible outcomes. Fig. 9B indicates a "corner solution". This corner solution is the point common to the frontier of X and a line joining the point e = (a,1-b) and the "bliss point" (1,1). It therefore closely resembles the bargaining solution of Kalai and Smorodinsky (1975), the difference lying in the definition of the bliss point. It is also worth noticing that n necessarily always lies on the line of slope U*/V* passing through e = (a,1-b). It is therefore a "proportional bargaining solution" in the sense described, for example, by Roth (1979).

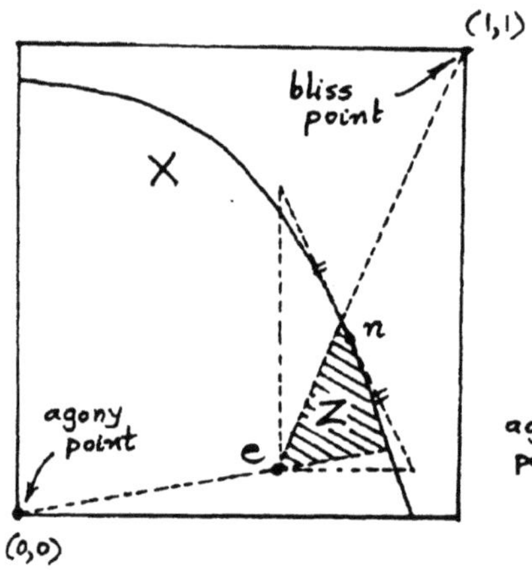

Figure 9A

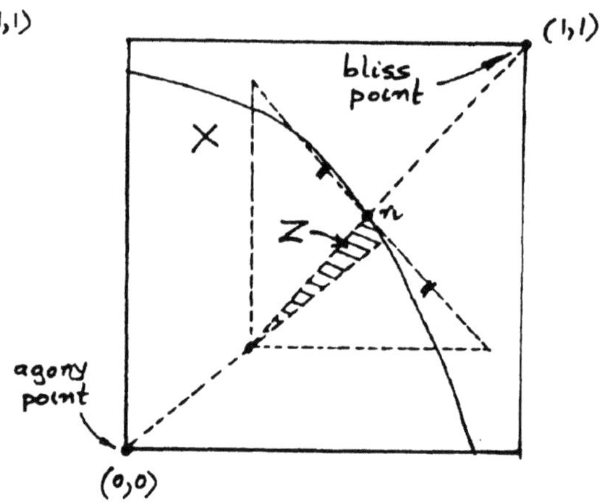

Figure 9B

None of these descriptions of the final outcome seem related to each other in any obvious way. Nor is their common relationship to the Rawlsian maximin principle at all evident. This illustrates a point which seems to me of considerable importance. There can be few, if any, amid the welter of mathematically formulated arbitration schemes and welfare criteria which compete for attention, that capture *nothing* of significance. But to evaluate what is significant in a particular scheme requires stepping back from the mathematics. In brief, all matters of any genuine importance in this connection are questions of interpretation.

3.3 No Commitment

Although the utility considerations of the previous subsection will remain applicable, it is now important to put its *static* attitude to the original position very firmly to one side. Commitment assumptions are to be entirely eliminated. As explained in Section 2, *time* then matters and it matters in many different ways. The chief reason is that, without commitment, no rational player will keep agreements unless he or she sees it to be in their current and future interest to do so. The endeavors of the past are a "sunk cost" and are treated as such. Unlike in the previous subsection, it follows that only utility pairs x corresponding to equilibria which are time-consistent, in an appropriate sense[36], belong in the feasible set X of Fig. 4A. (The objection that it is not necessarily irrational to be trusting or trustworthy (e.g. Sen 1987) was met at length in Section 2. In brief, the factual content of the objection is not denied, but it misses the point: which is that part of the aim of the analysis is to explain *why* the factual content is correct.) Time will also have a lesser, but still important, role to play in what goes on "in the original position". As in all bargaining, concern with the unproductive passage of time will have its part in determining the bargaining outcome. And so will the current state of society at the time the original position is envisaged as being evoked. This, of course, will change over time as the device of a hypothetical coordinating agreement is used repeatedly in steering society continuously from a primeval (but not necessarily "brutish") state-of-nature to a final utopian equilibrium.

As always in this paper, it is the interpretive questions which need the closest attention. What is meant by saying that a utility pair $x = (x_A, x_E)$ in the feasible set X of Fig. 4A corresponds to an equilibrium (or "state" in the terminology of Subsection 3.2)? It is important that x_A and x_E be interpreted as utility *flows*. Thus x_A and x_E record how much utility accrues, to Adam and Eve respectively, *per unit of time.* Fishburn and Rubinstein (1982) have provided the appropriate extension of the Von Neumann and Morgenstern theory, but the details will be no more relevant than the details of the latter theory were in Subsection 3.2. Admittedly, the interpretation of x_A and x_E as flows strains a formalism in which the set of feasible outcomes is envisaged as a set X in a two-dimensional space. A minimal requirement for this to make sense is that the universe outside the control of the players be construed as stationary. Such assumptions are forgivable only to the extent that they simplify the analysis.

What of the shape of the set X? In Subsection 3.1, the set X was assumed to be *convex* (and, tacitly, also closed, comprehensive and bounded above[37]). The standard justification for the convexity of X is that, if the points a and b

are both in X, then the midpoint t of the line segment ab must also lie in X, because t can be achieved by tossing a coin and implementing a if it shows heads and b if it shows tails. This is enough for X to be convex. No problem exists about this argument when commitment is possible. No problem exists either, without commitment, if the choice of random device[38] used to coordinate on an equilibrium is outside the control of the individuals whose behavior is to be coordinated (as in "sunspot equilibria"). Commitments to abide by the fall of the coin are then unnecessary, provided a and b are *equilibrium* outcomes, because it will be *optimal* for each player to do so, if he or she believes that the other plans to do likewise. But, what if nature has *not* provided a random device whose role as a coordination mechanism is beyond question? Imagine, for example, a dispute in a new nation about the side of the road on which to drive. Leftists argue that driving on the left leaves the right hand on the wheel when changing gear. Rightists doubtless have some rival argument. A conference decides to settle the dispute by tossing a coin. But why should the losers honor the agreement? They can produce another coin and argue for a further trial. If they do not behave in this way, it is because the act of honoring the agreement *is itself in equilibrium.* There may be various reasons why this should be the case. For example, repudiating the agreement may have consequences for their reputations: or, to be more reductionist, on the consequences to their future payoffs that would result from the behavior triggered in others by the repudiation.

Of course, as explained in Section 2, it is not necessary to argue that the parties to an agreement understand the mechanism that makes it equilibrium behavior to honor the agreement. It is enough that they operate some convention that generates equilibrium behavior. Their own explanation of their behavior may well be very different from that of a game theorist: but it does not follow that one of the two explanations must be wrong. The point here is *not* that the toss of the coin cannot serve as a coordinating device in such circumstances. Clearly it can. However, it cannot occupy this role alone. An elaborate system of checks and balances is needed to maintain the appropriate equilibrium. Admittedly, we are largely unaware of the necessary wheels turning, just as we are largely unconscious of what we are doing when we breathe. But, for a successful social animal, the actions required to keep society operational must necessarily assume the character of reflexes.

This point has been pursued, not so much because of its relevance to the shape of the set X, but because it will soon be necessary to consider an equivalent of the coin-tossing coordinating device *without* any back-up system of checks and balances. Its usefulness in this role will then be denied. As for the shape of X, the discussion simply shows that there is *not* a trivial argument for the convexity of X. Of course, there are other ways of mixing equilibria: like taking turns or sharing physical goods. But, such mixing does not have the same pleasant properties in respect of Von Neumann and Morgenstern utilities as probabilistic mixing.

Thus convexity necessarily has to appear as a *substantive* assumption rather than a near tautology. In many cases, it seems to me that it will be a *bad* assumption. However, to keep things simple, it will continue to be sustained in this paper, with the exception of an aside at the end of this subsection. In fact, X will be assumed to be *strictly* convex when necessary. Without such a convexity assumption the bargaining theory does not become impossible, but it does become more tiresome. (See paper 5 of Binmore/Dasgupta 1987.)

My preference would now be to proceed directly to a discussion of how hypothetical bargaining in the original position selects a point in X. However, game-theorists would then register a substantial lacuna in the discussion of the precise linkage between the strategies available to the players and the payoff pairs in the set X. Some comments on this point therefore follow, but those who know little game theory are invited to skip the next six paragraphs.

Recall from Section 2 that, in the absence of commitment assumptions, only behavior that is in equilibrium is to be regarded as admissible. And here the notion of equilibrium needs to be interpreted reasonably carefully. All agents will be assumed to be planning to optimize given their information and the plans of all other agents, *under all contingencies*, whether or not these contingencies are realized in equilibrium. Section 2 mentions Nash equilibrium and subgame-perfect equilibrium, and makes the point that the former is too weak a concept in the current context. Subgame-perfect equilibrium is the weakest of the various standard equilibrium concepts that captures what is necessary for the purposes of this paper. In what game are such subgame-perfect equilibria embedded? That is to say: what are the rules that constrain the players in making their decisions?

The position of this paper is that the players are to be understood as constrained *only* by physical and biological laws: that is to say, by "natural law" in the unsophisticated sense that this was understood in Section 1. Ethical conventions or legal statutes are *not* to be interpreted as constraints on the set of strategies from which players make their choices. The operation of the equilibrium chosen in the original position will require that "man-made laws" corresponding to such notions be honored because agents *choose* to honor them, *not* because it is impossible that matters could be otherwise.[39]

"Natural law", in this unsophisticated sense, is seen as being embodied in a *repeated* game over time. The fact that it is the *same* game, repeated over and over again, reflects the remark made earlier in this section that the underlying environment is to be assumed to be *stationary*. There would be little intrinsic difficulty in adapting the analysis to the case when more opportunities become available to society over time, so that the feasible set X grows steadily. However, the case when opportunities *contract* over time, so that X shrinks steadily, would not be so easy to analyze. Sharp conflicts would then arise if the current state-of-nature became non-viable, and it is not obvious how a bargaining analysis would proceed. The model also neglects the important possibility that the games to be played in the future are functions of *stochastic* events in the present. This seems to me to be an important omission, since I suspect that the reason why the device of the original position seems intuitively attractive to us is because the necessary considerations are similar to those which arise when *actual* bargaining takes place *before* an *actual* random event is realized.

The time interval between successive repetitions of the repeated game will be assumed to be very small so that time can be treated as a continuous variable. The players will not, of course, treat each occurrence of the repeated game as a new problem to be resolved *de novo*. Their strategic considerations, at any time, will need to take account, not only of what they know of what has happened in the past, but also of what they anticipate happening in the future. When a succession of repetitions of a game is considered as a whole, the result is sometimes called a *super-game*. The super-game to be considered here is a complicated object, and so are the strategies that can be used in it. In particular, the behavior induced by a strategy in the super-game will normally depend on the *history* of the game so far. It will therefore depend, among other things, on *time*, and hence so will the flows of utilities to the players which result from their strategy choices. To evaluate the outcome of a strategy pair, the players therefore need to evaluate a flow of utilities over time. In a manner familiar from the "permanent income hypothesis", it will be assumed that a player can always find an income flow which is constant over time that he or she regards as equivalent to any given income flow. It will be the values of these *constant* income flows which make up the payoff pairs in the feasible set X.

The preceding remarks are very general, but it is now proposed to introduce some assumptions designed to simplify the analysis. In particular, I want to strengthen the manner in which a point in X is to be interpreted. This requires some preliminary observations on how equilibria, especially those featuring apparently cooperative or altruistic behavior, are sustained. Abreu's (1986) formulation in terms of "punishment schedules" provides a simple basis for discussion. One looks at all pairs of income flows which can result from the use of (subgame-perfect) equilibrium strategies by both players. If one knows some of these pairs of equilibrium income flows, then one can construct others by using the known equilibria as prospective "punishments" which deter deviations from the behavior required in the equilibrium under construction. Such punishments are *credible* because only equilibrium behavior is ever called for. The subgame-perfect equilibria in the overlapping generations' game of Section 2 provide appropriate examples.

I propose to assume that, given any pair (u_A,u_E) of payoffs in the feasible set X, this pair of payoffs can be achieved *in the game which is repeated* (the "stage-game") if the players choose a suitable pair of actions in this game. Thus, if they choose these actions at *all* times in the super-game, they will achieve (u_A, u_E) as a *constant* income flow. In order for (u_A, u_E) to be a member of X, it is necessary that equilibrium strategies exist which generate a pair of income flows which is equivalent to the *constant* income flow (u_A, u_E). Now, consider the following strategies in the supergame. Always play the action necessary to generate the constant income flow (u_A, u_E) *until* someone deviates from this behavior. After a deviation, play the *equilibrium* strategy which generates income flows which are equivalent to a *constant* income flow of (u_A, u_E). No incentive to deviate from this prescribed play exists if the time lapse between a deviation and the response to the deviation is negligible. (It is for such reasons that time is to be treated as a continuous variable. I am aware that this creates a new technical difficulty to replace each of those which it resolves, but do not feel that these new difficulties are important in the current context.) The point of all this is that we can now construct equilibria in the super-game that are supported by punishment schedules which generate *constant* income flows over time.

The nature of the set X of feasible equilibria has now been discussed at length, along with the strategies which may be used in sustaining such equilibria. The next topic concerns the *selection* of equilibria from the set X.

Let us, for the moment, forget that the aim is to consider the implications of using the device of *hypothetical* bargaining behind a *veil of ignorance* as a coordinating device, and concentrate instead on what would be involved in allowing Adam and Eve *actually* to bargain about the selection of an equilibrium *without* their being blinkered by the informational constraints of the original position. Such bargaining would then need to be taken account of in the rules of the repeated "game-of-life" that Adam and Eve are conceived to play. What is proposed here is that they should be seen as playing a Rubinstein-type bargaining game of offer and counter-offer (Subsection 3.1) *simultaneously* with conducting their day-to-day affairs. An agreement at time t in the bargaining game is to be interpreted as an agreement on the equilibrium to be jointly operated in the future. Of course, in accordance with the no-commitment philosophy, such agreements will not *bind* the players. The agreements will be honored only to the extent that it is in their interests to honor them. A player is therefore always free to renegue on an agreement, whereupon bargaining about the future recommences.

Imagine Adam and Eve in the primitive state-of-nature s_0. This is to be thought of as an equilibrium generating the constant income flow pair d. By bargaining, they seek to agree on which, of all the equilibria in X, they should move to. Their feasible set is X and their disagreement point will be taken to be d. The understanding is that, while they are bargaining, they will continue to receive the income flows corresponding to the current state-of-nature. Thus d is their *impasse point* (in the terminology of Subsection 3.1). If they bargained forever without ever reaching an agreement or leaving to take up an outside option[40], the assumption is that they would still operate the conventions necessary to keep the current state-of-nature in being.[41] In Subsection 3.1, it was argued that, in such circumstances, it is appropriate to employ the Nash bargaining solution to locate a bargaining outcome n in the set X relative to the disagreement point d. But a gloss on this observation is now necessary.

Recall from Section 2 that, *without trust*, a problem may exist in *implementing* the agreement on the utility pair n. Following the observations of Section 2, it will be taken for granted that the various difficulties which exist on this question can be short-circuited, provided that a *continuous* path in X can be found which joins d and n and has the property that, for each point e on this path, the Nash bargaining solution for X, *relative to* e *as disagreement point*, remains n. The reason for this requirement is that the necessary adjustments, in getting from d to n, will need to be made little by little because of the entirely justified fear that, if such precautions are not taken, the other person will cheat. But once the adjustments have engendered a change of the current state-of-nature from d to e, then the whole situation is, in principle, open to *renegotiation*. If such renegotiation does not lead to a re-affirmation of the original agreement on n, then the original agreement on n would not be made, because the bargainers would predict that this agreement would *actually* result in the implementation of something else. Or, to put the same point a little differently, Adam and Eve would not include n in the feasible set because there would be no way of implementing it successfully, given their current location at d.

Notice that the points e on the continuous path joining d and n are not to be interpreted simply as way-stations on the route from the primitive state-of-nature to the final utopian state-of-nature. Each such point also corresponds to a punishment schedule that will be implemented should a player fail to carry out his or her part in moving society to the next way-station along the planned route.

On this side of the veil of ignorance, no problems at all arise in identifying a suitable continuous path from d to n. As illustrated in Fig. 10A, the straight line segment joining d and n will suffice. For each disagreement point e on this line, the Nash bargaining solution for X remains at n. The players can therefore get successfully from d to n by "trading utiles" at a fixed rate.[42] But, as will emerge shortly, matters are not quite so straightforward beyond the veil of ignorance.

The preceding paragraphs concern the implications of allowing *actual* bargaining to proceed simultaneously with the play of the repeated "game of life". But the object of the exercise is to examine the consequences of allowing the idea of *hypothetical* bargaining in the original position to be used as a coordinating device in selecting an equilibrium. Without commitment, of course, the device must be available for use *at any time*. The term "original position" is therefore somewhat misleading when used in the current context, since it invites thinking in terms of an agreement reached at some particular time in the far past. But here the word "original" simply indicates that players imagine themselves as passing behind a veil

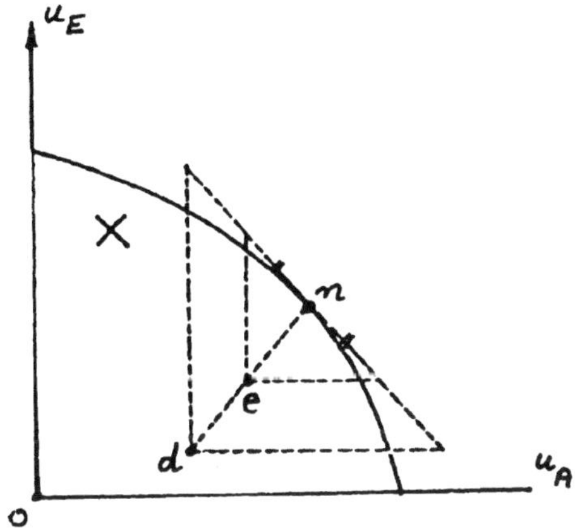

Figure 10A

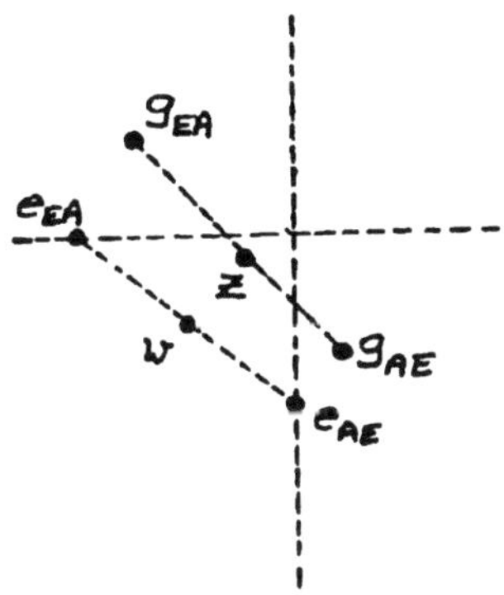

Figure 10B

of ignorance into a realm in which they no longer remember who they are in the real world. The players ask themselves what coordinating arrangements *would be* negotiated under *such fictional* circumstances and use the predictions they come up with to regulate their own *actual* behavior. If it is "common knowledge" that this coordinating device is in use, then no policing will be necessary to sustain the arrangement. Since only equilibrium behavior is advocated, each player will be optimizing given the behavior of the others. The analogy with "sunspot equilibria" is a natural one. As with "sunspot equilibria", no explanation is being offered about *how* such a common understanding might come to be established (although there is an evolutionary story that *could* be told in defence of something like the original position). And, again as with "sunspot equilibria", it is important that the coordinating device is not *itself* vulnerable to alteration through *actual* negotiation or otherwise.

It is important, if the original position is to be useful as a coordinating device, that the predictions the players make about what would be agreed under such circumstances be free from ambiguities. This remark is made because some authors seem to feel that, having perpetrated one fiction, one is then free to perpetrate other fictions governing, not only the informational circumstances of those in the original position, but also their *behavior* in the original position. They are assumed to act according to unorthodox principles of decision theory, or to be constrained by considerations of a moral character. My own view is that to proceed in this way is to lose the whole power of the approach. In any case, this paper will proceed as though players *really* believe that they *actually* can enter the original position at any time to renegotiate the coordination arrangements. Moreover, it will be assumed that the bargaining procedure followed in the original position is a common-or-garden procedure, with the bargainers acting as rational optimizers (given their beliefs) in the regular way. (*In equilibrium*, of course, rational players who believed that they really could enter the original position would never *actually* do so, because they would anticipate what would be agreed in advance and act upon the anticipated agreement *immediately*.) The "game of life" is therefore to be replaced by an expanded game in which players have the opportunity to bargain behind the veil of ignorance about how to coordinate in the future while simultaneously carrying on their day-to-day affairs. This expanded game will be analysed just as if the bargaining opportunities were actually available rather than being a hypothetical construct.

The bargaining in the original position is designed to coordinate the activities of the players in shifting society from a primeval state-of-nature through a continuous succession of intermediary states to a final utopian state. At any time and under any contingency, any player may call for a return to the original position for a renegotiation of the path to be followed in the future. But there is a cost involved in so doing. While bargaining in the original position proceeds, players receive only the income flows appropriate to the state which was current at the time the original position was entered. Progress towards the utopian state is therefore halted. This provides a disincentive for calling upon the original position unnecessarily. Of course, in the real world, a person cannot *actually* invoke the original position when someone is thought to have "cheated" by departing from the equilibrium behavior implicitly agreed upon. Instead, a punishment schedule is initiated and sustained until the deviation is rendered unprofitable. But, notice that, with the punishment schedules described earlier, the hypothetical invokation of the original position has the *same* effect as the initiation of such a punishment schedule.

What do players know in the original position? Certainly they are *not* to know their identities. Agent 1 will not know that he is Adam and agent 2 will not know that she is Eve. Instead, each will assign probability 1/2 to the two possible role assignments. All other information they have about the actual world will be assumed to remain available to them. Here, again, the object is to minimize on the fictions to be promulgated.

These informational assumptions mean that one must contemplate, not a single original position, but many "original positions", each of which corresponds to a possible history of events in the real world. Unless commitment assumptions are to be re-introduced via a side entrance, it is necessary that the bargaining between players 1 and 2 in any one of these many "original positions" is conducted *independently* of the bargaining in other "original positions". To be more precise, it will be assumed that, in *each* "original position", players 1 and 2 bargain their way to what they see as an optimal deal, taking the deals reached in other "original positions" as *given*. The situation envisaged is similar to that in a Nash equilibrium except that it is not *individual* deviations which must be ruled out as unprofitable, but *joint* deviations in any particular "original position".[43] As in Selten (1975), it sometimes eases the conceptual side of things to imagine each player split into a number of "agents": one for each of his or her potential decision points. The collection of such agents is then sometimes called a "team", following Marshak and Radner (1972). Each agent in a team has the *same* preferences as his or her team-mates but makes decisions *independently* of the others. Such a formulation makes it impossible to fall into the error of supposing that what a player plans to do at one decision point (i.e. under one contingency) can bind or commit the player to any particular action under another contingency. The formulation may also help to defuse Hare's (1975) complaint about Rawls' (1972) use of tenses. With agents bargaining independently for each possible state-of-nature, *when* the hypothetical bargaining takes place in real time ceases to be an issue.

Consider agents 1 and 2 in any one of the many possible "original positions". They know the sets X_{AE} and X_{EA}, and also the points e_{AE} and e_{EA}. These points represent the income flows associated with the current state-of-nature for the two possible role-assignments. The disagreement point for the two agents is therefore $w = (e_{AE} + e_{EA})/2$. Thus the two agents are faced with the situation illustrated in Fig. 5A, but with utilities re-interpreted as *flows*. But there is an important difference between the extension of Harsanyi's bargaining-with-commitment approach, as described in Subsection 3.2, and the bargaining that takes place in this new situation. This lies in the necessity of finding a continuous path from the disagreement point to the agreed outcome. The issue is complicated by the veil of ignorance and its associated trappings.

Suppose that agents 1 and 2 agree on a deal which they evaluate as equivalent to the utility pair x. This deal will only be genuinely feasible if there is a continuous sequence of adjustments which leads from the current state-of-nature to the utopian equilibrium corresponding to x in the real-world. Since the agents do not know the role assignments in the real-world, they therefore need to specify *two* paths: one from the current state-of-nature to the utopian equilibrium to be implemented when agent 1 is Adam and agent 2 is Eve, and one to be implemented when the roles are reversed. But an immediate incoherence arises unless the current state-of-nature is role-neutral (i.e. $e_{AE} = e_{EA}$, as defined in Subsection 3.2). This is explained below.

In Fig. 10B, e_{AE} is *not* role-neutral.[44] As a first step in implementing their deal, agents 1 and 2 agree to move from e_{AE} to g_{AE}, if agent 1 is Adam and agent 2 is Eve; and to move from e_{EA} to g_{EA}, if agent 1 is Eve and agent 2 is Adam. The result of this move is evaluated as being equivalent to $z = (g_{AE}+g_{EA})/2$ by agents 1 and 2. In the real-world, Adam now informs Eve that he has predicted what would have been agreed to in the original position, and, in consequence, action should be taken to move from e_{AE} to g_{AE}. But now a version of the slave-holder's dispute ensues (Subsection 3.2). A travel-stained Eve explains to a bright-eyed Adam about the realities of a trustless world. In the heroin-trading story of Section 2, for example, it *cannot* be optimal for the seller and the buyer to agree to settle the terms of trade on the toss of the coin, because it would not be optimal for the trader disadvantaged by the fall of the coin to honor the contract. To be more precise, if it *were* optimal in the original position to agree upon the utility pair z, then Eve would rather return to the original position than implement the terms of the deal supposedly reached in the original position. Adam would therefore be told that his account of the situation in the original position contains an inherent contradiction. He would also doubtless receive advice on how he could dispose of his hypothetical coin in a manner which would not be to his advantage.

It is this very simple point upon which my inverted pyramid balances. If it is rejected, the whole edifice collapses. But recall that Rawls (1972, p. 167) *would* reject the argument and, implicitly, Harsanyi also. However, perhaps Rawls (1972, p. 145) may be quoted against himself:

> "...they will not enter into agreements they know they cannot keep, or can do so only with great difficulty. Along with other considerations, they count the strains of commitment."

Later, Rawls (1972, p. 178) observes that "self-respect" may be counted upon to take up the strain. He follows this with various comments about the nature of self-respect. I do not see what he is getting at when he asserts that. "Self-respect is, not so much part of any rational plan of life, as the sense that one's plan is worth carrying out." This seems to beg all the relevant questions. The later assertion that "self-respect is reciprocally self-supporting" fits much more comfortably into my framework. Here the existence of an equilibrium is being proposed. But what is it that sustains this equilibrium?

Of course, within a fully fledged "natural law" setting, different ground-rules apply than those employed in this paper. However, my own feeling is that *any* argument which led to a defense of Adam's position would leave me looking very closely indeed at the "natural law" hypotheses on which it was based.

One immediate consequence of rejecting a hypothetical coin as a practical coordinating device, is that *no agreement at all is possible in the original position* unless the state-of-nature point is role-neutral. Thus, for any progress at all, it is necessary that the extended sympathy preferences attached to agents 1 and 2 must ensure that $e_{AE} = e_{EA}$. If the evolutionary defense of the existence of extended sympathy preferences is taken seriously, one must therefore hypothesize that

$e_{AE} = e_{EA}$: otherwise the existence of extended sympathy preferences would confer no evolutionary advantage.

A second immediate consequence is that any agreements can only be on a pair of states, s_{AE} and s_{EA}, with the property that both agent 1 and agent 2 are *indifferent* about which is implemented. Otherwise, there will always be someone who would prefer the hypothetical coin to be hypothetically tossed again.

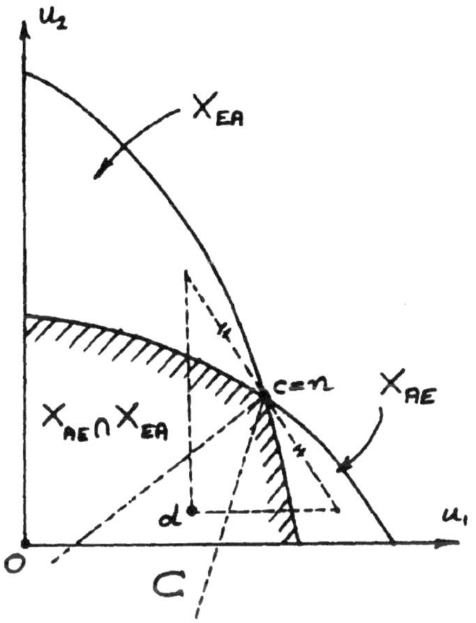

Figure 11A

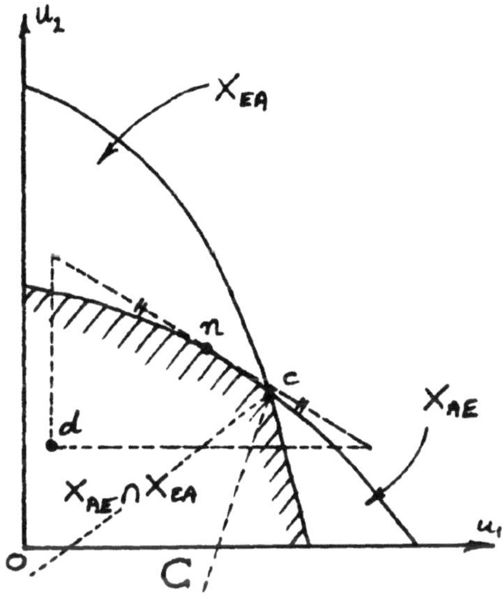

Figure 11B

Consider now the primeval "original position". The problem faced by the "primeval" agents is illustrated in Fig. 11A. Recall, from Subsection 3.2, that their extended sympathy preferences (and hence those attributed to Adam and Eve) are envisaged as being strategically determined. This allows an immediate justification for the primeval state-of-nature being assumed to be role-neutral. Then $d = e_{AE} = e_{EA}$ serves as the disagreement point for the primeval agents. Next, their feasible set must be determined. Since they are confined to deals which yield the *same* utility pair whatever the role assignment, the feasible set is $X_{AE} \cap X_{EA}$. That is to say, it is the set of all utility pairs which are *common* to X_{AE} and X_{EA}. It will be apparent from Fig. 11A that there is a region (labeled C for cone in the figure) with the property that, if d lies within C, then the Nash bargaining solution n for the feasible set $X_{AE} \cap X_{EA}$, relative to the disagreement point d, lies at the point c. This means that, if d lies in C, then the rational bargaining outcome is located at the cross-over point c. If d lies outside C, then the rational bargaining outcome n is either the same as the Nash bargaining solution of X_{AE} or the same as that of X_{EA}. (See Fig. 11B.)

However, the important point has already been made: namely that the feasible set is $X_{AE} \cap X_{EA}$. If the primeval agents have *symmetric* preferences, then Fig. 11A becomes symmetric, as illustrated in Fig. 8A (Fig. 8B is an important special case). The rational bargaining outcome (that is to say, the Nash bargaining solution of $X_{AE} \cap X_{EA}$ relative to the symmetrically placed disagreement point d^*) is therefore necessarily equal to c^* in Fig. 8A, as claimed in Subsection 3.2. This is just the Rawlsian maximin solution.

An *appendix* is concerned with identifying which of all the possible sets of symmetric preferences that can be attributed to players 1 and 2 are consistent with the principle of strategic choice introduced in Subsection 3.2. All that was shown there was that, if equilibria exist at all (and they do), then symmetric equilibria exist. Notice, however, that the content of the appendix is not necessary for the above defense of the Rawlsian maximin criterion.

It remains to confirm that c^* is actually achievable in a world without trust. Is it possible to get from d^* to c^* along a continuous path? The answer is that the straight-line segment joining d^* and c^* will suffice. But the fact that this is correct is not entirely obvious. As soon as the first step along this path is taken, a new state-of-nature is established. Appeals may then be made for a *review* of what should happen next. The appeal board will consist of the two agents in the "original position" corresponding to what is *now* the current state-of-nature. If the appeal-board agents are assigned the *same* extended sympathy preferences as the primeval agents, they will have no reason either to change the decisions made by the primeval agents nor cause to wish to change the preferences assigned to them. The reason is that symmetry makes all points on the line segment joining d^* and c^* role-neutral. But then, precisely the same arguments which applied to the agents in the primeval position apply also to the agents on the appeal board.

An objection that might be made to this line of argument is the following. No appeal board corresponding to a state which is *actually* visited while society moves from the primeval-state-of-nature to the final utopian state will contain an agent who would wish to report different extended sympathy preferences from the primeval agent on his team. However, agents on appeal boards which correspond to states which are *not* actually visited would typically *not* wish to report the primeval extended sympathy preferences, because these will not render *their* state-of-nature role-neutral. In brief, the preferences assigned to the agents *would* change *if* the equilibrium path from d^* to c^* *were* to be abandoned. Of course, the argument is that agents on the equilibrium path will not *wish* to leave the equilibrium path. However, a critic might observe that the argument identifying the feasible set with $X_{AE} \cap X_{EA}$ requires attributing the extended sympathy preferences of the *primeval* agents to Adam and Eve. Should not the feasibility of an outcome be evaluated instead by assigning to Adam and Eve the extended sympathy preferences of the agents on the *last* appeal board which reviewed the decision leading to that outcome? This consideration may seem excessively nice but there are game theorists who would argue this way (e.g. Friedman 1986, p. 114). I think such an argument is unsound in general. But in the particular case with which this paper is concerned, it seems enough simply to comment that, although various hypothetical agents may have various hypothetical preferences under various hypothetical circumstances, Adam and Eve are *actual* persons and hence can only have *one* set of extended sympathy preferences each.

This section concludes with a brief aside on the case when X is not convex and comprehensive. (The latter means that "free disposal" is assumed and so, if a certain pair of utilities is feasible, so is any pair which assigns the players smaller utilities.) As remarked earlier, these assumptions on X were introduced to simplify the analysis; not because they are thought to describe a necessary state of affairs. Without commitment, matters are more complicated than is often supposed. It is interesting that this aspect of the commitment problem seems to be generally recognized in philosophical discussions of Rawls' theory, albeit only implicitly. Emphasis is placed, as in Wolff (1977, p. 30-32), on the fact that the Rawlsian maximin criterion sometimes selects an outcome which assigns the players *unequal* utilities. It is then pointed out that the outcome selected is a Pareto-improvement of the best equal-split outcome. This phenomenon is lost when only strictly convex comprehensive X are considered because the best equal-split outcome then has no Pareto-improvements. To put it crudely, "maximining" is then equivalent to "equi-splitting".

No defense is to be made of the maximin criterion, as such, for feasible sets X which are not convex and comprehensive. Although I have not studied the question closely, I believe that the strategic choice of extended sympathy preferences will usually (and perhaps always) generate a situation in which the best equal-split point has no Pareto-improvements, even when X is not convex and comprehensive. But where this is not the case, it is not to be anticipated that the maximin criterion will emerge. My guess is that uniqueness will be lost and all Pareto-improvements of the best equal-split point will be admissible. I would still regard such a result as a vindication of Rawls but see that there is room for more than one opinion on this issue.

4. Cooperative Bargaining Solutions

In Subsection 3.2, it was noted that the Rawlsian maximin criterion, when translated from an extended sympathy preference framework into a personal preference framework, leads to a hybrid "cooperative bargaining solution" which mixes properties of the Nash bargaining solution with those of a "proportional bargaining solution" (Roth

1979). Since the "axioms" which characterize the former explicitly *forbid* inter-personal comparison of utilities in determining the bargaining outcome, while those which characterize the latter explicitly *require* such inter-personal comparison, it is not surprising that the extent to which the hybrid notion mixes the two bargaining solution depends on the strength of the *a priori* assumptions about the inter-personal comparison of utilities from which one begins.

In this section, it is proposed to look more closely at the *a priori* assumptions concerning inter-personal comparisons of utility introduced in Section 3, and to consider the consequences of varying these. A natural extension is to examine the consequences of allowing, not only extended sympathy preferences, but also personal preferences to be determined strategically. This examination will relate the ideas of the paper to the literature on "envy-free" allocations and the market mechanism (e.g. Thomson and Varian 1984).

In Section 3, it was taken for granted that Adam's and Eve's *personal* preferences are fixed and commonly known. This assumption will be maintained for the moment. Two benchmark states, h and H, were chosen to provide anchoring locations for their personal utility scales. It was assumed to be commonly understood that both Adam and Eve prefer the state H (heaven) to any state which might arise in practice and that both prefer any such state to the state h (hell). Their personal utilities are then normalized so that both attach utility 1 to state H and 0 to state h. Further assumptions were then made about their extended sympathy preferences (and hence about those of agents bargaining in the original position). These were assumptions about the relative well-being of Adam and Eve in the two states h and H. It is important to the analysis that these assumptions represent fixed and commonly known constraints on the extended sympathy preferences. The actual assumptions made in Subsection 3.2 were only one of many possible assumptions that might have been made. In what follows, the implications in respect of Adam's and Eve's final personal utilities will be explored for various alternative assumptions. Only assumptions which lead to *symmetric* extended sympathy preferences for those in the original position will be considered. This is not because the analysis fails in asymmetric cases (although it will no longer, in general, be true that the Rawlsian maximin will be implemented). It is simply because this seems the leading case of interest.

In Subsection 3.2, the parameter U^* is the utility commonly assigned to being Adam in heaven and the parameter $1-V^*$ is that commonly assigned to being Eve in hell. Matters are normalized so that the utility commonly assigned to being Adam in hell is 0 and that of being Eve in heaven is 1. The requirements that $U^* \geq 0$ and $V^* \geq 0$ will always be taken for granted. Further constraints, that $U^* \leq 1$ and $V^* \leq 1$, were made in Subsection 3.2, but now some alternatives will be considered.

Perhaps the simplest alternative requirement is that U^* and V^* be assigned fixed *a priori* values. Two cases must then be distinguished, depending on whether or not $aU^* + bV^* = 1$. Here a and $1-b$ are the personal utilities assigned by Adam and Eve respectively to the primitive state-of-nature s_0. If $aU^* + bV^* \neq 1$, then s_0 is not role-neutral and hence no progress towards a mutually better state can be achieved by the method of the paper. If $aU^* + bV^* = 1$, the utopian state finally achieved assigns Adam and Eve the personal utility pair u illustrated in Fig. 12A. It is therefore *a proportional bargaining solution* as discussed, for example, by Kalai (1977) and Roth (1979), following Raiffa (1953). In brief, it is the point u on the frontier of the feasible set X which lies on a line of given slope through the "disagreement point". In this case, the disagreement point is identified with the primitive state-of-nature point $e = (a, 1-b)$. Axiomatic defences of the notion are based on an axiom requiring "full inter-personal comparison of utilities". The slope of the line joining e and u is equal to U^*/V^*, and hence depends on how the relative well-being of Adam and Eve in the benchmark states h and H is evaluated.

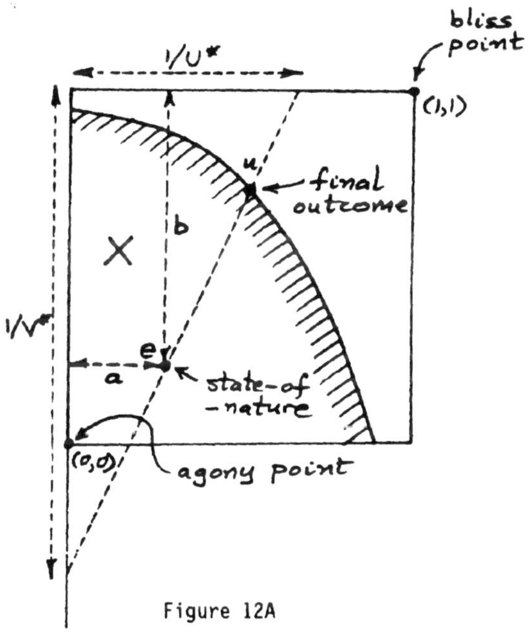

Figure 12A

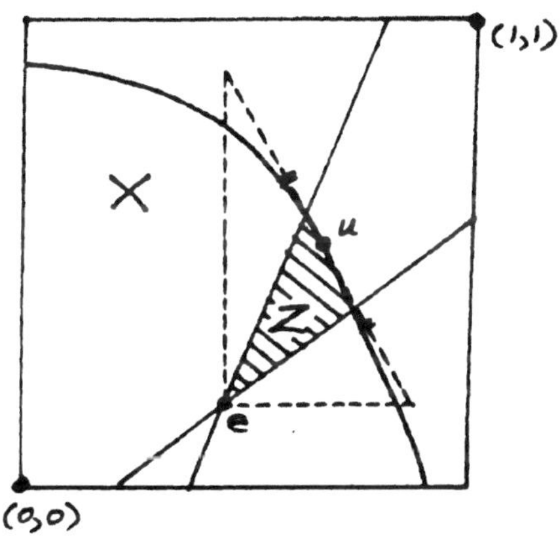

Figure 12B

Utilitarians, if they are willing to follow the argument this far, may care to note that a defense of their position can be mounted at this late stage. As long as the utilitarian outcome is a Pareto improvement on the primitive state-of-nature (which it may not be in Harsanyi's formulation), it can be implemented by choosing U^* and V^* to satisfy $aU^* + bV^* = 1$ and then arranging that the slope U^*/V^* takes an appropriate value. Fixing the parameters in this way engineers a commitment on the players' behavior that they cannot engineer for themselves. It may be thought that such a choice for U^* and V^* is too arbitrary to be considered seriously. However, if any credence is given to the quasi-evolutionary story offered in Subsection 3.2 about the origins of extended sympathy preferences, then this particular choice of U^* and V^* need not be seen as arbitrary. The argument throughout has been that a strategic choice of U^* and V^* *must* yield $aU^* + bV^* = 1$; otherwise the holding of extended sympathy preferences would convey no benefit. In selecting U^* and V^* subject to this constraint, mother Nature[45] is like a player in *Harsanyi's* original position, because her choice of U^* and V^* is not subject to revision. In essence, her choice *commits* Adam and Eve. She will therefore choose the slope U^*/V^* to implement a utilitarian outcome for precisely the reasons that Harsanyi describes.[46]

But such a defense of utilitarianism requires postulating a close relationship between the environment, that is to say the feasible set X and the current state-of-nature point e, and the parameters U^* and V^*. It is not hard to see how such a relationship could develop between these parameters and the current state-of-nature. Even amongst the disadvantaged the idea that we somehow all deserve our lot is quite remarkably pervasive. But it is difficult to see how the *detailed* structure of the perceived feasible set X would get written into our extended sympathy preferences without mediation by our critical faculties. Indeed, there are those who would maintain that we have such critical faculties largely *because* mediation in such matters is *necessary* for the survival of human societies. Without the capacity for such mediation, a society's coordinating arrangements would be unable to adapt quickly to external shocks. But none of this is to deny that society's accumulated experience about *past* environments may not appear in the form of constraints on the extent to which extended sympathy preferences are able to respond to external shocks. As in much else, there will always be a trade-off between the efficiency of the system in one particular environment and its capacity to adapt flexibly to changes in that environment. I have sought to capture this idea by envisaging the extended sympathy preferences of the individuals in society as being determined *strategically* subject to *constraints*. The strategic choice allows for flexibility: the constraints allow for close adaptation to a particular type of enviroment.

The particular constraints, $U^* \leq 1$ and $V^* \leq 1$, employed by way of example in Subsection 3.2, will serve to indicate the dangers of inflexibility. These constraints make it *impossible* to render the state-of-nature role-neutral when $a + b < 1$. A society subject to these constraints and operating a coordinating system of the type discussed in this paper, would therefore be unable to respond at all if shocked from a situation with $a + b > 1$ to one with $a + b < 1$.

On the other hand, without constraints at all on the strategic choice of extended sympathy preferences, the coordinating system of this paper has no ethical benefits whatsoever to offer to society. The analysis of the *appendix* then goes through without it being necessary to make allowance for the possible existence of "corner solutions". The conclusion is then that the point u of Fig. 12A will simply be the Nash bargaining solution for X relative to the disagreement point e. That is to say, the result is the same as would be obtained as a consequence of bargaining (hypothetical or otherwise) *without* a veil of ignorance. This observation echoes a point which is often made (e.g. Harsanyi 1977, p. 55, Hammond (1976b)) about inter-personal comparison of utilities being essential as a basis for ethical behavior. My view differs from that commonly held on this issue only in that I do not feel comfortable without an explanation of the origin of the inter-personal comparisons to be made.

Tying down U* and V* totally, leads to a "proportional bargaining solution". Leaving them completely unconstrained leads to the Nash bargaining solution. However, it has been argued that the interesting cases are the intermediate ones in which U* and V* are *partially* constrained. The constraints studied in Subsection 3.2 are only one of many that might be imposed. If the constraints take the form of restricting U* and V* to fixed intervals, then a generalization of the situation depicted in Figs. 9A and 9B is obtained. The generalization is shown in Fig. 12B. The constraints determine two rays through the primitive state-of-nature point e. The final outcome point u is then the Nash bargaining solution for the feasible set Z consisting of all points in X between the two rays.[47] The result is what might be called a "hybrid bargaining solution" in that it incorporates features both of "proportional bargaining solutions" and of the Nash bargaining solution.

Such a conclusion may seem satisfactory for the two-person case. But a major difficulty has been finessed. Such "cooperative bargaining solutions" take for granted that *personal* preferences are given. But much of what has been said about extended sympathy preferences also applies to personal preferences. In particular, when people interact it will often be in their individual interests to misrepresent their preferences. But if a person systematically chooses to reveal, by his or her behavior, a certain set of preferences, then, from an operational point of view, these *are* the person's preferences. It is not all evident, indeed, that a person will be readily able to distinguish between his or her "true" preferences and those which he or she has chosen to reveal over a long period. It is a cliché that we tend to become what we pretend to be. Such considerations, which seem to me inescapable in this context, raise the question of how the conclusions of this section are affected if *personal* preferences, as well as extended sympathy preferences, become the object of strategic choice.[48]

The following discussion will be confined to the case in which what has to be decided consists simply of the distribution of physical commodities between the two persons involved. Two subcases will be distinguished. In the first, everything available for distribution will be assumed to be already part either of the first person's endowment or of the second person's endowment. In the second, neither person will be assumed to hold any initial property rights in what is to be distributed.

The two subcases are treated in papers 9 and 10 of Binmore/Dasgupta (1987). In paper 10, it is shown that, if only mild constraints are placed on the personal preferences that are reported, then, almost regardless of the cooperative solution concept used, the final outcome in the first subcase will be a *Walrasian equilibrium*[49] relative to the initial endowment point. (See also Hurwicz 1979, and Schmeidler 1980). In paper 9, under similar conditions, Sobel shows that the final outcome in the second subcase will also be a *Walrasian equilibrium*, but *not* relative to the initial endowment. Instead, each commodity is divided equally between the two persons involved and the resulting equal-split allocation is treated as the endowment point from which the Walrasian equilibrium is calculated. It is interesting that this is the *same* conclusion that others have derived from axiomatic definitions of "equitable" or "envy-free" allocations. (See Thomson and Varian (1984) for a survey.)

It is currently fashionable to defend, and even to advocate, the market mechanism in an *ethical* context, and conclusions like those mentioned above may seem to provide some support for such views. Varian (1975), for example, makes some sense of Nozick's (1974) observations on these matters by using the "envy-free" criterion as the standard for a "fair" allocation.[50] My own view is that the idea of a Walrasian equilibrium has much the same status, when personal preferences may be strategically chosen, as the Nash bargaining solution has when they are fixed: that is to say, its use maximizes the ability of the system to respond quickly and flexibly to external shocks, but at the cost of denying society the ethical benefits that a more rigid distributive scheme could make available.

Consider, for example, the following situation. A benefactor presents two bottles of vermouth and twenty bottles of gin to Adam and Eve on condition that they agree on how to split the donation between them. Adam mixes martinis only in the ratio of one part of vermouth to 100 parts of gin. Eve mixes martinis only in the ratio of one part of vermouth to 10 parts of gin. Neither has any use for gin or vermouth beyond mixing such martinis. An equal split of the donation gives 1 bottle of vermouth and 10 bottles of gin to each. Any trade makes Eve worse off and so no trade will take place. That is to say, the equal-split endowment is already the Walrasian equilibrium. But one can hardly say that it is "fair" in any meaningful sense. It is true that Adam does not envy Eve's allocation and that Eve does not envy Adam's, but Eve is able to enjoy 10 times as many Martinis as Adam

Figs. 13A and 13B are of the familiar Edgeworth box. The indifference curves have been chosen so that, in both cases, a diagram showing the feasible set in terms of utility pairs will be symmetric. Fig. 13A is meant to illustrate Subcase 2 with all of one commodity originally owned by Adam and all of the other by Eve. Fig. 13B illustrates Subcase 2 and so the initial endowment is placed at the equal-split allocation.

The purpose of drawing these figures is not to decry the market mechanism as an efficient, cheap and flexible distributive system: it is simply to press the point home that it operates independently of ethical considerations. A society that relies exclusively on the market mechanism for distributive purposes therefore cuts itself off from the benefits to be derived from ethically based interactions. To gain such benefits, within the framework described here, it would be necessary to impose substantive constraints on the strategic choice of personal preferences as well as on the extended sympathy preferences.

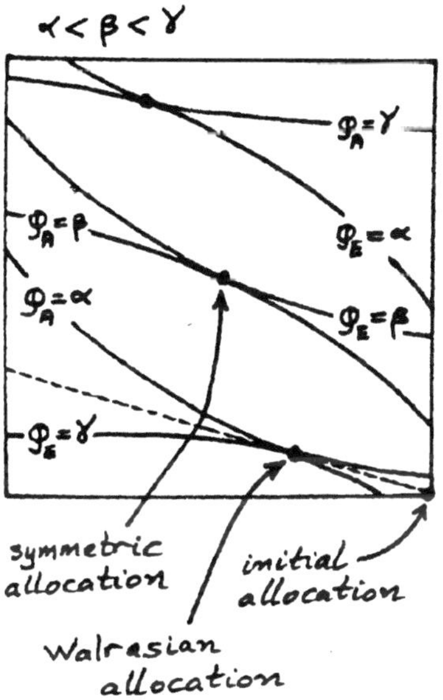

Figure 13A

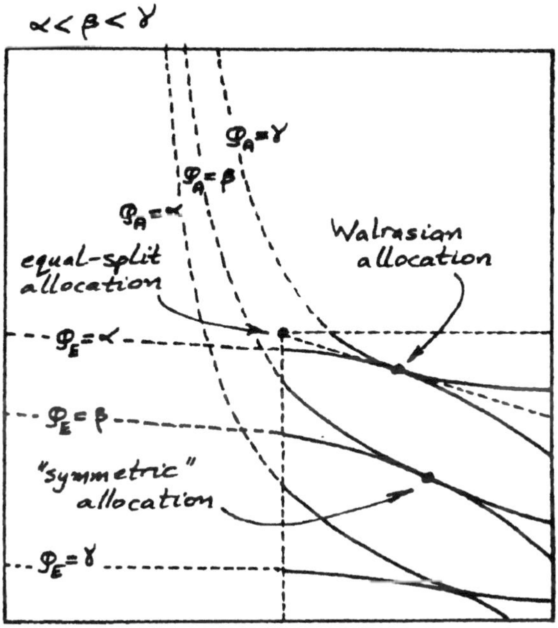

Figure 13B

5. Who is a Citizen?

In all the preceding analysis, only two-person societies have been considered. Of course, much of what has been said is immediately relevant to interactions between two *classes* of individuals.[51] In this section, it is proposed to comment on the difficulties which arise when more than two persons or classes are considered. I believe these difficulties to be more deep-rooted than usually seems to be assumed. This is not meant to imply that one cannot extend the preceding two-person analysis to the n-person case and so obtain a theory parallel to that of Harsanyi and Rawls.

Technical problems would need to be finessed[52], but not in a manner likely to cause consternation. However, I believe that such a straightforward extension would neglect various questions of importance that I do not see how to incorporate adequately into the framework of this paper.[53] One of these questions concerns society's membership criteria. That is to say: who is my neighbor?

Any theory worth its salt should give an endogenous answer to this question. It is worth quoting Sen (1970, p. 141) at length on this point:

> "A half-jocular, half-serious objection to the criterion of fairness of Rawls and others runs like this: Why confine placing oneself in the position of other human beings only, why not other animals also? Is the biological line so sharply drawn? What this line of attack misses is the fact that Rawls is crystallizing an idea of fairness that our value system does seem to have, rather than constructing a rule of fairness based on some notion of biological symmetry. Revolutions do take place demanding equitable treatment of human beings in a manner they do not demanding equality of animals. 'If I were in his shoes' is relevant to a moral argument in a manner that 'If I were in its paws' is not. Our ethical system may have, as is sometimes claimed, had a biological origin, but what is involved here is the *use* of these systems and not a *manufacture* of it on some kind of biological logic. The jest half of the objection is, thus, more interesting than the serious half."

This quote is interesting, not only because it raises the question of who or what is to be counted as a citizen, but because it raises it in a (socio-)biological context. Clarification on the latter issue is therefore necessary before the former can be addressed.

I certainly do believe that the place to look for the origins of the ethical systems we operate is in our evolutionary history: not only in the evolutionary history of *homo sapiens* as a species, but also in the evolutionary history of human culture. But I do not, for one moment, believe that what has *actually* evolved is universally describable in terms of rational coordination mediated by hypothetical bargaining in the original position. *Sometimes*, of course, such a description may fit the facts reasonably closely, and it is not hard to think of socio-biological arguments why this should be so. But, for much of what we face in modern society, it seems obvious that things have moved too fast for there to have been time for any kind of close adaptation to our current environment to have evolved, and that many of our difficulties arise from operating outmoded systems adapted to some other environment.

My concern is therefore not with the *use* of existing ethical systems (as Sen attributes to Rawls), but with the *manufacture* of *ideal* systems. The purpose is largely *rhetorical*. Those, for example, who argue that the market mechanism should reign supreme in every aspect of our lives need to be opposed.[54] But it is hard to win a public debate when one's own view is complex and that of the opponent is simple. Among other things, the model of this paper can be seen as providing a *counter-example* for use against the opposition in such a debate. For this purpose, it has the virtue of being constructed according to the "logic" of the opposition. In particular, a cynical view is taken of human motivations, in that self-interest is always assumed to be triumphant in the long-run. When pressed on how agents determine what is in their best interests, defenders of this view, as in this paper, fall back on an evolutionary explanation. Moreover, the opposition is much concerned with *legitimacy*. By and large, the current state-of-nature is somehow seen as legitimized by history and attention concentrates on the legitimacy of the *process* by means of which future states are achieved. This view, also, is built into the assumptions of the paper. But, as has been shown, the model of the paper demonstrates that these hypotheses are *not* inconsistent with the operation of a coordination mechanism which operates according to Rawls' maximin criterion for social justice. This is of some importance because it seems to me that no utopian scheme founded on a denial of one or other of the hypotheses has any prospect of being implemented successfully. That is to say, the conservatives are right to this limited extent. There are few who are sufficiently clear-sighted not to regard whatever power and privilege they may hold as only just and right; and evolution ensures that the successors, even of the high-minded, will soon be corrupted. It is true that it would be more pleasant to live in a different jungle, but this is the jungle in which an ethical system must survive.

The point of this digression on aims is to make it clear that I am not entitled to Sen's defense of Rawls on the citizenship question. The role of animals is not at issue for this paper, although I agree with those who maintain that a properly general discussion of social justice would need to treat the role of animals seriously. The reason is that coordination by means of hypothetical bargaining in the original position requires a capacity for imagination and analysis of which an animal is not capable. This is also true of the senile, the desperately ill, the feeble minded and of young children. The assertion that such groups, by their very nature, are *necessarily* excluded from citizenship, is liable to misunderstanding. A citizen, for the purposes of this paper, is an individual who participates in sustaining society by coordinating his or her behavior with that of other citizens according to a commonly understood convention. That studied here is the device of hypothetical bargaining in the original position. But to assert that a human being (or an animal) is not a citizen in this sense, is not to argue that he or she should be, or necessarily will be, plundered, exploited, neglected or ignored. For example, in the overlapping-generations' model of Section 2, the old are not technically citizens. They do nothing to sustain society in equilibrium: they simply consume. Nevertheless, society acts in a manner which ensures their comfort and ease. Of course, in this example, the welfare of at least one old person figures explicitly in the personal preferences of every young citizen. But, even when the citizens have no direct interest in the welfare of non-citizens, it does not follow that no *modus vivendi* can exist. Some other convention may be operated between citizens and various classes of non-citizens, as with the convention which operates between shepherds and sheepdogs.

Having excluded those who, for whatever reason, lack the necessary intelligence or imaginative powers from those to be counted as citizens, it is natural to argue that the hypothetical bargaining in the original position should then be supposed to encompass all those who *do* have the necessary intelligence and powers of imagination to serve as a citizen. Such a view would serve well enough if it were true that society could be seen as a monolithic organization without subsidiary structure, actual or potential, above the level of the individual. But this is neither the way things are; nor the likely outcome of any utopian reasoning which adequately takes account of the nature of the human animal. Society, as currently constituted, consists of a complex system of interlocking organizational hierarchies which coordinate behavior at many different levels. An attempt to describe the current state-of-nature which took no account of this reality would clearly be futile. So would any attempt to describe the potential states to which society might aspire. Not only do the subsocieties into which society-as-a-whole is split exist, they are essential to its efficient operation.

However, the existence of subsocieties necessarily requires that distinctions be made between insiders and outsiders. This is *not* to argue that the irrational "dehumanizing" criteria currently used by society to "blackball" various classes of unfortunates are inevitable or legitimate. It is simply to recognize, as a matter of logical necessity, that everybody cannot be insiders in everything. But it does not follow that, because outsiders must exist, then they must necessarily be seen as implacable opponents. Indeed, two individuals may well be fellow-insiders within an organization devoted to one issue while simultaneously belonging to rival organizations concerned with orthogonal questions. However, some caution is perhaps appropriate before xenophobic behavior is unreservedly categorized as irrational. Is it not possible that, under certain constraints on what is possible, an ideal society might not consist of two exclusive classes of individuals? Insiders within each class would coordinate their behavior relative to each other according to a hypothetical deal reached in the original position by all members of that class. Behavior between members of different classes would be coordinated by a hypothetical deal reached in the original position by a representative insider and a representative outsider. The point here is that an individual's extended sympathy preferences would depend on whether he or she were inter-acting with an insider or an outsider. Under such circumstances, the question of who is to be treated as a citizen is clearly not properly posed. One needs to ask: who is a citizen *in what?*

A very simple example, in which nearly all the difficulties are evaded, will now be offered with a view to indicating that various approaches are not viable. Suppose that Adam and Eve are joined by Ichabod, who is entirely useless, both to himself and to others. Instead of Fig. 4A, it is then necessary to contemplate Fig. 14A. For simplicity, the state-of-nature point is located at the origin. Fig. 14B replaces Fig. 4B.

Players 1, 2 and 3 are Adam, Eve and Ichabod in the original position but unaware of their identities. They are assumed to have fixed, identical extended sympathy preferences, according to which the state-of-nature is role-neutral and located at the origin. Notice that the feasible set in the original position is *not*

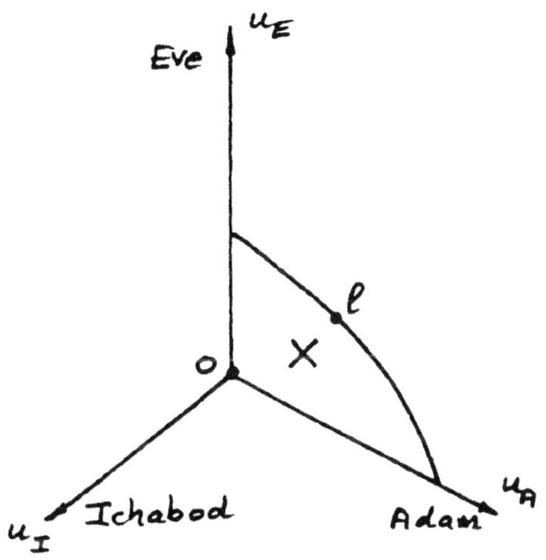

Figure 14A

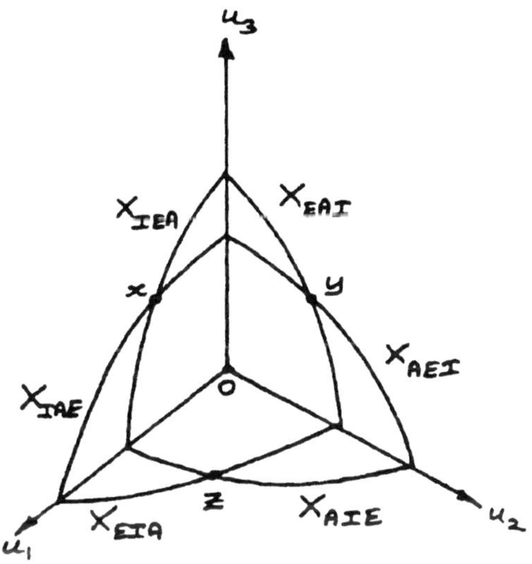

Figure 14B

the intersection of the six sets obtained from X_{AEI} by permuting the indices. The reason is that, although Ichabod has no reason to honor the fall of the hypothetical coin, his failure to do so affects nothing, and is therefore irrelevant. A viable deal, in the original position, is therefore for x to be implemented if player 1 turns out to be Ichabod, y if player 2 turns out to be Ichabod, and z if it is player 3. The result of this deal is labeled ℓ (for leximin) in Fig. 14A.

Next, suppose that Ichabod has some small use as a hewer of wood and drawer of water, while matters remain the same in respect of Adam and Eve. Fig. 15A illustrates the new situation. The set indicates what is available if all three act together. The set X continues to be what is available to Adam and Eve if Ichabod fails to cooperate. The Harsanyi utilitarian point h and the Rawlsian maximin point r are indicated. Neither seems very relevant given that Adam and Eve can get ℓ without Ichabod. Some Pareto improvement m of ℓ is presumably what should result from an analysis which takes proper account of the strategic realities.

In Fig. 15B, Ichabod has a comparable status with Adam and Eve. The figure serves to illustrate two points. The first is that the notion of the core, as proposed in this general context by Howe and Roemer (1981), is not adequate in general, since, for the situation illustrated, the core is empty. Although some of the ideas about coalition formation incorporated in such notions from cooperative game theory must be relevant, it is hard to believe that they can be directly applicable without much rethinking. The second point is that there does not seem a great deal that the device of hypothetical bargaining in the original position can do about a situation with as much built-in conflict as Fig. 15B. The general arguments

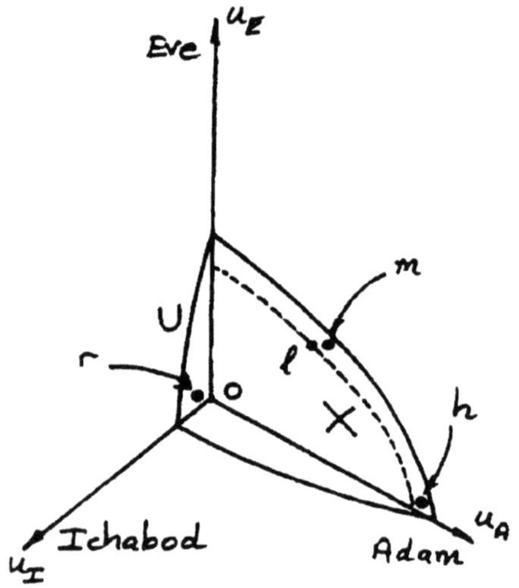

Figure 15A

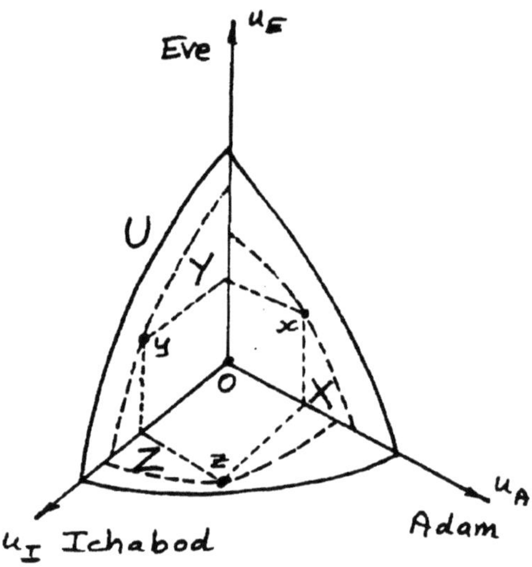

Figure 15B

given in Binmore (1983) in favor of the conclusion that a bargaining analysis should lead either to the outcome x and the exclusion of Ichabod, or to y and the exclusion of Eve, or to z and the exclusion of Adam, would seem to apply whether or not the bargaining takes place behind a veil of ignorance.

6. Conclusion

It is widely believed that to model man as a selfish optimizer is to exclude from consideration a wide range of moral intuitions that introspection suggests are actually important in governing real-world behavior. Often discussion proceeds as though it is *only* when individually sub-optimal actions are chosen that morality becomes manifest. In so far as such views are buttressed with examples, these seldom reflect the complexity of the real-world situation from which they are abstracted. In particular, nearly all real-world decisions involve on-going relationships with other people that are ignored in the over-simplified, formal models[55] which are normally held to embody all that is relevant to the decision problem. A more realistic view of the nature of most human inter-action allows game theory to be seen as a vehicle for giving expression to Hume's opinion that ethical codes of conduct exist (or should exist) for the purpose of coordinating the behavior of selfish, but enlightened, individuals. This paper has sought to explore what can be done in this direction, albeit in a very abstract manner and for the case of only two persons.

At one level, the paper can be seen as naively constructive. From this perspective, the device of hypothetical bargaining in the original position is perceived as a primitive "ethical axiom" from which all other ethical behavior is to be *derived* by formal reasoning. It seems to me that there are no acceptable half-way houses if this approach is to be taken seriously. It is cheating to sneak hidden ethical premises into the analysis of behavior in the original position or elsewhere. In this paper, behavior is therefore always assumed to be individually optimal, given the specified constraints. Moreover, a serious approach means working out in detail precisely what would be necessary *if it really were possible* to retire into the original position in order to reach coordinating agreements. One is not excused, by the fact that what goes on in the original position is hypothetical, from the duty to ensure that what is imagined to occur should actually make proper sense *if the hypothesis were realized.* The paper claims that, if these requirements are relentlessly pursued to the bitter end, then the Rawlsian maximin criterion will result from an orthodox game-theoretic analysis, without the need to follow Rawls in denouncing orthodox Bayesian decision theory. The maximin criterion emerges within a framework of extended sympathy preferences. If the conclusion is rewritten in terms of the individuals' personal preferences, results consistent with certain ideas from cooperative game theory are obtained. Implementing the maximin criterion in terms of personal preferences turns out, under certain conditions, to be equivalent to using a hybrid notion which crosses the Nash bargaining solution with the idea of a proportional bargaining solution. The extent to which each appears in the cross depends on the extent to which inter-personal comparison is constrained. With *no* inter-personal comparison, the Nash bargaining solution appears: with *full* inter-personal comparison, a proportional bargaining solution. The interesting cases are the intermediate ones.

Aside from the naively constructive interpretation, the paper can also be read as rejoinder to the extravagant claims of conservative intellectuals for the unique virtues of the market mechanism.[56] On the basis of hypotheses to which no Chicago economist has any right to object[57], it is shown that it can be entirely rational to act according to ethical principles uncaptured by the market mechnism. It may be thought wimpish to offer an intellectually respectable defense for a middle road between two extremes. But that is the direction in which the analysis points. Distributive mechanisms which are maximally flexible in the face of change pay no attention to inter-personal comparisons of utilities. Mechanisms which devote full attention to inter-personal comparison are unable to adapt easily to change. The middle road is to use a flexible system subject to checks and balances derived from partial inter-personal comparisons.

References

D. Abreu (1986). On the theory of infinitely repeated games with discounting. Harvard University.

Anscombe, F. and R. Aumann (1963). A definition of subjective probability. Annals of Math. Stats., **34**: 199-205.

Arrow, K. (1963). Social Choice and Individual Values. Newhaven: Yale University Press.

Arrow, K. (1978). Extended sympathy and the possibility of social choice. Philosophia, **7**: 223-237.

Aumann, R. (1981). Survey of repeated games. In: Essays in Game Theory and Mathematical Economics in Honor of Oskar Morgenstern. Wissenschaftsverlag, Bibliographisches Institut, Mannheim/Wien/Zürich.

Aumann, R. (1987). Correlated equilibrium as an expression of Bayesian rationality. Econometrica, **55**: 1-18.

Aumann, R. and S. Sorin (1985). Bounded rationality forces cooperation. Strasbourg University.

Axelrod, R. (1984). The Evolution of Cooperation. New York: Basic Books.

Barker, E. (1947). Social Contract: Essays by Locke, Hume and Rousseau. Oxford: Oxford University Press.

Binmore, K. (1983). Bargaining and coalitions. ICERD discussion paper 83/71, L.S.E. Also (1985) in: A. Roth, (ed.), Game Theoretic Models of Bargaining, Cambridge: Cambridge University Press.

Binmore, K. (1984). Game Theory and the Social Contract, Part I. ICERD discussion paper 84/108, London School of Economics.

Binmore, K.(1987a/b). Modeling rational players I/II. Forthcoming in J. Econ. and Philosophy (different issues).

Binmore, K. and P. Dasgupta (1987). Economics of Bargaining. Oxford: Basil Blackwell.

Binmore, K., A. Rubinstein and A. Wolinsky (1986). The Nash bargaining solution in economic modelling. Rand J. Econ. **17**: 176-187.

Binmore, K. A. Shaked and J. Sutton (1985). An outside option experiment. ICERD discussion paper 85/124, London School of Economics.

Braithwaite, R. (1955). Theory of Games as a tool for the moral philosopher. Cambridge: Cambridge University.

Buchanan, J. (1976). A Hobbsian interpretation of the Rawlsian difference principle. Kyklos **29**: 1.

Buchanan, J. (1987). Towards the simple economics of natural liberty. Kyklos **40**: 1.

Calvo, G. (1978). Some notes on time inconsistency and Rawls' maximin criterion. R. Econ. Studies, **45**.

Costa, G. (1987). The over-lapping generations economy as a noncooperative game. Bergamo: Quarto convegno di teoria dei giochi e applicazioni: 155-165.

Cross, J. (1965). The Economics of Bargaining. New York: Basic Books.

D'Aspremont, C. and L. Gevers (1977). Equity and the informational basis of collective choice. Review of Economic Studies, **44**: 199-209.

Deschamps, R. and L. Gevers (1978). Leximin and utilitarian rules: a joint characterization. J. Econ. The., **17**: 143-163.

Farrell, J. and E. Maskin (1987). Renegotiation in repeated games. Harvard University.

Friedman, J. (1986). Game theory with applications to economics. Oxford: Oxford University Press.

Gauthier, D. (1986). Morals by Agreement. Oxford: Clarendon Press.

Gevers, L. (1979). On interpersonal comparability and social welfare orderings. Econometrica, **47**: 75-89.

Hammond, P. (1976a). Equity, Arrow's condition and Rawls' difference principle. Econometrica **44**: 793-804.

Hammond, P. (1976b). Why ethical measures of inequality need interpersonal comparisons. Theory and Decision, **7**: 263-274.

Hare, R. (1975). Rawls' theory of justice. In: N. Daniels (ed.), Reading Rawls, Oxford: Blackwell.

Harsanyi, R. (1953). Cardinal utility in welfare economics and in the theory of risk-taking. Journal of Political Economy, **61**: 434-435.

Harsanyi, R. (1955), Cardinal welfare, individualistic ethics, and interpersonal comparisons of utility. Journal of Political Economy, **63**: 309-321.

Harsanyi, J. (1958). Ethics in terms of hypothetical imperatives. Mind **47**: 305-316.

Harsanyi, J. (1977). Rational behavior and bargaining equilibrium in games and social situations. Cambridge: Cambridge University Press.

Howe, R. and J. Roemer (1981). Rawlsian justice as the core of a game. Amer. Econ. R., **71**: 880-895.

Hume, D. (1789). A Treatise of Human Nature. (ed. Mossner, 1969), Harmondsworth: Penguin.

Hurwicz, L. (1979). On allocations attainable through Nash equilibria. J. Econ. Th., **21**: 140-65.

Isbell, J. (1960). A modification of Harsanyi's bargaining model. Bull. Amer. Math. Soc., **66**: 70-73.

Kalai, E. (1977). Proportional solutions to bargaining situations = intepersonal utility comparisons. Econometrica, **45**: 1623-1630.

Kalai, E. and M. Smorodinsky (1975). Other solutions to Nash's bargaining problem. Econometrica, **45**: 1623-1630.

Kolm, S. (1974). Sur les consequences Economiques des principes de justice et de justice practique. Revue d'Economie Politique, **84**: 80-107.

Lewis, D. (1969). Convention: a Philosophical Study. Cambridge: Harvard University Press.

Luce, R. and H. Raiffa (1957). Games and Decisions. New York: Wiley.

Marschak, J. and R. Radner (1972). Economic Theory of Teams. New Haven: Yale University Press.

Myerson. R. (1977). Two-person bargaining and comparable utility. Econometrica, **45**: 1631-1637.

Nash, J. (1950). The bargaining problem. Econometrica, **18**: 155-162.

Nash, J. (1951). Non-cooperative games. Annals of Math., **54**: 286-295.

Nozick, R. (1974). Anarchy, State and Utopia. New York: Basic Books.

Peters, H. and E. van Damme (1987). A characterization of the Nash bargaining solution not using IIA. Discussion paper No. A-117, Bonn-University: Sonderforschungsbereich 303.

Raiffa, H. (1953). Arbitration schemes for generalized two-person games. In: Kahn and Tucker (eds.), Contributions to the Theory of Games II. Princeton.

Rawls, J. (1958). Justice and fairness. Philosophical Review, **57**.

Rawls, J. (1968). Distributive justice; some addenda. Natural Law Forum, **13**.

Rawls, J. (1972). A theory of justice. Oxford: Oxford University Press.

Rawls, J. (1976). Some reasons for the maximin criterion. Amer. Econ. R., **59**: 141-146.

Roberts, K. (1980a). Possibility theorems with interpersonally comparable welfare levels. Review Econ. Studies, **47**: 409-420.

Roberts, K. (1980b). Interpersonal comparability and social choice theory. Review Econ. Studies, **47**: 421-439.

Rodriguez, A. (1981). Rawls' maximin criterion and time consistency: a generalisation. R. Econ. Studies, **48**.

Roth, A. (1979). Axiomatic Models of Bargaining. Lecture Notes in Economics and Mathematical Systems 170, Berlin: Springer Verlag.

Rousseau, J.-J. (1763). Du Contract Social. English translation, M. Cranston (1968). The Social Contract. Harmondsworth: Penguin.

Rubinstein, A. (1982). Perfect equilibrium in a bargaining model. Econometrica, 50: 97-100.

Savage, L. (1951). The Foundations of Statistics. New York: Wiley.

Schelling, T. (1960). The Strategy of Conflict. Cambridge: Harvard University Press.

Schmeidler, D. (1980). Walrasian analysis via strategic outcome functions. Econometrica, **48**: 1585-93.

Schmeidler, D. and K. Vind (1972). Fair net trades. Econometrica, **40**: 637-642.

Selten, R. (1975). Re-examination of the perfectness concept for equilibrium in extensive

Sen, A. (1987). On Ethics and Economics, Oxford: Basil Blackwell.

Sen, A. (1970). Collective Choice and Social Welfare. San Francisco: Holden Day.

Sen. A. (1976). Welfare inequalities and Rawlsian axiomatics.. Theory and Decision, **7**: 243-262.

Shubik, M. (1981). Society, land, love or money. J. Econ. Behavior and Organization, **2**: 359-385.

Springer, P. (1987). In defence of animals. Oxford: Blackwell.

Suppes, P. (1966). Some formal models of grading principles. Synthese **6**: 284-306.

J. Tinbergen (1957). Welfare economics and income distribution. Amer. Econ. R., Papers and Proceedings,, **47**: 490-503.

Varian, H. (1974). Equity, envy and efficiency.. J. Econ. Theory, **9**: 63-91.

Von Neumann J. and O. Morgenstern (1944). The Theory of Games and Economic Behavior. Princeton: Princeton University Press.

Wolff, P. (1966), A Refutation of Rawls' Theorem of Justice. Journ. of Philosophy, **63**: 179-190.

Wolff, Peter (1977). Understanding Rawls. Princeton: Princeton University Press.

Yaari, M. (1981), Rawls, Edgeworth, Shapley, Nash: theories of distributive justice re-examined. J. Econ. The., **24**: 1-39.

Zeuthen, F. (1930). Problems of monopoly and economic welfare. London: Routhledge and Kegan Paul.

Footnotes

[1]For example, Howe and Roemer (1981), Yaari (1981), Gauthier (1982).

[2]In this paper, a convention is not to be understood in the technical sense proposed by Lewis (1969), although his general views are very close to those expressed here. In brief, he gives three reasons for distinguishing between a convention and a social contract. The first is that a social contract might not select an equilibrium, but this possibility is explicitly rejected here. The second is that conventions should select only from the set of what he defines to be "coordination equilibria", but here no reason is seen for singling out such a special class of equilibria. A convention in this paper will simply be a rule for selecting one equilibrium from the set of all possible equilibria. The third reason has to do with common knowledge requirements which seem to me of only secondary relevance in this context.

[3]It is not easy to compare Gauthier's (1986) approach. His aims and attitudes are almost identical to mine and so our differences on matters of execution may not be immediately apparent. But these run very deep. For example, the sentence which occasions this footnote is the second sentence of his book, the first having been written by Hume. To pin down the differences more closely, it is necessary to refer to the five "core conceptions" that he delineates in an overview of his theory (1986, pp. 13-16). I agree wholeheartedly with the fifth, which emphasizes the role of bargaining in the original position. I also agree with the preliminary observation of his first conception: namely, that the *analysis* of what is moral needs to take place in a "morally free zone", although why this term should be thought to apply to "perfectly competitive markets" is not clear to me. In choosing to study equilibria in a price-mediated distributive mechanism, one is making a value-judgement about the equilibria of other distributive mechanisms. Nor do I sympathize with Gauthier's second conception in which he invents an idiosyncratic definition of a "rational" bargaining outcome. My complaint is similar to that made later (Subsection 3.2) about Rawls' *fiat* use of the maximin criterion. To be convincing as a piece of rhetoric, an argument ought to take an *orthodox* line on such matters. His third conception allows for unenforced commitment, and this is the major assumption found in the literature on these matters which I wish to avoid. His fourth conception is rather technical, but some similar requirement emerges as a necessity in my approach also, although the difference in interpretations will not make the similarity easy to recognize. In spite of all these differences, Gauthier's (1986) approach remains very close in spirit to mine.

[4]Neither uses the term "game" in this context. In Harsanyi's case, this is presumably because his assumptions reduce the situation very rapidly to an essentially one-player situation. In Rawls' case (1972, p. 145), it is apparently because he sees the word "game" as a synonym for "zero sum game". Similarly, in other places, he uses the word "bargaining" in a much narrower sense than is common in game theory.

[5]A technical aside may avert later misunderstandings. With the simplifying convexity assumptions made in the text, the maximin criterion reduces to an equal-splitting convention. This does not trivialize the criterion, because the paper seeks, among other things, to treat the question of what it is that gets split equally without evading questions of the inter-personal comparison of utilities. The use of the maximin criterion in the absence of convexity assumptions is not to be defended. Although I have not considered this case closely, it seems unlikely that the approach adopted in this paper would lead to such a neat conclusion. However, I would regard any criterion which led to some Pareto-improvement of all equal-split outcomes as a vindication of the spirit of Rawls' position (as actually expressed in Rawls (1958)). Some brief discussion of the non-convex case appears at the end of Subsection 3.3.

[6]We follow Hare (1975) in characterizing Rawls' argument in this way in spite of Rawls' protestation to the contrary.

[7]The Hobbesian view that such a failure to agree results in a return to a brutish "state of nature" clearly requires some re-evaluation in the current context.

[8]The device of the original position seems to lose much of its force if such value judgements are required in predicting the bargaining outcome, as in Gauthier (1986).

Only hypotheses about *individual* optimizing behavior are required in this paper.

[9]So much nonsense is now extent on the "Prisoners' Dilemma" that its usefulness as an example has become much attenuated. In any case, it remains controversial as to what game theory prescribes as "rational" in the finitely repeated game (Binmore 1987a).

[10]A punishment may, of course, consist of withholding a carrot as well as administering a stick.

[11]The conventional use of the term *status quo* here may be a little confusing to those who know some bargaining theory, where the term refers to what will result if coordination attempts *fail* rather than the result when they have succeeded.

[12]Lewis (1969, p. 89) makes relevant comments on this point but my approach is not consistent in matters of detail with his, in particular, I would argue that the fact that there might be "losers" in Hobbes' state-of-nature, who regret the actions they took which led to their lack of success, does not imply that his state-of-nature is not in equilibrium. Even granting the static viewpoint implicit in such an assertion, the issue is not whether the Hobbesian brutes regret their actions after they have observed the consequences, but whether their actions seemed optimal given the information they had when the decisions were made.

[13]In making explicit reference to some surrogate of a state-of-nature against which proposed improvements are to be evaluated, this paper deviates, for example, from Rawls (1972) and also from Harsanyi (1977). But not, I believe, in any important way. It is true that Rawls has been criticized (Wolff 1966) for neglecting state-of-nature questions, but the symmetry to be found in his original position would seem to make it clear what the appropriate standard of comparison is to be.

[14]The game also has Nash equilibria in mixed strategies. These require players to use a random device in selecting which pure strategy to play. Correlated equilibria (Auman 1987) may also be relevant in such circumstances.

[15]It is, for example, the Nash bargaining solution.

[16]Or, to be more accurate, so that *control* over the money and drug changes hands continuously. Even in apparently instantaneous exchanges, what is presumably happening in practice is that the probabilities of being able to walk off with various commodity bundles are changing very rapidly over time.

[17]In some cases there may be various candidates for the disagreement outcome without its being obvious which should be selected (Binmore/Rubinstein/Wolinsky 1986).

[18]One may criticize these axioms but, if one wishes to follow Rawls (1972) in rejecting expected utility maximization, then it is necessary to deny one of the axioms. One of the attractions of the current application is that the context can be seen as a "closed universe" situation (Binmore 1987, Section 6).

[19]A more popular characterization is to describe n as the point in X at which the "Nash product" $(u_1-d_1)(u_2-d_2)$ is maximized.

[20]A Nash equilibrium (Section 2) is quite a different concept from the Nash bargaining solution. Without *some* uncertainty, any point on the frontier of X corresponds to a Nash equilibrium of the demand game provided it gives both players at least their disagreement payoff. (See paper 3 in Binmore/Dasgupta 1987).

[21]If it does not matter *when* we agree, then it does not matter *if* we agree (Cross 1965).

[22]Unless, perhaps, information is asymmetric.

[23]In an earlier version of the paper (Binmore 1984), a "fair bargaining game, based on the principle of divide-and-choose, was used in the "original position" to obtain a *unique* outcome. But, at that time, I was mistakenly taking a Hobbesian view of the "state-of-nature".

[24]Although the symmetry of the original position obscures some of the differences.

[25]In Binmore (1984) the set X was drawn differently and reference made to Tiresias. But this joke was in poor taste.

[26]Harsanyi restricts the choice set to be the intersection of T with the line $u_1 = u_2$. In the immediate context, this leaves the conclusion unaffected.

[27]Note that the Nash product $(u_1-d_1)(u_2-d_2)$ of Footnote 20 is maximized at the same point in X as the Nash product $(B_1u_1-B_1d_1)(B_2u_2-B_2d_2)$.

[28]Following Harsanyi, "intra-personal" is used to indicate a subjective comparison of utility scales which is internal to one specific individual. "Inter-personal" indicates an objective comparison in that it is common to all individuals.

[29]Although this "doctrine" is associated by game theorists with Harsanyi, a similar notion has a seemingly independent existence in social choice theories which use extended sympathy ideas. For example, the doctrine is built into the work of Hammond (1976a) and those with similar concerns. Hammond traces the idea back to Tinbergen (1957) via Kolm (1974).

[30]The set X consists of all utility pairs $(\phi_A(s),\phi_E(s))$, where s is any relevant state.

[31]And therefore "in their *own* shoes" if they so choose. Hence the reference to self-consciousness. I have written at length elsewhere (Binmore 1987b) about the significance of this viewpoint to equilibrium selection in the theory of games.

[32]The point in distinguishing between coordination *within* kin-groups and *across* kin-groups is that the game paradigm will not, in general, be appropriate in the former case. For example, coordination between identical twins should be treated as a *one*-player game, from the biological viewpoint, because there is only *one* set of genes. It is instructive to note that Harsanyi's own formulation of his theory, as opposed to my bowdlerized version, reduces the situation to an essentially *one*-player problem. But, of course, in the context of the current discussion, the Harsanyi doctrine could only apply *within* a kin-group.

[33]Why does this line lie below the main diagonal? Recall that arbitrary simplifying assumptions were made about what things are like in heaven and hell. It is the same asymmetry which requires that a distinction be made between the cases $a+b<1$ and $a+b\leq 1$. The asymmetry appears in the restriction that U_i and V_i do not exceed 1.

[34]Provided any Nash equilibria exist at all.

[35]Assuming that it is granted that a coin exists somewhere in the universe for which *objective* probabilities are given (Aumann/Anscombe 1963, Harsanyi 1977). Diamond 1967 makes an honest attempt to argue against the "sure thing principle" in this context, but cannot be said to be very convincing.

[36]Subgame perfect equilibria were mentioned in Section 2 but the treatment which follows is too abstract for the precise definition of the equilibrium concept employed to matter a great deal.

[37]Closed means that its boundary points are members of the set. Comprehensive means that unlimited "free disposal" is permitted (i.e. utility can be freely thrown away). Bounded above means that neither player can exceed a certain fixed utility level no matter what.

[38]More complicated coordinating devices than coins are envisaged in Aumann's (1987) notion of a correlated equilibrium, but it would add very little to take these into account.

[39]An aside on "rights" and "entitlements" is called for. Rawls (1972) includes "liberty" among his primary goods and somehow manipulates his analysis of the original position so that the maximin principle is applied to this particular good before the others. I do not properly understand even how "liberty" can be thought of as a "good" in a quasi-economic sense. Goods have to do with preferences, while liberty relates to constraints on strategy choices.

In any case, my treatment consigns the subject of rights and obligations to a secondary level of analysis, in that their discussion requires an examination of the means by which equilibria are *implemented*. This is an issue abstracted away in my approach. Nevertheless, some general observations may be helpful:

(1) My approach makes it meaningless to talk about some categories of "rights". It takes for granted that players will optimize and is concerned only with *coordinating* optimal play. Thus, for example, if one attaches a sufficiently high value to one's own survival, then it will never be optimal to acquiesce in one's murder. The Hobbesian "right to self-defense" is therefore not even on the agenda as being part of a coordinating convention.

(2) Some rights would have very little value if everyone were rational. We value such rights because we see them as providing some protection against the irrationalities of others. Such considerations, although admittedly very important, cannot be captured by an analysis of the type attempted in this paper. But, some care is necessary before classifying most rights into this category. For example, in a rational world, presumably governments would not be tyrannical and citizens would not rise up in rebellion. However, perhaps the government is not tyrannical because, *if it were*, the citizens *would* rise up in rebellion: hence the "right to free assembly" and the "right to bear arms".

(3) This last point suggest a *definition* of a "right" in terms of out-of-equilibrium behavior. A right is a category of behavior whose adoption *never* leads to a punishment strategy being adopted in the original position. An obligation is a category of behavior which, if neglected, *always* leads to a punishment strategy being agreed to in the original position.

[40]Here this means abandoning the attempt to found a society and opting instead for a feral existence.

[41]I know that I am guilty of over-simplification here. In principle, the course the failed negotiations took might serve as a re-coordinating device. However, I shall be on firmer ground when I make the same assertion for what goes on beyond the veil of ignorance.

[42]Peters and van Damme (1987) characterize the Nash bargaining solution in this way. Reference should also be made to the early work of Zeuthen (1930) which came at the Nash bargaining solution from this direction.

[43]Although unusual, the idea is not original. Harsanyi (1977, p. 226) refers to a similar situation as a "bargaining equilibrium". As with much else in the paper, the idea is much more attractive beyond the veil of ignorance than it is in general.

[44]It cannot be that $e_{AE} \le e_{EA}$. This inequality requires that $U_1 a \le 1 - V_1 b$ and $1 - V_2 b \le U_2 a$, and the latter is inconsistent with the requirements that $0 \le U_2 \le 1$, $0 \le V_2 \le 1$ and $a+b<1$. Similarly, it cannot be that $e_{EA} \le e_{AE}$.

[45]Invoked here with the usual non-teleological understanding.

[46]Here it is Harsanyi's (1977) "ideal observer" formulation to which appeal is being made, rather than to the bargaining formulation of Subsection 3.2.

[47]Sometimes the points "between the two rays" will be those "below" e rather than those "above" as illustrated. In the former case, u will then just be e.

[48]In Subsection 3.2, the strategic choice of extended sympathy preferences is made behind the veil of ignorance. For personal preferences, however, the interesting case would seem to be when the strategic choice is made with no veil of ignorance.

[49]That is to say, the allocation will be that which results when trading takes place at market-clearing prices.

[50]Although he goes to some trouble not to commit himself personally to any particular view.

[51]Provided the individuals within each class are reasonably homogeneous in respect of the property that distinguishes the two classes. Thus Adam may serve as a representative of the male sex and Eve of the female. Or, in the overlapping generations' model of Section 2, a representative mother and a representative daughter might be selected.

[52]In particular, one would have to confine attention to *stationary* equilibria in the Rubinstein bargaining model described in Subsection 3.1.

[53]The original version of this paper (Binmore 1984) was to have a second part on these issues, but this will never now appear.

[54]As do those who see no role for markets at all. But the latter do not currently seem to be in fashion.

[55]For example, we often have to play real-life versions of the much discussed "Prisoners' Dilemma", and we seem to be equipped with "moral intuitions" for such purposes. But these real-life games are not the one-shot, never-to-be-repeated game against an anonymous opponent that appears in textbooks. Those who deduce the appropriate moral intuitions for the real-life game from an examination of the textbook game are simply applying an incorrect analysis to the wrong game.

[56]This is written in the United Kingdom at a time when the newspapers are agog with scandal about under-provision in the National Health Service. Simultaneously, it is possible for a respected political journalist (Brian Walden in the Sunday Times) to write, "The National Health Service is not part of the open market, so it has no rational mechanism for establishing priorities."

[57]It is not an objection that the coordination is not achieved, in practice, via hypothetical bargaining in the original position. It is enough that an equilibrium exist which is *as-if* this were the case.

Appendix

Subsection 3.3 describes the model advocated in this paper as appropriate for hypothetical bargaining behind the veil of ignorance. The bargainers are assumed to be equipped with extended sympathy preferences, and the outcome of the hypothetical deal reached in the original position is expressed in terms of these extended sympathy preferences. For the analysis of Subsection 3.3 to be of interest, it is necessary that the primitive state-of-nature be *role-neutral* relative to the extended sympathy preferences attributed to the bargainers. This assumption is justified by supposing that extended sympathy preferences are determined *strategically*. Subsection 3.2 makes other claimes about the consequences of supposing that extended sympathy preferences are chosen strategically. In particular, it is claimed that *symmetric* equilibria lead to the "hybrid bargaining solution" illustrated in Figs. 9A and 9B, when matters are expressed in terms of Adam's and Eve's *personal* preferences. The purpose of this appendix is to discuss the technicalities of the *symmetric* case with a view to justifying the various claims made in the text.

To review the situation to be examined: players 1 and 2 are in the original position, behind the veil of ignorance. They are assumed to have *symmetric* extended sympathy utility functions Φ_1^* and Φ_2^*. More precisely, it is assumed that

$$\Phi_1^*(A,s) = \Phi_2^*(A,s) = U^*\Phi_A(s)$$

$$\Phi_1^*(E,s) = \Phi_2^*(E,s) = V^*(\Phi_E(s)-1) + 1.$$

The primitive state-of-nature is denoted by s_0. As in Subsection 3.2, $a = \Phi_A(s_0)$ and $1-b = \Phi_E(s_0)$, where $a + b > 1$. It is assumed that

$$aU^* + bV^* = 1, \tag{1}$$

so that the primeval state-of-nature is role-neutral.

The strategic problem is to determine the values of U^* and V^* from the range [0,1] which would make it unprofitable for either player 1 or player 2 to report different values, u or v, from the range [0,1] for these parameters, provided the other player reports U^* and V^* truthfully. The restriction of the choice of parameters to the range [0,1] represents an underlying understanding about the inter-personal comparison of utilities. (Discussion of this issue, and of alternative assumptions, appears in Section 4.) In order to determine what is or is not profitable for a player in reporting his or her extended sympathy preferences, it is necessary to know how a player's *actual* expected utility is affected by the report that he or she chooses to make. For this purpose, the analysis of Subsection 3.3 is required.

The relevant information is summarized in Figs. 11A and 11B. The extended sympathy utility functions Φ_1 and Φ_2 that get reported, determine the sets

$$X_{AE} = \{(\Phi_1(A,s),\Phi_2(E,s)) : s \in S\}$$

and

$$X_{EA} = \{(\Phi_1(E,s),\Phi_2(A,s)) : s \in S\}.$$

Subsection 3.3 comments on the shape of the set

$$X = \{(\Phi_A(s),\Phi_E(s)) : s \in S\}.$$

This set, and hence X_{AE} and X_{EA}, are assumed to be closed, bounded above, comprehensive and *strictly* convex. The primitive state-of-nature s_0 is role-neutral, relative to Φ_1 and Φ_2, if and only if $\Phi_1(A,s_0) = \Phi_1(E,s_0)$ and $\Phi_2(A,s_0) = \Phi_2(E,s_0)$). The point d in Figs. 11A and 11B is the pair of utilities commonly assigned to s_0 by both players in the case when s_0 is role-neutral. That is to say, $d = (\Phi_1(A,s_0),\Phi_2(E,s_0)) = (\Phi_1(E,s_0),\Phi\ (A,s_0))$. The conclusion from Subsection 3.3 needed in this appendix is that, if s_0 is role-neutral, then the result of bargaining in the original position will be the Nash bargaining solution n for the set $X_{AE} \cap X_{EA}$ relative to the disagreement point d. If the Nash bargaining solution of either X_{AE} or X_{EA}, relative to d, happens to be a member of X_{AE} or X_{EA}, then this will necessarily be the Nash bargaining solution of $X_{AE} \cap X_{EA}$ relative to d. (This follows from Nash's (1950) "independence of irrelevant alternatives".) Fig. 11B illustrates this situation. Otherwise, the Nash bargaining solution of $X_{AE} \cap X_{EA}$ relative to d lies at a cross-over point c, as illustrated in Fig. 11A. A cross-over point is defined to be a point which is common to the Pareto-frontiers of both X_{AE} and X_{EA}.

To operate the agreement on the Nash bargaining solution n, the players need to implement one state $s = s_{AE}$ if player 1 turns out to be Adam and player 2 to be Eve; and to implement another state $t = s_{EA}$ if the roles are reversed. The requirements on the states s and t are that

$$n = (\Phi_1(A,s),\Phi_2(E,s)) = (\Phi_1(E,t),\Phi_2(A,t)).$$

The above conclusions relate to the case when the reported Φ_1 and Φ_2 render s_0 role-neutral. If not, then no improvement on the primitive state-of-nature s_0 will be achieved by bargaining in the original position.

Consider first the case when players 1 and 2 both truthfully report U* and V* so that $\Phi_1 = \Phi_1^*$ and $\Phi_2 = \Phi_2^*$ in the above analysis. Then $d = d^*$ and $X_{AE} \cap X_{EA}$ are symmetric, and hence s_0 is the Nash bargaining solution n. (In Figs. 8A and 8B, $n = c^* = r$.) It follows that n corresponds to a role-neutral state s^*. That is to say, the agreement in the original position will result in the *same* state s^* being implemented, regardless of the role-assignment.

Next consider the case in which player 1 deviates from reporting U^* and V^* truthfully, and instead reports u and v. Player 2 is assumed to continue to report U^* and V^* as before. It is then necessary to study what happens when $\Phi_2 = \Phi_2^*$ but Φ_1 is given by

$$\Phi_1(A,s) = u\Phi_A(s)$$

$$\Phi_1(E,s) = v(\Phi_E(s)-1) + 1.$$

Suppose that bargaining in the original position leads to the implementation of σ, when the role assignment is AE, and τ, when the role-assignment is EA. The expected payoff to player 1 is then equal to

$$\{\Phi_1^*(A,\sigma) + \Phi_2^*(E,\tau)\}/2.$$

Notice three things. The first is that, although player 1 reports Φ_1, his actual preferences continue to be represented by Φ_1^*. The second point is simply that both players regard the role-assignments AE and EA as equi-probable. The third point is that, if the primitive state-of-nature s_0 is not role-neutral with respect to Φ_1 and Φ_2^*, then $\sigma = \tau = s_0$.

It is now possible to state, in precise terms, the requirement that it be unprofitable for player 1 to deviate from truthfully reporting U^* and V^*. The requirement is that

$$\{\Phi_1^*(A,\sigma) + \Phi_1^*(E,\tau)\}/2 \le \{\Phi_1^*(A,s^*) + \Phi_1^*(E,s^*)\}/2. \qquad (2)$$

(Notice that a deviation which does not leave s_0 role neutral is never profitable and so this possibility will be ignored from now on.) Inequality (2) may be rewritten as

$$U^*\Phi_A(\sigma) + V^*(\Phi_E(\tau)-1) + 1 \le U^*\Phi_A(s^*) + V^*(\Phi_E(s^*)-1) + 1,$$

which reduces to

$$U^*(\Phi_A(\sigma) - \Phi_A(s^*)) \le V^*(\Phi_E(s^*) - \Phi_E(\tau)). \qquad (3)$$

To make use of this inequality, it is necessary to formalize some of what is known about σ, τ and s^*. Note first that, since s^* is role-neutral,

$\Phi_1(A,s) = \Phi_1(E,s^*)$, and so

$$U^*\Phi_A(s^*) = V^*(\Phi_E(s^*)-1) + 1. \tag{4}$$

(This is to be compared with the requirement that the primeval state-of-nature s_0 be role-neutral: namely

$$U\Phi_A(s_0) = V(\Phi_E(s_0)-1) + 1),$$

which is expressed more succinctly as Equation (1).) The states σ and τ are associated with similar equations via the requirements that $\Phi_1(A,\sigma) = \Phi_1(E,\tau)$ and $\Phi_2^*(E,\sigma) = \Phi_2^*(A,\tau)$. The equations are

$$u\Phi_A(\sigma) = v(\Phi_E(\tau)-1) + 1 \tag{5}$$

$$U^*\Phi_A(\tau) = V^*(\Phi_E(\sigma)-1) + 1. \tag{6}$$

Since only deviations to states in which s_0 remains role-neutral are to be considered, an analog of Eq. (1) is also required, namely

$$au + bv = 1. \tag{7}$$

Necessary Conditions on U^* and V^*

Figs. 8A and 8B illustrate the case when both players report U^* and V^* truthfully, so that a *symmetric* configuration results. The Nash bargaining solution then lies at the *symmetric* cross-over point c^*. (This follows from Nash's (1950) axioms of "Pareto efficiency" and "symmetry".) Much of the interest of the paper derives from the fact that c^* can also be characterized as the Rawlsian maximin point r, but this is not significant here. Fig. 8B illustrates the case in which c^* is not only a cross-over point, but is also the Nash bargaining solution of X_{AE} (and hence also of X_{EA}, by symmetry). This occurs when $u_1 + u_2 = c_1^* + c_2^*$ is a supporting line of the convex set X_{AE}. This observation is used in defining extremal values of U^* and V^* below.

The parameters U^* and V^* (chosen from the interval [0,1] will be said to be *extremal* if and only if one or more of the following three criteria is satisfied:

(i) $U^* = 1$;

(ii) $V^* = 1$:

(iii) the line $u_1 + u_2 = c_1^* + c_2^*$ supports X_{AE}.

In the following proposition, "arbitrary close" means that, given any $\varepsilon > 0$, u and v can be found with $|u-U^*| < \varepsilon$ and $|v-V^*| < \varepsilon$. In view of the evolutionary argument offered in Subsection 3.2, the fact that non-extremal U^* and V^* are *locally* unstable (as well as globally unstable) is worthy of note.

Proposition 1. *Unless U^* and V^* are extremal, arbitrarily close u and v exist, which make it profitable for one player to deviate by reporting u and v, provided that the other continues to report U^* and V^*.*

Proof. Consider Figs. 16A and 16B. The axes show Adam's and Eve's *personal* utilities. In particular, the set X is defined by $X = \{(\Phi_A(s),\Phi_E(s)) : s \in S\}$. (The figures are therefore comparable with Figs. 9A and 9B, but *not* with Figs. 8A, 8B, 11A or 11B in which the axes show utilities for players 1 and 2 that are derived from Adam's and Eve's *extended sympathy preferences*). Points are labeled with the state to which they correspond.

In Fig. 16A, the line ℓ drawn through the points labeled s_0 and s^* registers the validity of Eqs. (1) and (4). The broken line m drawn through the point labeled s_0 registers that u and v are to be assumed to satisfy Eq. (7).

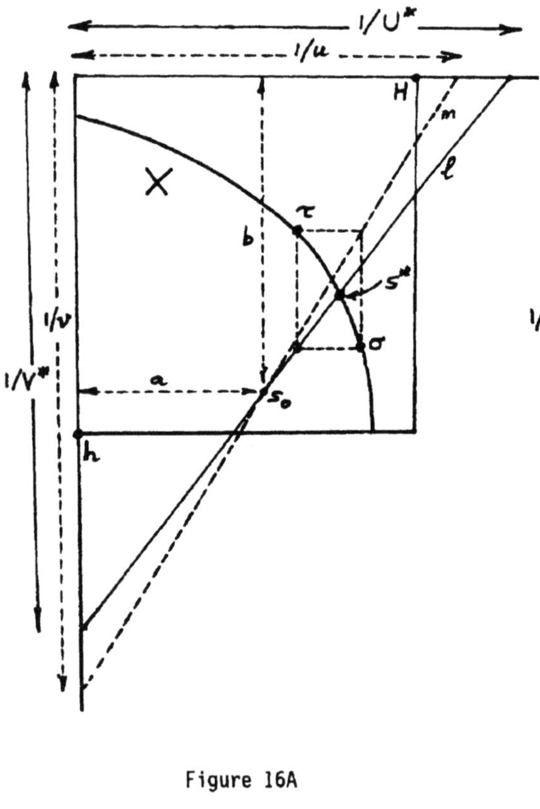

Figure 16A

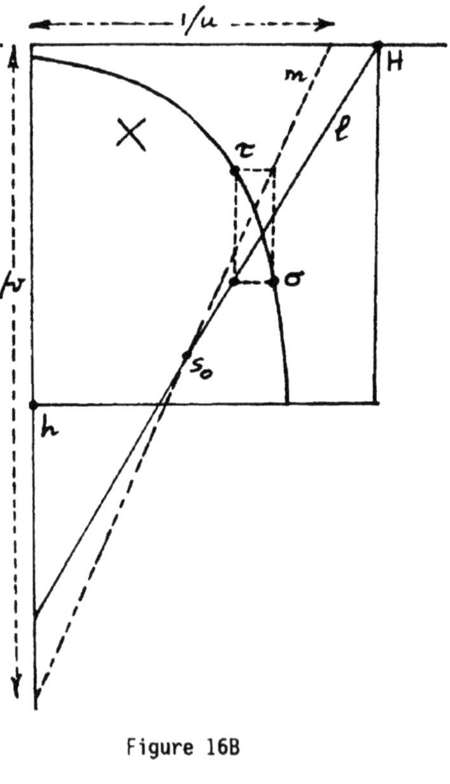

Figure 16B

The point labeled H (for heaven) is the "bliss point" (1,1) and that labeled h (for hell) is the "agony point" (0,0).

Proceed, for the moment, as though items (i) and (ii) of the definition of "extremal" are absent. This allows attention to be concentrated on item (iii), if this fails to hold, then the situation illustrated in Fig. 8A applies. The important point is that a *small* deviation by player 1 or player 2, from reporting U^* and V^* to reporting u and v, leaves the Nash bargaining solution n at a cross-over point, as illustrated in Fig. 11A. (If (iii) holds, as illustrated in Fig. 8B, then such a small deviation may generate the situation illustrated in Fig. 11B, in which the Nash bargaining solution n is *not* a cross-over point.)

The last consideration explains why the points labeled σ and τ in Fig. 16A are both placed on the Pareto-frontier of the set X. (This is justified, for the case when U^* and V^* do not satisfy (iii) and u and v represent a small deviation, by the fact that $n = c$. In Fig. 11A, lies on the Pareto-frontier of *both* X_{AE} *and* X_{EA}.) Their location is otherwise determined by the necessity of satisfying Eqs. (6) and (7). The former requires that $(\Phi_A(\sigma),\Phi_E(\tau))$ lie on the line m, and the latter that $(\Phi_A(\tau),\Phi_E(\sigma))$ lie on the line ℓ.

Some algebra is now required. Subtract Eq. (6) from Eq. (4). Then

$$U^*(\Phi_A(s^*)-\Phi_A(\tau)) = V^*(\Phi_E(s^*)-\Phi_E(\sigma)). \tag{8}$$

This has to be compared with the requirement (3), which expresses the condition that it is unprofitable for player 1 to deviate. Because σ lies on the Pareto-frontier of X, $\Phi_A(\sigma))$ cannot be increased without decreasing $\Phi_E(\sigma)$, and similarly for τ. (Recall that strict convexity is assumed.) It is then easy to deduce the following equivalence from Eq. (8):

$$\sigma >_A s^* \Leftrightarrow \sigma <_E s^* \Leftrightarrow s^* >_A \tau \Leftrightarrow s^* <_E \tau. \tag{9}$$

If any one of these strict preferences hold, then the equilibrium condition (3) *fails* because the left-hand side will be positive and the right-hand side negative. (Conversely, if any of the strict preferences fail, then (3) necessarily holds.)

It follows that, if *any* small deviation by player 1 is permissible, then the equilibrium condition (3) cannot hold, when (iii) does not apply. Fig. 16A illustrates an appropriate destabilizing deviation. But it is not true that any small deviation is allowed. If $U^* = 1$, deviations with $u > 1$ are not permitted, and similarly if $V^* = 1$. This explains the necessity for conditions (i) and (ii) in the definition of "extremal". (Fig. 16B illustrates a case in which $U^* = 1$ and destabilizing deviations by player 1 violate the constraint $u \geq 1$.)

It was only necessary to consider deviations by player 1 in the above argument. Deviations by player 2 may be treated simply by interchanging the symbols σ and τ in Eqs. (1) - (9).

The "Hybrid" Bargaining Solution

Section 3.2 makes a claim about the personal utilities that Adam and Eve obtain in a symmetric equilibrium of the type studied in this appendix. The claim is illustrated in Figs. 9A and 9B. Since $(\Phi_A(s^*),\Phi_E(s^*))$ lies on the ray of slope U^*/V^* which passes through the point $e = (a,1-b) = (\Phi_A(s_0),\Phi_E(s_0))$, the "corner solution" illustrated in Fig. 9B corresponds to the case $U^* = 1$. The alternative "corner solution" corresponds to the case $V^* = 1$. This explains how the claim arises when U^* and V^* are extremal in consequence of items (i) or (ii) of the definition. It remains to examine the relevance of item (iii).

Proposition 2. *A necessary condition for item (iii) is that* $(\Phi_A(s^*),\Phi_E(s^*))$ *be the Nash bargaining solution of* X, *relative to the disagreement point* $e = (a,1-b) = (\Phi_A(s_0)\Phi_E(s_0))$.

Proof. Item (iii) implies that c^* of Fig. 8B is the Nash bargaining solution of X_{AE}, relative to the disagreement point d^*. The strictly increasing, affine transformation

$$(u_1u_2) \rightarrow (U^*u_1,\ V^*(u_2-1) + 1)$$

maps X onto X_{AE}, e onto d^* and $(\Phi_A(s^*),\Phi_E(s^*)$ onto c^*. Since the Nash bargaining solution is invariant under such transformations (by Nash's (1950) "invariance" action), Proposition 2 follows.

Sufficiency Conditions

Refer to Figs. 9A and 9B. If n is the Nash bargaining solution for Z, relative to the disagreement point e, then the line through e and n determines parameters U^* and V^*, as indicated in Fig. 16A. For a symmetric equilibrium, it is necessary and sufficient that U^* and V^* be determined in this way. One then has that $n = (\Phi_A(s^*),\Phi_E(s^*))$.

This characterization is not essential to the text and so only one aspect of the proof will be mentioned here. Suppose that the requirement for U^* and V^* to lie in [0,1] were absent (so that Z in the previous paragraph would be the same as X). After proving a converse of Proposition 2, it would then be necessary to demonstrate that a deviation by player 1 in the configuration of Fig. 8B would not be to his advantage. Such a deviation either replaces Fig. 8B by a configuration like Fig. 11A or else by a configuration like Fig. 11B. The former case can be dealt with by reversing the argument of Proposition 1. The latter case requires a new argument. The important point is that, if player 1 engineers a deviation which leads from one Fig. 11B configuration to another, then player 2's payoff remains unchanged. The reason is that a deviation by player 1 generates a strictly increasing, affine transformation of his own utility scale while leaving that of player 2 fixed. The result therefore follows from Nash's (1950) "invariance axiom". Since player 2's payoff is unchanged, the same must be true of player 1's actual payoff since the original outcome is Pareto-efficient.

Akira Okada and Hartmut Kliemt

Anarchy and Agreement - A Game Theoretic Analysis of Some Aspects of Contractarianism

I. Introduction

Our paper is an exercise in *"the logic of collective action"*. It analyzes some difficulties that arise if we take seriously the fundamental ethical premise of contractarianism, namely, that nothing short of free consent of all individuals concerned can justify an institutional order. The discussion is based on an extensive game of consensual choice of rules or institutions in prisoner's dilemma situations. In this game the fundamental distinction between collective consent under the unanimity rule -- a collective act in a predefined group -- and private consent of individuals -- an act that precedes group formation -- shows up in the process of institution creation. We think that an explicit game theoretic formulation of these two forms of unanimity may be helpful for a better understanding of the scope and limits of contractarian theories of social institutions: In a purely private decision rational players of a set N of concerned individuals who share a common interest decide whether or not they become members of a group S. Only after the group has been constituted its members may make decisions according to some collective decision rule. The decision of whether or not a rational player should join S depends on the predictable results of the game for a member or for an outsider of S. If group membership is purely voluntary well known arguments suggest that S=N will hold good only under quite specific circumstances. Some of the relevant circumstances are explicitly modeled subsequently (II.-IV.). We describe a general model although, in the present paper, our explicit analysis is restrained to the special case of three interacting individuals and a numerically specified prisoner´s dilemma situation. This is sufficient to illustrate game theoretically some more philosophical points. After relating our non-cooperative game theoretic analysis to Olson's (cf. 1965) only implicitly game theoretic one (V.1.) it is indicated that circumstances favorable to individually rational group formation may not be fully compatible with contractarian ethical premises as those of Nozick (cf. 1974) (V.2.) because individualistic contractarian and collectivistic non-contractarian versions of the unanimity principle are incompatible in general (V.3.).

II. The game theoretic model

Our elementary game theoretic model is designed to study some of the problems that might arise if rational individuals try to overcome an n-person prisoner's dilemma situation by the formation of collective agencies.

Let $N = \{1, 2, ..., n\}$ be the set of all players. Simultaneously every player $i \in N$ can choose to take one of two actions that are conveniently dubbed C_i (cooperation) and D_i (defection). The payoff for each player depends on his action $a_i \in \{C_i, D_i\}$ and also on the number h_{-i} of other individuals who cooperate. The payoff function of player $i \in N$ is given by

$f_i(a_i, h_{-i})$; where $h_{-i} = |\{ j \in N \mid j \neq i, a_j = C_j\}|$.

For every $i \in N$

(i) $f_i(C_i, h_{-i}) < f_i(D_i, h_{-i})$, for $h_{-i} = 0, 1, 2, ..., n-1$,

(ii) $f_i(C_i, n-1) > f_i(D_i, 0)$.

Condition (i) implies that every player $i \in N$ has a dominant strategy D_i. He or she will be better off playing this strategy regardless of the value of h_{-i} or of what the other players do. But, from condition (ii) it follows that the result of playing the dominant strategies will be Pareto-inferior to playing the cooperative and dominated strategies. This is common knowledge among all players and thus they have a common interest or good reason to "hope" that the situation may be changed in a way that would allow them to reach a Pareto-efficient outcome in a rational manner.[1] Therefore they might consider along the usual lines of contractarian argument to form groups that put into operation an agency that will enforce agreements among the members of the group.

The process of group formation may be characterized as a constitutional game Γ. Γ comprises several stages the last of which is the underlying prisoner's dilemma game played under the constitutional rules that are the outcome of

(i) negotiations among all players in N about the formation of a group with an enforcement agency, and

(ii) negotiations for collective action within the group formed.

More precisely the extensive form of Γ consists of the following four stages.

[1] We are alluding here to Hobbes' distinction between obligations to act and obligations to hope that something will become true. The former arise "in foro externo" while the latter hold good only "in foro interno"; cf. for this important though not always noticed distinction Hobbes 1651/1968, 215.

(1) The participation decision

All players in N decide simultaneously whether or not they participate in group formation bargaining. Formally, every player $i \in N$ simultaneously may either select $d_i=1$ (in) or $d_i=0$ (out). For any decision vector $d=(d_1, ..., d_n)$ the set of participants is determined by

$$S := \{ i \in N \mid d_i=1 \}.$$

The players outside the group S do not participate in any further negotiations. They will only "reappear" in the final stage of the game.

The individuals in the group S intend to leave the institution free "anarchic" state of affairs as far as their relationships *among each other* are concerned. To that purpose they enter negotiations about installing a kind of government, a legal staff, or a police that is going to enforce contracts among members of S. The negotiation process is assumed to be subject to the rule of the second stage.

(2) Negotiating for the installation of an enforcement agency and the intensity of punishment

In this stage, all members of the group S negotiate about whether or not they should install the police within S, and also about how much punishment power should be bestowed on the police. Formally, every player $i \in S$ selects $q_i \in \mathbb{R}_+$ which is interpreted as the amount or quantity of a penalty that can maximally be imposed on deviators from contractual agreements among the members of S. Let $s:=|S|$, for any combination $(q_i)_{i \in S}$ selected by members of S, the amount q of the penalty is determined by the unanimity rule:

$$q(q_1, ..., q_s) = \begin{cases} q \text{ if } q = q_i & \text{for all } i \in S \text{ and } s \geq 2 \\ 0 & \text{otherwise} \end{cases} .$$

If all members of the group S agree to install the police then it will enforce unanimous agreements of players in S.[2] In the group S that can make use of the services of the police force a collective decision can be made about which kinds of action the police should enforce. Following the main thrust of contractarian argu-

---

2 Only now a collectivity S has been constituted that is governed by the constitutional rule of unanimous agreement. Thus the two phenomena of joining or not a collectivity (a purely individual decision) and of consenting or dissenting to its collective actions (a genuinely collective decision) are kept apart. Collective decision and collective action becomes possible only after the relevant collectivity is determined by *individual consent*. Contrary to those "contractarian" approaches that are ultimately based on consent *under* the unanimity rule in our model it is not exogenously determined to whom the collective choice principle applies.

ment we will assume that the unanimity principle as a collective decision rule governs this choice too (at least on the highest level of the agreement) and thus the bargaining process involved (cf. for instance Buchanan and Tullock 1962). This is explicitly stated in the rule characterizing the next stage of the game.

(3) Negotiating for collective action

All payers in the group S negotiate about cooperation. Formally, every player $i \in S$ may simultaneously select either $\underline{a}_i = C_i$ or $\underline{a}_i = D_i$. Agreement on cooperation is reached only if $\underline{a}_i = C_i$ for all $i \in S$, otherwise no collective action to enforce cooperation in the final stage of the game may be taken.

(4) Taking actions

In the final stage all players in N actually play the prisoner's dilemma game. Possibly diverging from his declaration "$\underline{a}_i$" in the negotiation for collective action stage every player simultaneously selects either $a_i = C_i$ or $a_i = D_i$. This, in turn will determine the final payoff $F_i(a_1, ..., a_n)$ for each player i in N.

Again let $h_{-i} = |\{ j \in N \mid j \neq i,\ a_j = C_j \}|$.

(i) If unanimous agreement is reached in S then

$$\forall\ i \in S,\ F_i(a_1, ..., a_n) := \begin{cases} f_i(a_i, h_{-i}) & \text{if } a_i = C_i \\ f_i(a_i, h_{-i}) - q & \text{if } a_i = D_i \end{cases}$$

$$\forall\ i \notin S,\ F_i(a_1, ..., a_n) := \quad f_i(a_i, h_{-i}).$$

(ii) When unanimous agreement in S is not reached

$$\forall\ i \in N,\ F_i(a_1, ..., a_n) := \quad f_i(a_i, h_{-i})\ .$$

At each stage of the game Γ every player makes his decision with complete knowledge of the outcomes of all past stages of the game.

Given this description of the general game we may analyze its strategies and solution. A pure strategy σ_i for every player $i \in N$ is a function that assigns a choice to each of his decision nodes in Γ. Given any pure strategy combination $\sigma = (\sigma_1, ..., \sigma_n)$ of the n players the play of the game Γ is uniquely determined and so are the payoffs to the players. A game played at one of the stages of Γ is called a stage game of Γ. For every player the payoff in each stage game depends on what is going to happen in all future stage games of Γ. For a pure strategy combination $\sigma = (\sigma_1, ..., \sigma_n)$, the concept of a σ-stage game of Γ is defined as the stage game that arises from the assumption that all future stages are played according to σ.

The non-cooperative solution concept that we are going to employ in our discussion of Γ is that of a subgame perfect equilibrium point with the "undominatedness" property. This solution concept is defined as follows.

(A) Subgame perfectness: The strategy combination $\sigma^* = (\sigma^*_1, ..., \sigma^*_n)$ of Γ is called a *subgame perfect equilibrium point* if and only if it induces a Nash equilibrium on every σ^*-stage game of Γ.

(B) Undominatedness: A subgame perfect equilibrium point $\sigma^* = (\sigma^*_1, ..., \sigma^*_n)$ of Γ is called *undominated* if and only if the Nash equilibrium induced by σ^* on every σ^*-stage game of Γ does not entail any (weakly) dominated strategies for any of the players.

For a more precise definition of the concept of a subgame perfect equilibrium point and a detailed account of its properties the reader is referred to Selten (1975). As indicated in the introduction we are presently interested only in illustrating and making more precise a basic philosophical point with game theoretic tools. Therefore we need not analyze in a general way the class of games characterized by the assumptions of our game theoretic model. For illustrative purposes it suffices to discuss one numerical example because all philosophically relevant aspects show up already in a three person prisoner's dilemma game with a very simple numerical structure (cf. also sec. IV below).

III. A simple example of a three person game

III.1. The example

We consider a 3-person prisoner's dilemma game with a set of players $N=\{1, 2, 3\}$ satisfying for $i = 1, 2, 3$ and $h_{-i} = 0, 1, 2$

$f_i(C_i, h_{-i}) := 3h_{-i} + 3$, and $f_i(D_i, h_{-i}) := 3h_{-i} + 5$;

where h_{-i} is again the number of other individuals except of i who cooperate. The payoff matrix of this prisoner's dilemma game is

2		C		D	
3		C	D	C	D
1	C	9, 9, 9	6, 6, 11	6, 11, 6	3, 8, 8
	D	11, 6, 6	8, 3, 8	8, 8, 3	5, 5, 5

Table 1 payoffs of the original prisoner's dilemma

The ***participation decision stage*** of our game Γ is represented by the following extensive form:

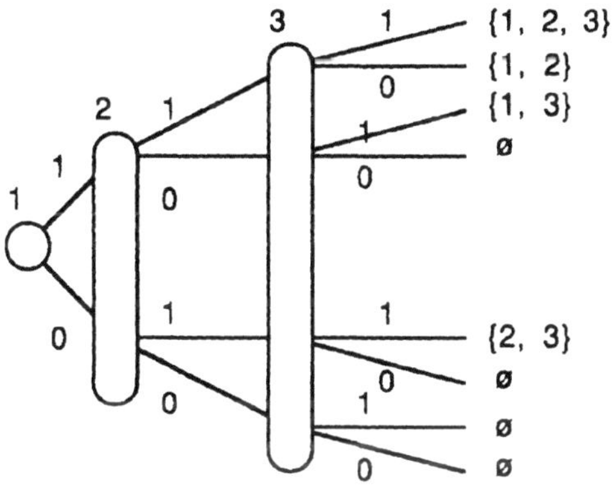

Figure 1 tree of participation decision stage

When the empty set "ø" shows up at the end of one of the pathes through the tree this indicates that no group of at least two individuals has formed. In this case, no collective decisions can be taken within a group and the game proceeds directly to the final stage, i.e., the action decision stage. If at least one group S with at least two members has formed collective decisions in that group become viable The game proceeds to the second stage, i.e., the ***negotiation stage for installing the police and the intensity of punishment***.

In this second stage of Γ each member of the group S has infinitely many choice in selecting the intensity q of punishment by the police. The situation for the 3 person group is roughly described by the picture below. If each of the player chooses a value of q along a vertical choice line, then there are infinitely many instances of agreement and disagreement. The agreement lines are horizonta whereas the disagreement lines are not:

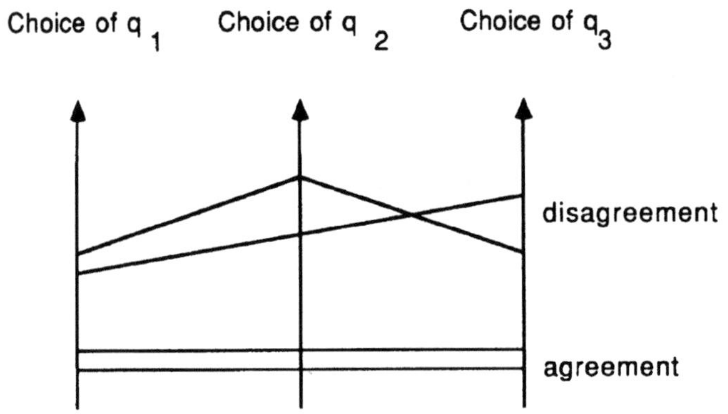

Figure 2 agreement and disagreement on intensity of punishment

If a horizontal line results from these choices and thus unanimous agreement about q is reached, players can enter the negotiation stage about whether the police as their agent should enforce some sort of collective agreement on actions in the underlying game. The ***negotiation stage for collective action*** when S={1, 2, 3} is represented by the extensive form below

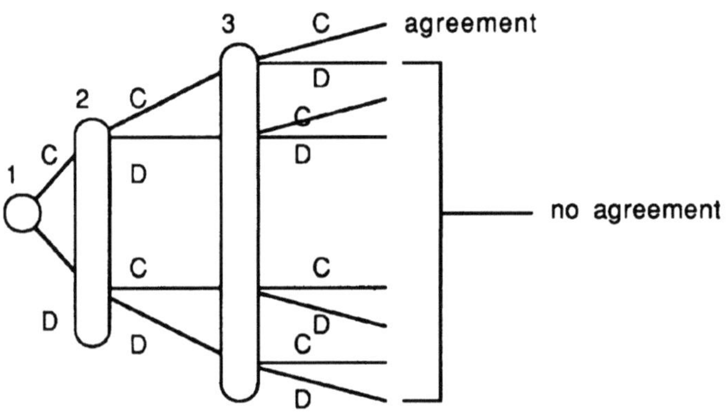

Figure 3 tree of negotiation stage of collective action

When a two person subgroup S of N has formed the extensive form of this third stage of the game Γ can be given in a similar manner.

If agreement is reached in S = {1, 2, 3}, then at the final ***action decision stage*** of Γ players select their actions in the prisoner's dilemma game under the "institutions" stemming from their own agreement. They can still deviate from cooperation, but if they do, a penalty is imposed by the police. If all members of S have unanimously agreed on installing the police in their group the police is authorized by the consent of every player to take action against deviations from the actions that have been unanimously agreed on. The extensive form of the action decision stage of Γ under police operation is then given by

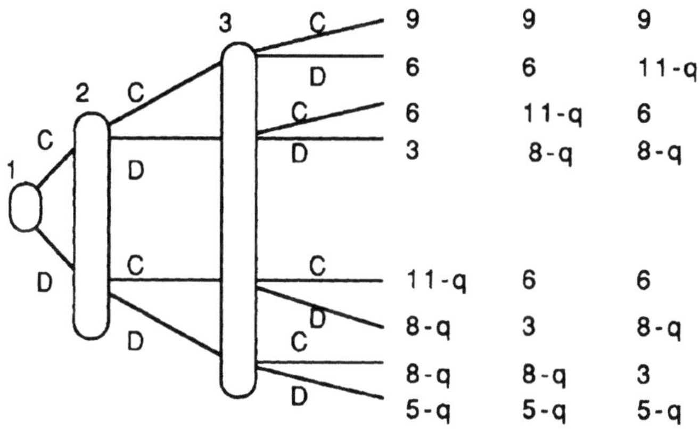

Figure 4 extensive form of action decision stage under agreement in {1, 2, 3}

If agreement is reached in a two-person group S of N, say S = {1, 2}, the police has been authorized by consent only to take action towards those individuals who were part of both agreements that of installing the police and that of taking collective action. The extensive form of the final stage of Γ is given then by

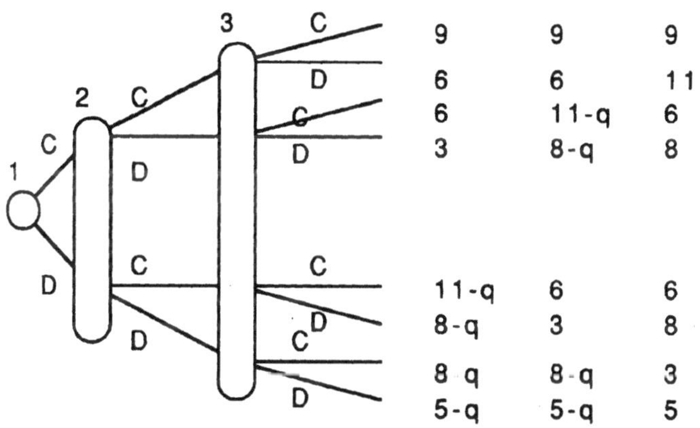

Figure 5 extensive form of action decision stage under agreement in {1, 2}

If player 3 -- as presently assumed -- is not a member of S the police cannot legitimately impose a penalty on him should he choose not to cooperate. At least his own prior consent does not authorize such an act.

If no agreement is reached, all players in N select their actions in the original ("institution free") prisoner's dilemma game. Therefore, the extensive form of the final stage is given by

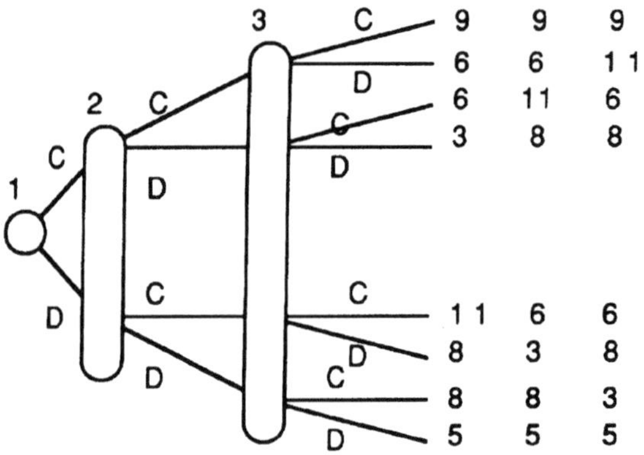

Figure 6 extensive form of action decision stage without agreement

After we have introduced and illustrated the rules of our example we now analyze whether or not the outcome of our example will be S=N and also which value of q rational players would choose.

III.2 Analyzing the example

We will characterize a subgame perfect equilibrium point of the example with the "undominatedness" property. The analysis is based on "backward induction". First we seek Nash equilibria of the final stage of the game which satisfy the "undominatedness" property. Next, we seek Nash equilibria of the third stage game under the assumption that the final stage game is played according to the Nash equilibria already obtained. This procedure ends when the game at the participation decision stage is solved.

In a first step we will assume that some form of agreement on group formation is reached between the players. We will further distinguish between those three cases in which a proper subgroup S of N of at least two individuals is formed, (i), on the one hand and the case that the group S = {1, 2, 3} forms, (ii), on the other hand. In a final step we will drop the initial assumption that a group has formed and solve the game for the participation decision stage (iii). After this we will be in a position to answer the question whether or not one and if so which of the groups S will form.

(i) Since the game is symmetric it suffices that we focus on the case S = {1, 2}.

1) The payoff matrix of the *action decision stage* in which every player chooses either $a_i=C_i$ or $a_i=D_i$ is (when no agreement was reached either on installing the police or on collective action we have q=0),

	C	D		C	D
C	9 9 9	6 11-q 6	C	6 6 11	3 8-q 8
D	11-q 6 6	8-q 8-q 3	D	8-q 3 8	5-q 5-q 5
	C			D	

Table 2 payoff matrix of the action decision stage for S={1, 2}

The Nash equilibria depend on the value of q. The following table which gives all pure strategy equilibria shows the dependency,

Value of q	Corresponding Nash equilibrium	
	strategy combination	payoffs
q>2	(C, C, D)	(6, 6, 11)
q=2	(C, C, D) (D, C, D) (C, D, D) (D, D, D)	(6, 6, 11) (6, 3, 8) (3, 6, 8) (3, 3, 5)
0≤q<2	(D, D, D)	(5-q, 5-q, 5)

Table 3 dependency of pure strategy equilibria on q

2) The payoff matrix for the *negotiation stage for collective action* in which every player in the group {1, 2} chooses either $\underline{a}_i=C_i$ or $\underline{a}_i=D_i$ is

	C	D
C	x_1 x_2	5 5
D	5 5	5 5

Table 4 payoff matrix of the negotiation stage for collective action with S={1, 2}

where (x_1, x_2) is the payoff of a Nash equilibrium determined in 1). As the Nash equilibria of the final stage depend on q the same holds good for the Nash equilibria of the next to last stage. The equilibrium strategy combinations ($\underline{a}_1, \underline{a}_2$) of that stage and the payoff combinations eventually resulting from these choices are shown in the following table

Value of q	(x_1, x_2)	strategy combin. ($\underline{a}_1, \underline{a}_2$)	payoff combination (α_1, α_2)
$q > 2$	(6, 6)	(C, C)	(6, 6)
$q = 2$	(6, 6) (6, 3) (3, 6) (3, 3)	(C, C) (C, D) (D, C) (D, D)	(6, 6) (5, 5) (5, 5) (5, 5)
$0 < q < 2$	(5-q, 5-q)	(D, D)	(5, 5)

Table 5 equilibrium strategy combinations and payoffs of negotiation for collective action stage with S={1, 2}

Only if $(x_1, x_2) = (6, 6)$ can ($\underline{a}_1, \underline{a}_2$) = (C, C) be in equilibrium. Otherwise at least one player i=1, 2 is better off after announcing $\underline{a}_i$ = D. For, then, unanimity will not prevail at the negotiation for collective action stage. q will be set to zero which implies that (D, D, D) is the only equilibrium of the action decision stage with payoffs (α_1, α_2) = (5, 5).

3) The payoff function for the *negotiation stage for installing the police and the intensity of punishment* in {1, 2} is for i=1, 2,

$$f_i(q_1, q_2) = \begin{cases} 6 & \text{if } q_1 = q_2 > 2 \\ \alpha_i & \text{if } q_1 = q_2 = 2 \\ 5 & \text{if } 0 \le q_1 = q_2 < 2 \text{ or } q_1 \ne q_2 > 2 \end{cases}$$

where

$$\alpha_i = \begin{cases} 6 & \text{if } (x_1, x_2) = (6, 6) \\ 5 & \text{if } (x_1, x_2) \in \{(6, 3), (3, 6), (3, 3)\} \end{cases}$$

Considering the values of α_i obviously the (undominated) Nash equilibria in this negotiation stage are characterized by:

$$q_1=q_2 \geq 2 \quad \text{if} \quad \alpha_i = 6 \quad \text{and} \quad q_1=q_2 > 2 \quad \text{if} \quad \alpha_i = 5.$$

In equilibrium individuals will choose the same value of q which is greater or equal to 2. Pursuing the undominated subgame perfect equilibrium strategy will then lead to a payoff of (6, 6) for players 1 and 2 respectively.

(ii) The case of S = {1, 2, 3} can be analyzed basically along the same lines as the three cases that were discussed in (i).

1) The payoff matrix of the *action decision stage* is

C

	C	D
C	9, 9, 9	6, 11-q, 6
D	11-q, 6, 6	8-q, 8-q, 3

D

	C	D
C	6, 6, 11-q	3, 8-q, 8-q
D	5-q, 3, 8-q	5-q, 5-q, 5-q

Table 6 payoff matrix of the action decision stage with S={1, 2, 3}

where again q=0 when either no agreement on installing the police and the intensity of punishment or no agreement on collective action was reached. In the latter cases the payoff in the last stage of the game will be (5, 5, 5). As before the equilibrium outcomes of the collective decisions depend on the equilibrium outcomes of the action decision stage and thus on the value of q. The Nash equilibria of the action decision stage are listed below.

Value of q	strategy combin	payoff combination
q > 2	(C, C, C)	(9, 9, 9)
	(C, C, C)	(9, 9, 9)
	(C, C, D)	(6, 6, 9)
	(C, D, C)	(6, 9, 6)
q = 2	(D, C, C)	(9, 6, 6)

	(D, D, C) (D, C, D) (C, D, D) (D, D, D)	(6, 6, 3) (6, 3, 6) (3, 6, 6) (3, 3, 3)
$0 < q < 2$	(D, D, D)	(5-q, 5-q, 5-q)

Table 7 Nash equilibria of the action decision stage with S={1, 2, 3}

2) The payoff matrix of the *negotiation stage for collective action* in the group {1, 2, 3} is

C

	C	D
C	x_1, x_2, x_3	5, 5, 5
D	5, 5, 5	5, 5, 5

D

	C	D
C	5, 5, 5	5, 5, 5
D	5, 5, 5	5, 5, 5

Table 8 payoff matrix of the negotiation stage for collective action in S={1, 2, 3}

where (x_1, x_2, x_3) is the payoff of a Nash equilibrium determined in 1).

The payoffs at the negotiation for collective action stage depend on (x_1, x_2, x_3) and thus on q. Basically three cases must be distinguished:

If $q > 2$ then only one undominated equilibrium exists at the last or action decision stage. All players $i = 1, 2, 3$ should announce $\underline{a}_i = C$ at the next to last stage. Because deterrence is strong enough the combination of strategic announcements ($\underline{a}_1, \underline{a}_2, \underline{a}_3$) = (C, C, C) will lead to the expected payoff combination ($\alpha_1, \alpha_2, \alpha_3$) = (9, 9, 9) at the negotiation for collective action stage.

If $q < 2$ there is again exactly one undominated equilibrium at the last stage. No player should agree on collective action at the next to last stage. The equilibrium announcements ($\underline{a}_1, \underline{a}_2, \underline{a}_3$) = (D, D, D) will lead to a payoff of (5, 5, 5).

If q = 2 then all strategy combinations (a_1, a_2, a_3) of the last stage are equilibria. If the payoff vector of such an equilibrium fulfills (x_1, x_2, x_3) ≥ (5, 5, 5) then the announcements ($\underline{a}_1, \underline{a}_2, \underline{a}_3$) = (C, C, C) will be in equilibrium. The payoffs resulting from this will fulfill the relationship ($\alpha_1, \alpha_2, \alpha_3$) > (5, 5, 5). If, however, there is a player i with $x_i < 5$ this player should announce $\underline{a}_i$ = D; i.e. player i should veto collective action. Then the payoffs will be ($\alpha_1, \alpha_2, \alpha_3$) = (5, 5, 5).

The following table sums up the discussion of the three cases:

Value of q	(x_1, x_2, x_3)	Strategy combin. ($\underline{a}_1, \underline{a}_2, \underline{a}_3$)	Payoff combination ($\alpha_1, \alpha_2, \alpha_3$)
q > 2	(9, 9, 9)	(C, C, C)	(9, 9, 9)
	(9, 9, 9)	(C, C, C)	(9, 9, 9)
	(9, 6, 6)	(C, C, C)	(9, 6, 6)
	(6, 9, 6)	(C, C, C)	(6, 9, 6)
q = 2	(6, 6, 9)	(C, C, C)	(6, 6, 9)
	(6, 6, 3)	(C, C, D)	(5, 5, 5)
	(6, 3, 6)	(C, D, C)	(5, 5, 5)
	(3, 6, 6)	(D, C, C)	(5, 5, 5)
	(3, 3, 3)	(D, D, D)	(5, 5, 5)
0 < q < 2	(5-q, 5-q, 5-q)	(D, D, D)	(5, 5, 5)

Table 9 equilibrium strategy combinations and payoffs of negotiation for collective action stage with S={1, 2, 3}

3) The payoff functions for the stage of the *negotiations for installing the police and the intensity of punishment* depend on the choice of q and on the equilibrium selected if q=2. In general we have

$$\forall\, i \in S,\ f_i(q_1, q_2, q_3) = \begin{cases} 9 & \text{if } q = q_1 = q_2 = q_3 > 2 \\ \alpha_i & \text{if } q = q_1 = q_2 = q_3 = 2 \\ 5 & \text{if } q = q_1 = q_2 = q_3 < 2 \text{ or not } q_1 = q_2 = q_3 \end{cases}$$

where α_i is the payoff to player i if the Nash equilibrium of the negotiation for collective action stage leads to a payoff combination ($\alpha_1, \alpha_2, \alpha_3$) for q=2. As there are five different combinations ($\alpha_1, \alpha_2, \alpha_3$) the payoff functions may be grouped in five classes each comprising one function for each of the three players. The class related to ($\alpha_1, \alpha_2, \alpha_3$) = (6, 9, 6) for instance comprises

$$f_1(q_1, q_2, q_3) = \begin{cases} 9 & \text{if } q = q_1 = q_2 = q_3 > 2 \\ 6 & \text{if } q = q_1 = q_2 = q_3 = 2 \\ 5 & \text{if } q = q_1 = q_2 = q_3 < 2 \text{ or not } q_1 = q_2 = q_3 \end{cases}$$

$$f_2(q_1, q_2, q_3) = \begin{cases} 9 & \text{if } q = q_1 = q_2 = q_3 > 2 \\ 9 & \text{if } q = q_1 = q_2 = q_3 = 2 \\ 5 & \text{if } q = q_1 = q_2 = q_3 < 2 \text{ or not } q_1 = q_2 = q_3 \end{cases}$$

$$f_3(q_1, q_2, q_3) = \begin{cases} 9 & \text{if } q = q_1 = q_2 = q_3 > 2 \\ 6 & \text{if } q = q_1 = q_2 = q_3 = 2 \\ 5 & \text{if } q = q_1 = q_2 = q_3 < 2 \text{ or not } q_1 = q_2 = q_3 \end{cases}$$

The payoffs of the Nash equilibria that result from the payoff functions f_i at the stage of the negotiations for installing the police are shown in the following table,

	$(\alpha_1, \alpha_2, \alpha_3)=$	(9, 9, 9)	(9, 6, 6)	(6, 9, 6)	(6, 6, 9)	(5, 5, 5)
q =	$q_1=q_2=q_3>2$	(9, 9, 9)	(9, 9, 9)	(9, 9, 9)	(9, 9, 9)	(9, 9, 9)
	$q_1=q_2=q_3=2$	(9, 9, 9)	(9, 6, 6)	(6, 9, 6)	(6, 6, 9)	(5, 5, 5)
	$q_1=q_2=q_3<2$ or $\forall$ i,j: i≠j -> $q_i \neq q_j$	(5, 5, 5)	(5, 5, 5)	(5, 5, 5)	(5, 5, 5)	(5, 5, 5)

Table 10 Nash equilibria of the negotiations for installing the police in S={1, 2, 3}

(iii) The *stage of the participation decision* can be characterized by a table showing the possible combinations of joining a group (d_i=1) or not joining it (d_i=0).

	1	0
1	*	6, 11, 6
0	11, 6, 6	5, 5, 5
	1	

	1	0
1	6, 6, 11	5, 5, 5
0	5, 5, 5	5, 5, 5
	0	

Table 11 payoff matrix of the participation decision stage

For "*" one of the five equilibrium payoff vectors

(9, 9, 9), (9, 6, 6), (6, 9, 6), (6, 6, 9), (5, 5, 5)

of the subgame played by S={1, 2, 3} if all players simultaneously chose the "in" option may be substituted. It can immediately be seen that players in N will not choose to form a group S for collective action which would comprise all players. None of the possible substitutions would lead to an equilibrium. In an equilibrium of the game Γ either one of the three two person groups or no group at all will form. The group S=N will not be constituted in equilibrium. -- Before we turn to the philosophical insights that may be gained from this result a short remark on the more general case may be in order.

IV. On generalizing the analysis and its results

In a subsequent paper that builds on the premises and the ideas of the present one, one of the authors (Okada) using a somewhat stronger solution concept based on payoff dominance among equilibrium points demonstrates that for any group N the number $|S|$ of individuals participating in group formation will be determined as the value s^*, $2 \leq s^* \leq |N|$, such that $f_i(C, s^*\text{-}2) < f_i(D, 0) < f_i(C, s^*\text{-}1)$. Thus, in a symmetric N-person game of the type analyzed in the present paper there is in general a well defined value for the number of players who in equilibrium will participate voluntarily in a group S providing some public good. (In the example above we have $s^*=2$.) This value closely resembles the minimum requirement "n" of other cooperators that plays such a crucial role in Michael Taylor's well known analysis of the N-person prisoner's dilemma supergame (cf. the "revised discussion" in 1987, 82 ff.). If in Taylor's model the number of co-operating players in equilibrium must exactly equal n+1 (neither n nor n+2 will do), so we have $s^*=|S|$ in our model and its generalization (and neither s^*-1 nor s^*+1 will do). However, in Taylor's analysis the parameter n is exogenously fixed whereas in the present model s^* is endogenously determined. Again, in our model costless enforcement services are exogenously provided should individuals agree to use them whereas Taylor's analysis relies on endogenous policing based on contingent strategies of the players.

The two models may be viewed as complementary. Though it might be of some general interest to pursue the relationship between them somewhat further this will not be possible here. There are other links to arguments proposed in recent discussions of general social and political theory which may be of still greater interest. To these we will now finally turn.

V. Some interpretations and consequences

V.1. A short look at Olson's logic of collective action

As has been indicated in the introduction our model is an exercise in the logic of collective action. In his book with the same title Mancur Olson (1965) revived the political wisdom of some of the British and Scottish Moralists (cf. Raphael 1969 and Schneider 1967 for an overview and in particular Hume 1948, 97 ff. -- book III, part II, sec. VII of the *treatise*). He insisted that the behavior of collectivities should be explained as the outcome of individually rational behavior and the incentives operating on individuals. Against the bulk of political theory of that time he pointed out that rational individuals with a common interest will not necessarily have sufficient incentives to act on behalf of their common interest. There may be latent interest groups that will not be able to act in a concerted way.

The analysis of our model reproduces Olson's basic general claim.[3] In equilibrium S may be empty or it may be a proper subset of N. In the latter case players in the complementary set $N \setminus S$ will "exploit" those in S even though all players are identical and there is no Olsonian exploitation of the more intensely interested by less interested players. Further, our model shows explicitly that this may hold

[3] It is, of course, no surprise that non-cooperative game theoretic analysis of group formation parallels Olson's theory. For, his theory in the last resort boils down to stating that coalition formation cannot be taken for granted but has to be modeled explicitly.

good *even if enforcement institutions are provided at no costs.* Of course, there are opportunity costs of group membership. But these opportunity costs are not the same thing as membership fees and other direct costs that one might intuitively think of.

In any case it should be clear that our model is compatible with Olson's views. On the other hand, we are not interested in the logic of collective action as such but rather in its application to some basic principles of contractarianism.

V.2. The model and Nozick's invisible hand explanation of the emergence of the state

The modern theorist who most consequently endorses a (Lockean) contractarian view of the foundations of collective institutions is presumably Robert Nozick (cf. 1974). According to his basic ethical premise the application of *fundamental coercive power*, i.e. power that is not based on the prior consent of those who fall victim to it, can never be justified. All collective institutions that legitimately may apply power, then, must emerge from a process of private contracting -- and consensual transference of indiviual rights -- that starts in an institution free setting. This setting closely resembles the one that is sketched in our model.

Nozick imagines an anarchic state of nature in which individuals interact with one another under the rules of natural law. He concedes to the anarchist that in the best anarchic state that we can reasonably hope for individuals by and large would play by the rules without any institutional provision for their enforcement. However, as it is not reasonable to assume that the state of nature is a state of prestabilized harmony, there will be occasional violations of individual rights even in the best state of nature that we can reasonably hope for. Rational individuals will seek a remedy for these risks of the state of nature. Natural law provides clearance to do this. For, individuals have a natural right to punish violations of their rights. But, without courts and without "a sword" enforcing the findings of the courts there will be no reliable (due) process of rule or rights' enforcement. This deficiency of the state of nature will attract "protective agencies". These agencies will offer contracts to provide rights' protection, rule enforcement, and dispute settlement. In the execution of punishment they will act as agents to whom the individuals as principals have merely ceded their natural right to punish. Therefore, the protective agencies operate within the limits of natural law.

The protective agencies are nothing but private enterprises or firms. In the provision of protective services there will be economies of scale that can be exploited by these firms. Together with the specific character of the good that is provided by the agencies there even will be a proclivity to form dominant protective agencies over certain territories. These dominant agencies are almost states. Thus, as if guided by an invisible hand rational individuals will transform anarchy without violating any of the normative premises of anarchism into something akin to a state (cf. Nozick 1974, chap. 2).

This is the core of Nozick's quite famous invisible hand explanation of the state. Nozick is not very explicit about the game theoretic nature of the process involved. But, even from the rough sketch of his invisible hand explanation given before it should be obvious that a more explicit formulation of his model in game theoretic terms would come quite close to the model of the present paper. From this point of view the model proposed here might be regarded as a first step to an explicit game theoretic reformulation of Nozick's argument. On the other hand it might seem doubtful whether such a formulation would actually be of any help in solving the crucial problem of justifying a monopolistic claim of protective agencies.

Nozick wants to show that a monopolistic claim of a dominant protective agency *can* be legitimate (cf. 1974, chap. 3-6). In his alleged refutation of anarchism he argues that every individual has a natural right -- before a collectivity is founded -- that any rights' disputes between himself and others will be settled only according to procedures that he himself accepts as appropriate or secure. Such a right, Nozick assumes, by its very nature implies the right to exert an externality on those who do not accept the same procedure. If everybody is endowed with this right, then individual rights will not be compossible. To put it more simply in Kantian terms it is not possible that "everybody follows his own mind" if we do want a public order.[4]

This quite elementary argument is completely independent of the invisible hand explanation. It justifies the application of fundamental coercive power "the short way" so to say. The monopolistic claim is not based on individual consent. Therefore the argument violates Nozick's own contractarian premises. After the dust has settled his model of the gradual emergence of protective agencies seems to be superfluous. It does not contribute to the justification of the monopolistic claim of the state. If the application of fundamental coercive power is never justified then we must stick to procedures of institution formation as those in our model. However, our model indicates that processes of consensual institution creation may

---

[4] Kant in his *metaphysics of morals* (cf. esp. §§ 42-44) insists that we are under a moral obligation to accept some form of a public or collective monopoly. This obligation exists prior and independently of individual consent and so do the rights of the relevant collectivity. Such a legitimation of collective authority may or may not be valid. We need not enter this discussion here: the legitimation is evidently not compatible with the basic ethical norm of contractarianism.

not even come close to dominant protective agencies. In particular it will not be necessarily true, as Nozick seems to imply, that the application of fundametal coercive power will be concentrated on some "stubborn" outsiders with strange preferences after the invisible hand has done its work.

Again, even if his alleged contractarian justification of the state itself should be a failure Nozick evidently could and would maintain that his model is important in understanding the proper scope and limits of state action. Contractarianism not only deals with the justified foundation of the institution of the state but also with what the state may do after it was founded. Here, Nozick would argue that the state should operate basically in the same way as the enforcement institution of our model. This institution or protective state will offer its services at no cost and thus facilitate collective action in the provision of some collective good or other for some group or other. But it will never act itself as a productive state nor will it levy taxes for common purposes. Levying taxes without prior consent of each individual taxpayer would involve the application of fundamental coercive power. Therefore the provision of collective goods may not be based on taxes. No collective good -- except for the enforcement of rights -- may be provided in a compulsory way. Though Nozick does not deny that there are public goods he denies that the state may ever act on behalf of the individual interests involved. Even if severe prisoner's dilemmas must be overcome production must be left to the free contracting of individuals.

As some of the more radical libertarians Nozick seems to think that the problems involved in public goods' provision mostly stem from direct costs of transactions. If we design the institution of contract such that time and effort of contracting are minimized then we may be quite optimistic that those public goods that are really worthwhile for individuals will be provided in a quasi-anarchic way without the intervention of government institutions. Our model sheds some light on this optimism. Even if there are no direct transaction costs and even if there is a perfect enforcement institution there may still be some "generalized hold out problem" as the one we sketched in our model. Even under conditions of zero transaction costs the invisible hand may be attached to an arm too weak for the provision of public goods.

Though we think that our model has some relevance for the assessment of contractarian ideas we do not want to overstate its implications. There are other conceivable ways of successive contracting than the one we analyzed. There are many ways to further collective action on a purely voluntary basis. Game theoretic analysis itself can be helpful in the process of designing different contracting games. Still, it seems to be quite clear that our model captures in a "natural way" some essential features of group formation under conditions of (quasi-)anarchic, purely private contracting: In our model fundamental coercive power is never applied. It even grants -- what otherwise might be hard to swallow -- that in a state of nature there are already enforcement agencies offering their services. These services facilitate contracting between rational agents but they raise no monopolistic claim excluding competing agencies or -- what basically amounts to the same thing -- make membership compulsory. Therefore their existence can indeed be compatible with the essential premises of contractarian state of nature theory. They not only offer their services at no costs, they also refrain from imposing them on any individual who has not given his prior consent.

Our model goes as far as one can go within the limits to collective action that are set by the basic ethical norm of contractarianism. Contracting is not spurred by any form of public compulsion even if it is derived from such sources as the application of equity law.[5] Unanimity of all individuals is the sole basis of collective action. But, as may be illustrated by our model too, even the term "unanimity" itself should be interpreted quite carefully and restrictively in the context of contractarianism. To this we now turn by way of some final remarks.

V.3. Contractarian and non-contractarian varieties of unanimity[6]

Purely private agreement as the foundation of collective action is widely spread. In a free society almost every firm, club, or corporate actor will be based on free contracting. Contrary to this the political realm seems to be differently structured. It is governed by institutions that are founded without prior consent of its members. As a matter of fact membership is mostly compulsory in the realm of politics. But there are exceptions to this.

The U.S. constitution for example was ratified by "the unanimous assent of the several States that are parties to it ..." (cf. Federalist 39). Actually any nine of the thirteen states of that time could form the union and put the constitution into operation (cf. Federalist 43). This closely parallels the premises of our model. S may be a proper subset of N. Collective action may be taken if any (or any number beyond a certain threshold) have agreed to act. One might wonder whether and, if

---

[5] E. Ostrom (1988) provides many illuminating real world illustrations for a successful management of common pool resources without direct state intervention. She presents convincing empirical evidence that the process of free contracting may carry us a long way and perhaps further than we suggest with our model. On the other hand, she also points to such influences as *the shadow of the court*. In any case, it would be interesting to examine her real world examples and to find out whether or not there are -- as in the case of the shadow of the court -- slight deviations in the direction of compulsory group membership or not. Though this would be definitely worthwhile we cannot do it within the space limits of the present paper.

[6] Cf. on this also V. Ostrom (1986, esp. chap. 3).

so, how these realistic varieties of contractarian unanimity could be transferred from the realm of inter-state to that of inter-individual relationships.

First, to most scholars of modern political theory it seems to be almost self-evident that collective action is legitimate if all individuals concerned agree to it. Taking unanimity as a *sufficient* justificatory reason for collective action seems indeed quite unobjectionable. Though one may point out such possibilities as manipulated and badly informed agreement the latter objections may be raised against the legitimacy of any act of private consent too. Therefore we may savely put them aside for our discussion of the differences between private and collective consent and assume that unanimous agreement in general is a sufficient normative reason to justify collective action.

Second, some scholars at least implicitly seem to assume that unanimity as a necessary condition of collective action is an unproblematic norm too. However, usually a monopolistic claim is associated with this norm. Action may be taken *only* if all agree and no individual or subgroup of individuals may act on behalf of an individual or common interest of the subgroup. In this strong version unanimity as a necessary condition of action requires that no action -- neither private nor collective -- from a certain realm of actions is taken unless all individuals concerned have agreed. Thus, on the one hand, unanimity as a rule of collective choice bestows veto power on every individual. On the other hand, the monopolistic claim implied by unanimity as a necessary condition of action basically amounts to the same thing as compulsory group membership because individuals or subgroups of individuals may not anymore act on their own in certain ways. As has been discussed at length by Buchanan and Tullock in their *calculus of consent* (cf. 1962) veto power under this strong interpretation of unanimity cuts both ways: It provides an individual veto against collective acts and at the same time a collective veto against individual acts.

The strong version of unanimity as a necessary condition of action clearly violates the principle of individual consent. It presupposes a claim to a collective monopoly on the execution of certain acts. This claim is not justified by prior purely individual consent. The strong version of the unanimity norm and the principle of individual consent as fundamental justificatory principles are incompatible[7] and the difficulties of contractarianism mostly stem from this simple fact.

[7] It seems that the distinction was not always well understood in political philosophy. For instance Rousseau says: "Had there been no original compact, why, unless the choice were unanimous, should the minority ever have agreed to accept the decision of the majority? What right have the hundred who desire a master to vote for the ten who do not? The institution of the franchise is, in itself, a form of compact, and assumes that, at least once in its operation, complete unanimity existed." (1971, 179) What is "complete unanimity"? If we interpret him very friendly then Rousseau -- regardless of the otherwise somewhat collectivistic undercurrent of his approach -- seems to suggest here that there is a difference be-

However, unanimity (or any less demanding collective rule) as a necessary condition of legitimate collective action can be introduced in a less demanding way. At the outset every individual must be allowed to opt in or out as far as a group organizing collective actions of a certain class is concerned. Further, any subgroup of the class of all individuals must be allowed to form. Unanimity of *all* individuals in N is not a necessary condition for collective action then. But, after joining a group S the consent of every group member may well be a necessary condition of collective action of *that* group.[8] These features are congruent with the central features of our model. It therefore captures the essence of any introduction of collective institutions that is acceptable from a strictly contractarian point of view.

(We would like to express our gratitude for helpful comments to Roy Gardner, Werner Güth, Elinor Ostrom, and Jennifer Robaek.)

tween an original compact and the unanimity principle. He also seems to imply that even if there is no original compact unanimity still may serve as a legitimation of collective action. A thoroughgoing individualistic contractarian could agree with both tenets though he certainly would insist that a strong version of unanimity that raises a monopolistic claim and thus gives precedence to collective over individual authority is unacceptable.

[8] Introducing the exit option may roughly amount to the same thing as consensual group formation; cf. Platon's dialogue *Criton* and Hume's critical essay *Of the original contract* (1948, 356 ff.).

References

Buchanan, J.M. and Tullock, G. 1962: The Calculus of Consent. Ann Arbor.

Hobbes, Th. 1651/1968: Leviathan. Harmondsworth.

Hume, D. 1948: Moral and Political Philosophy (ed. Aiken). New York and London.

Nozick, R. 1974: Anarchy, State, and Utopia. New York.

Olson, M. 1965: The Logic of Collective Action. Cambridge, Mass.

Ostrom, E. 1988: The Commons and Collective Action. mimeo.

Ostrom, V. 1986: The Political Theory of a Compound Republic. Lincoln and London.

Raphael, D.-D. (ed.) 1969: British Moralists. Oxford.

Selten, R. 1975: Reexamination of the Perfectness Concept for Equilibrium Points in Extensive Games; in: International Journal of Game Theory, Vol. 4, 25 ff.

Schneider, L. (ed.) 1967: The Scottish Moralists on Human Nature and Society. Chicago and London.

Taylor, M. 1987: The Possibility of Cooperation. Cambridge.

The Federalist Papers by Alexander Hamilton, James Madison and John Jay. Edited by G. Wills. Toronto et. al. 1982.

IRRIGATION INSTITUTIONS AND THE GAMES IRRIGATORS PLAY: RULE ENFORCEMENT WITHOUT GUARDS

by

Franz Weissing and Elinor Ostrom

Abstract

This is the first of three efforts examining how irrigation institutions affect equilibrium stealing and enforcement rates. In this chapter, we examine rule–following and rule–enforcement rates of behavior adopted by irrigators on systems where rules are self–enforced rather than enforced by formal guards. To do this, we assume that irrigators rotate into the position of a turntaker. When in the position of a turntaker, they choose between taking a legal amount of water and taking more water than authorized (stealing). The other irrigators are turnwaiters who must decide whether to expend resources to monitor the behavior of the turntaker or not. In all our models, we find no combination of parameters where the rate of stealing by the turntaker falls to zero. In other words, there is always some stealing going on.

We give a complete equilibrium analysis for the situation where all irrigators have the same payoffs, monitoring efficiencies, and norms of behavior. Then, we examine how stealing and monitoring rates are affected by changes in parameters including: number of irrigators, cost of monitoring, detection probabilities, relative benefits of stealing, losses felt when stealing occurs, and the reward for successful discovery of a stealing event. Our final analysis addresses how the introduction of asymmetries in payoffs, monitoring efficiencies, and/or norms of behavior affects the distribution of equilibrium outcomes. One consequence of asymmetry is that there is no longer a unique Nash equilibrium. We find both paradoxical and non–paradoxical equilibria. In the paradoxical equilibria, just those turnwaiters monitor less who are inherently in a better position to monitor. In the non–paradoxical equilibria, one class of turnwaiters sticks to a pure strategy while the turntaker adjusts his stealing rate to the cost–benefit ratio of the others who are monitoring at an intermediate rate.

Several of our results are counterintuitive at first sight and reflect the interactive nature of the situation and the interdependence of the players. Our findings also contribute to a more general understanding of the relative weight of 'primary' and 'secondary' interaction effects in a mixed–strategy context.

We are both much appreciative of the editing of Patty Dalecki and her preparation of the drawings in this paper. We also wish to thank Reinhard Selten and George Tsebelis for their helpful comments on an earlier draft. Elinor Ostrom wishes to appreciatively acknowledge the support given her research by the National Science Foundation in the form of Grant Number 8921884.

1. Introduction

This paper represents the collaborative work of a game theorist and a policy analyst in an attempt to examine several questions that go to the heart of the way individuals achieve social order. The capacity of any individual to undertake long-term, productive activities that affect and are affected by the actions of other individuals depends upon gaining a minimal level of predictability among those involved. No one could successfully drive to work if the behavior of other drivers were not relatively predictable. No one could operate a store, if potential consumers did not purchase, rather than steal, the commodities offered to the public. No one would make any investments other than those that would be made by a solitary individual in an isolated setting (the classic Robinson Crusoe situation).

One of the ways that individuals achieve predictability in social arrangements is to agree to follow a set of normative prescriptions about what they must, must not, or may do. Agreeing to a set of prescriptions is relatively easy. Actually following those prescriptions over time when temptations arise offering potentially high payoffs, is not at all easy. In natural settings, individuals follow agreed upon prescriptions to a greater or lesser extent depending on enforcement levels. A frequent assumption made by policy analysts and game theorists is that enforcement is external to the situation under analysis. That assumption allows the theorist to examine what would happen if a particular set of prescriptions were enforced so effectively that all participants followed them without fail.

If one wants to understand, however, the parameters that affect whether and how much enforcement actually occurs and how that, in turn, affects behavior in conformance with legal prescriptions, one can no longer rely on the assumption of external enforcement. One must make the question of enforcement and its consequences endogenous to the analysis. That is the central task of our effort. This paper is the first of a three-part series of papers on this topic.

This is such a large and complex task that one could easily be overwhelmed by the number of variables potentially involved and the multiplicity of ways that individuals could relate to one another. Consequently, we examine this broad question within what at first glance appears to be a very limited context – the games that irrigators play. The world created in our models is that of a set of irrigators sharing the same canal. Each irrigator rotates into the position of a turntaker, who must decide upon whether to steal water (take more than is prescribed) or not to steal water (take the prescribed amount). When not a turntaker, each irrigator is a turnwaiter who must decide whether to monitor the turntaker or not. The large question of the enforceability of legal prescriptions is addressed by creating a specific type of situation where stealing and monitoring can occur and asking what combinations of variables affect the level of stealing and monitoring at equilibrium in that situation. As the reader will see, even in this quite simple environment, the analysis soon becomes quite complex. The analysis becomes tractable by providing this very specific context while it would soon become intractable by trying to answer the more general questions without a more focused analysis.

The particular context we have chosen to model is also a topic of considerable relevance for policy analysts in many countries. As we discuss in Section 2, the levels of stealing water that occur on many irrigation systems throughout the world threaten the viability of these systems and the level of agricultural productivity to be derived from these systems. Consequently, the more narrow focus of this chapter also enables us to use game theory to address policy questions of importance to the organization of irrigation systems.

Since this is an interdisciplinary effort, the chapter attempts to communicate with at least two types of readers. To do this requires somewhat more redundancy than if the chapter were addressed to either game theorists or policy analysts. Section 2 is addressed to substantive problems and will be somewhat more relevant to the policy analyst than to the game theorist. The formal models are introduced in Section 3 in a manner designed to make them understandable for policy analysts not trained in game theory. Game theorists can read Section 3 rapidly and move onto Section 4 where the models developed in this chapter are presented formally. Policy analysts will want to read through Section 4, but the major conclusions and implications are all presented in Section 5 in a non–formal manner.

In this part of the effort, we include models without formal guards and thus explore parameters that affect self–monitoring and its results. In the second part of this effort, we add the position of a formal guard and examine the results in terms of monitoring and stealing levels under various scenarios describing typical types of irrigation systems with guards. In a future effort, we will analyse how such games might change over time as feedback about the results of past behavior is taken into account.

2. The Nature of the Problem

The questions of why rules are followed and when and how rules are enforced are critical for policy analysts and game theorists.[1] These questions are, however, addressed more by assumption than by analysis. Policy analysts frequently predict the likely results of a proposed policy by assuming that affected individuals will comply with proposed rules and that law enforcement agents will enforce these rules effectively. Consequently, governmental policies are adopted that depend for their implementation on widespread behavior that conforms with these rules and effective rule enforcement. Game–theoretical solutions are similarly based on an assumption that the players in a game meticulously follow the rules of the game and cannot break these rules. This assumption is bolstered by a further assumption that the rules of the game are rigorously enforced by agents that are outside the game being analysed. By making these assumptions game theorists focus directly on the question of what equilibria are present when rules are followed and the strategies of all players at equilibria.[2]

Policies that are adopted without an analysis of whether the rules will be followed and how they will be enforced frequently fail when they are neither effectively enforced nor followed (See Pressman and Wildavsky, 1973). Contemporary irrigation policies, particularly in developing countries, are an example of policies adopted on the assumption of rule compliance and enforcement. Immense sums have been invested in constructing large-scale irrigation systems in many developing countries during the past 25 years in attempts to increase agricultural yield and reduce poverty. Predictions concerning increased yields and reduced poverty are based on assumptions that rules allocating water in an efficient manner will be adopted, followed, and enforced.

The policy of constructing major irrigation projects based on these assumptions has been uniformly disappointing. The amount of land actually devoted to irrigation has been substantially less than projected. The value of the crops produced has not met expectations. Agricultural yields obtained after project construction have been lower or more variable, in some cases, than prior to the project.[3] Evaluation teams asked to assess the performance of projects after a decade of operation have found low or even negative benefit/cost ratios. Water supplies are unreliable. Farmers follow less than optimal cropping and water-use patterns and engage in notoriously illegal activities. Public officials are unwilling or unable to enforce rules related to water allocation or resource mobilization. The level of maintenance has been shockingly low.[4]

In contrast, farmers in many settings have constructed their own irrigation systems, devised their own rules for allocating water and mobilizing resources, followed their own rules without external enforcers by self-monitoring and self-enforcing these rules, and maintained their systems in effective working order (Tang, 1989). From an outsider's perspective, the incentive for a farmer to steal water rather than follow water allocation rules would appear to be the dominant strategy on the modern, large-scale, irrigation projects as well as on farmer-owned and managed systems. The lack of effective, external rule enforcement should produce similar patterns of massive rule infractions on both types of systems. The behavioral patterns found in empirical settings, however, show that a substantial difference exists between irrigation systems where rules are devised by the participants themselves and self-monitored and self-enforced, as contrasted to systems where rules are devised by external officials and monitored and enforced by external officials. This is the central puzzle to which this paper is addressed.

Consequently, this paper is an analysis of a particular policy problem as well as an effort to explore a question of central interest to game theorists more generally. By overtly exploring how various parameters affect the level of stealing that occurs on an irrigation system, we can make some policy conclusions of relevance to the design of irrigation policies in many countries. By overtly examining stealing and guarding behavior using noncooperative game theory, we present results of interest to game theorists in general who wish to explore factors affecting the level of and results from rule enforcement rather than assuming that effective rule enforcement is exogenous to a particular game. Before we turn to our game–theoretical analysis, however, we briefly discuss: (1) rule following and rule enforcement on government–owned irrigation systems, (2) rule following and rule enforcement on farmer–owned systems, (3) the efforts to organize farmers' associations, and (4) the theoretical significance of studying self–organizing irrigation systems.

2.1 Rule Following and Rule Enforcement on Government–Owned Irrigation Systems

Large–scale, government–owned irrigation systems throughout the world are frequently plagued by massive problems of illegal water diversions and unwillingness of farmers either to pay water fees or contribute other resources to operate and maintain these systems. Five years after the completion of the Mahaweli project in Sri Lanka, for example, a study found that only one half of the farmers served received water through legally authorized outlets from the canal (Corey, 1986). The other half obtained water through illegal diversions or drainage out of other fields. Instead of following regular rotation systems, farmers blocked and unblocked the ditches and outlets trying to get more than their authorized shares. At times, upstream irrigators obtained the full flow of an irrigation canal. The following incident illustrates the problem.

> In one case, an unauthorized breach was observed to be taking the entire supply of water from a ditch. The downstream farmer said he was not able to obtain water to irrigate his paddies even though he had appealed to the farm leader. When asked why he did not close the breach himself, he said he was afraid of being assaulted by the man who had made the breach. When the farm leader was asked why he permitted this situation to exist ... he said he was afraid to take further action on his own initiative for fear of being "hammered" by the offending farmer (Corey, 1986).

Farmers attempting to obtain more water than their allotted share pursue a variety of strategies. "Common practices include constructing illegal outlets, breaking padlocks, drawing off water at night, and bribing, threatening, or otherwise in some way inducing officials to issue more water" (Chambers, 1980: 43). After reviewing the performance of several irrigation projects in Sri Lanka, Harriss (1984: 322) reports that "gates are missing, structures damaged, channels tapped by encroachers and others." Harriss asked two young Technical Assistants why they did not prevent some of the more blatant offenses. They replied "that they were afraid to because of the fear of being assaulted" (ibid.). Fear is backed up by a sense of futility since little effort is devoted by local police or court systems to sanctioning those who are charged with breaking irrigation rules. Some irrigation systems have been described as "hydrologic nightmares" given the level of stealing and physical violence that characterizes them (Uphoff, 1985).

The dynamics of these hydrologic nightmares frequently lead to mutually re–enforcing tendencies. As farmers increase their illegal diversions, the reliability of the water supply for the entire system is adversely affected. As the reliability of the water supply decreases, the individual payoff from stealing water increases. Henry Hart (1978) calls this pattern a "syndrome of anarchy" affecting both the farmers and the officials involved. Robert Wade describes this syndrome in the following way:

> The farmers lack the confidence that if they respect the restraints on entry to the irrigation commons (refrain from stealing the water, breaking the structures, bribing the officials) they will nevertheless get the expected amount on time. The officials lack the confidence that if they work conscientiously to deliver the water on time the farmers will refrain from breaking the rules. It is a syndrome in the sense that the behavior of each party tends to confirm the negative expectations held by the other (Wade, 1990: 4).

2.2 Rule Following and Rule Enforcement on Farmer–Owned Irrigation Systems

Not all irrigation systems, however, are characterized by high levels of illegal water diversions and the syndrome of anarchy. When farmers themselves organize, govern, and manage irrigation systems, they frequently adopt rules for allocating water (and maintenance responsibilities) that they follow at a high rate. Often, it is the farmers who monitor conformance to rules and sanction those who deviate. Some of these farmer–organized, irrigation systems have survived for centuries. Since these self–organized systems appear to cope with the problem of illegal water diversions more successfully than recently constructed government–owned systems, one can learn a great deal from studying them.

The particular rules used by self–organized irrigation systems vary substantially from system to system. Some have devised intricate allocation schemes dividing the flow of water in a channel into time intervals that are then allocated to farmers based on historic rights, amount of land owned, location on the channel, water needed for specific crops, or on the rental or purchase of water rights.[5] Others have constructed physical dividers or weirs that automatically allocate water to different channels or farmer intakes according to pre–established formulas.

In some systems, little specialization of labor is involved. All irrigators operate the water works, divide the water, participate in maintenance, monitor each others' behavior, and distribute rewards and punishment to one another. Most self–organized systems, however, do develop specialized positions such as ditch tenders, group leaders, and record keepers, even when these are not full–time positions and the farmers rotate into these positions on a temporary basis. Many of these systems operate for long periods of time without relying extensively on outside authorities to impose sanctions on those who break internal rules of the system and the level of rule conformance is high.[6]

And yet, in many of these systems, the potential gain that one irrigator can obtain by diverting water illegally is substantial. An irrigator can achieve higher private returns by taking water when or where it is most valuable, rather than when or where it is authorized by allocation rules. In the short run, the total loss resulting from one irrigator taking more water than allocated by the rules is simply the amount of water lost to the others. The total loss is spread across all others receiving water from the same source. The individual loss borne by each of the other irrigators may thus be quite small. If many irrigators succumb to the temptation to cheat, however, the predictability of the water supply for all farmers decreases over time. As the predictability of the water supply declines, the long–term agricultural yield achieved by the farmers as a group can decline precipitously.[7]

Given the substantial temptations to break allocational rules that continue to face farmers, learning how self–organized systems maintain high rates of rule conformance without relying on external enforcers is important both for its immediate policy implications and for its theoretical significance. In recent years, policy makers in many countries have tried to create farmer organizations on government–owned and managed systems in the hope that such organizations would help end the anarchy that exists on these systems. Many of these efforts have failed due to a lack of understanding of the structure of incentives that farmers establish when they create their own organizations.

2.3 The Effort to Organize Farmers' Associations

Officials of national governments and donor agencies have known for some time that the overemphasis on the engineering aspects of many irrigation projects built in the 1960s and 1970s and their underemphasis of the importance of farmer organization was largely responsible for many of the earlier failures. Consequently, it has become almost fashionable to advocate the importance of establishing farmers' associations. The Asian Development Bank, for example, was among the early advocates of creating farmers' associations as part of the policy process related to new irrigation projects.

> The success of an irrigation project depends largely on the active participation and cooperation of individual farmers. Therefore, a group such as a farmers' association should be organized, preferably at the farmers' initiative or if necessary, with initial government assistance, to help in attaining the objectives of the irrigation project. Irrigation technicians alone cannot satisfactorily operate and maintain the system (Asian Development Bank, 1973: 50).

Most policy documents written since the mid–1970s have stressed the importance of stimulating farmers' associations related to any investment made in irrigation facilities.[8] Knowing that farmer organization is important is not equivalent, however, to knowing how to create or enhance the operation of farmers' associations that generate a high level of rule conformance.

The effort to develop farmers' associations has frequently been located in central government agencies where officials draft the official by–laws for all farmer associations that will be recognized in a particular country. This design is viewed as a predetermined 'blueprint' for how farmers will organize themselves. On some projects, officials have overlooked pre–existing but unrecognized farmer associations and have recognized only those organizations that have followed the official blueprint.[9] On other projects where organizational efforts have been made, the farmers meet and elect the officials they are requested to elect, but that is all that happens.[10] Farmers have then resisted efforts to develop procedures to allocate water and refuse to participate in the maintenance of field canals. Farmers are then perceived by officials as being difficult, irresponsible, and irrational. The failure of the projects to meet predicted benefit levels is frequently blamed on the farmers rather than on engineering design or the lack of effective institutional development.[11] Official efforts to organize farmers have failed more frequently than they have succeeded.

In addition to the general acceptance of the view that farmer organization is important to the success of any irrigation project, recent studies have found that modest investments in smaller–scale irrigation systems, particularly those that are owned and managed by the farmers themselves, tend to yield higher economic returns than investments in large–scale, government–operated, systems. Not only is the cost of investing in the physical systems lower, but the increased level of reliability of water deliveries achieved in these systems when operating well allows farmers to make more productive investments of their own labor and capital inputs.[12] When donors or national irrigation agencies have tried to help already established locally organized systems, however, their interventions have sometimes led to the collapse of the farmer organization that operated the systems prior to the attempt to improve them.[13] A concrete lined, but non–functional irrigation system represents an investment that leads to a reduction in agricultural yield after the investment rather than an increase.

Experience in the design and operation of irrigation projects has demonstrated a pervasive lack of knowledge about why and how farmers organize, devise rules for allocating water, monitor each other's behavior, and maintain their field channels. What is known is that the highest economic returns can be made by investing in physical and organizational systems that are difficult to establish and can be destroyed easily. Learning more about how those self–organized systems that achieve high levels of rule conformance and resulting system reliability operate could have substantial payoff if these lessons could be applied in future investments in irrigation to enhance agricultural yields.

2.4 Theoretical Significance of Studying Self–Organizing Irrigation Systems

Turning to the theoretical significance of studying how self–organized irrigation systems cope with the temptation to divert water illegally, we confront the difference between the assumptions made in most theories of social organization (at least those based explicitly or implicitly on non–cooperative game theory) that individuals follow the rules of a game because external enforcers monitor their behavior and impose sanctions on them for violations. It is usually presumed that without external enforcers, individuals cannot make credible commitments to one another to follow a set of rules where substantial temptations exist to break these rules in the future.

Rather profound social dilemmas potentially exist in almost all self–organized irrigation systems. Looking at these systems from the outside, it appears that stealing water would be the dominant strategy for all irrigators, given the lack of external enforcement of the rules. Why should any irrigator refrain from stealing if he is not forced to comply by the threat of strong sanctions by external officials? If external officials do not enforce rules, why should the irrigators themselves engage in self–monitoring and self–enforcement? After all, each act of monitoring and enforcement is itself costly for the individual who monitors and enforces. Is there not just as strong an incentive to shirk in regard to monitoring and enforcement activities, while hoping that others will undertake these costly actions, as there is to cheat on the rules themselves? How self–organized systems continue to operate for long periods of time, dependant primarily on the monitoring and sanctioning activities of the irrigators themselves, is a major question of interest not only to policy analysts but also to scholars of social order more generally.

The question of how a set of individuals would engage in mutual monitoring of conformance to a set of their own rules is not easily addressed within the confines of most current theories of collective action. In fact, the usual theoretical prediction is that they will not do so. The presumption that individuals will not monitor a set of rules, even if they devise them, is summarized by Jan Elster in the context of a labor union:

> Before a union can force or induce workers to join it must overcome a free–rider problem in the first place. To assume that the incentives are offered in a decentralized way, by mutual monitoring, gives rise to a second–order free–rider problem. Why, for instance, should a rational, selfish worker ostracize or otherwise punish those who don't join the union? What's in it for him? True, it may be better for all members if all punish non–members than if none do, but for each member it may be even better to remain passive. Punishment almost invariably is costly to the punisher, while the benefits from punishment are diffusely distributed over the members. It is, in fact, a public good: To provide it, one would need second–order selective incentives which would, however, run into a third–order free–rider problem (Elster, 1989: 40–41).[14]

Similar questions can be asked of an irrigator. Why should a farmer spend any time helping to devise a set of allocation rules? Once they are devised, they will be made available to all farmers using the same system whether or not the farmer participated in the costly process of negotiating a set of rules. Why should a farmer spend any time watching that others do not steal water? By remaining passive, the farmer can use that time to increase private yields and can free ride of any monitoring provided by others. These questions are specific versions of very general questions that are at the heart of social organization. Good theoretical answers to these questions are not easily available.

3. The Application of Game–Theoretical Concepts in a Positive Analysis

We address the questions posed in Section 2 by developing a series of game–theoretic models. We concentrate in this chapter on the question of whether self–monitoring can be stably maintained within a population of irrigators and whether self–enforcement of a set of rules can effectively deter irrigators from stealing. In a related paper, we will examine how the situation is changed if official monitoring and enforcement mechanisms are introduced.

This is the first effort, to our knowledge, to use game–theoretic models to analyse strategic behavior and equilibrium outcomes in differently structured irrigation systems.[15] We shall start, therefore, by clarifying why we use this approach. Since game theory is frequently used in normative, rather than positive analyses, we discuss how we interpret game–theoretical concepts in an explanatory endeavor.

3.1 The Use of Formal Models

Even in simple social systems, the number of variables that simultaneously affect individual behavior is relatively large. In addition, these variables are complexly interrelated. Most of the variables have multiple effects on individual behavior, and each aspect of behavior is influenced by the interaction of several variables. When one tries to explain behavior in such situations, purely verbal argumentation can lead to very general insights. Obtaining precise qualitative or quantitative conclusions, however, crucially depends on the exact configuration of key variables. Developing formal models where the basic variables and their configuration are well–specified is the only way to arrive at precise and detailed predictions of human behavior in complex social systems.

Consider, for example, the question of what happens to the rates at which farmers divert water illegally (steal water) when the number of farmers using the same irrigation system is increased. One effect of growth in the number of farmers is an increase in the number of individuals who can monitor each others' behavior. This should lead to an increase in the rate of detecting any stealing that occurs. On the other hand, given the expectation that the detection rate will rise, each farmer is more likely to wait for the other farmers to expend resources in monitoring activities. Further, the loss resulting from a stealing event is spread across a larger group, and the harm to any one farmer is reduced. A firm conclusion about the overall effects of a change in the number of farmers requires a method of combining these separate effects in a systematic way.

In some instances, the consequences of a change in a variable seem obvious. Obvious primary effects are often, however, confounded with secondary effects that can easily be overlooked when relying on verbal reasoning. We will investigate several instances where the combination of primary and secondary effects lead to conclusions that are counterintuitive at first sight, but that prove to be quite reasonable after closer inspection. To take a specific example, consider a change in the costs of monitoring. Other things being equal, one expects that increasing the resources that one farmer must expend to monitor whether another farmer steals water or not, should decrease the rate with which farmers monitor each others' behavior. Decreasing the rate of monitoring, however, leads to an increase in the incentive to steal water. A higher stealing rate, in turn, induces an increase in the rate of monitoring. Within our models, the primary and secondary effects offset each other. In fact, we shall arrive at the counterintuitive result that a change in the costs of monitoring has no net effect on the equilibrium rate of monitoring.[16]

Even in those cases where the qualitative effects of a change in a variable are unambiguous, more specific information about the strength of this effect is often desirable. Consider a community of irrigators who are faced with the problem of how to reduce stealing. An increase in the detection efficiency of monitoring farmers, a more severe punishment imposed on cheaters, and the employment of an external guard all appear to be appropriate means to reduce the incentive to steal water. Which device, however, is the most effective? This question cannot be addressed without a formal analysis that compares the costs and benefits of accomplishing objectives in a system of complexly interrelated variables.

The conclusions of formal models are derived in a series of logical steps from a set of basic assumptions. The key variables and their interaction have to be made explicit as does the derivation procedure. The conclusions of formal models are thus more transparent to the reader than those that are derived by verbal reasoning. Formal analysis has, however, its price in a certain lack of realism. No real-world system of some complexity can be described adequately by some assumptions about a few key variables and their interrelationships. In modeling self-organized irrigation institutions, therefore, we do not attempt to model specific real-world systems even though in-depth descriptions of the rules followed and strategic behavior in specific locations have inspired our work.[17] Regularities observed across many systems have been particularly important in the way we have formulated basic assumptions about the structure of the games we model. We hope we have singled out some key aspects for understanding what outcomes are likely to occur in self-organized irrigation systems and why.

We have tried to maintain generality by treating all parameters as variables that are assigned specific values only in rare occasions. Solutions will be given as functions of these underlying parameters. In addition to its generality, this approach has the advantage of providing specific information on how the outcomes are affected by the parameters. In fact, we have gone some length in using separate parameters in order to keep our models as general as possible. When we examine more complex relationships, however, we are sometimes forced to concentrate on more specific parameter constellations.

3.2 The Players and Their Strategies

In this section we illustrate our general modeling approach with an overview of how we translate a simplified description of an interactive, social situation into a game structure that presents the underlying strategic structure of the situation. Such a structure specifies (see Selten, 1975, 1983) (1) which decisions have to be made by whom in what order; (2) the information a player has, whenever it is his turn to move; (3) the choices available to a player, when he has to move; (4) all chance events and the probabilities associated with them; and (5) the final outcomes at the end of the game. In 'extensive form', a game can be represented graphically by a labeled tree (see, e.g., Figure 1). For simplicity (and slightly abusing established terminology), the corresponding representation of a game model will be called a 'game tree'.

We provide an overview of our first model without justifying in this section why we make particular assumptions. The purpose is to enable the reader, who is not a game theorist, to become familiar with how we use game trees to capture the structure of a situation. More detailed descriptions of the models in this chapter are presented in Section 4.

Let us take as an example a situation where two farmers use the same irrigation ditch as their main supply of water to irrigate their crops. Water is scarce. Each farmer can always use more water than he receives to increase agricultural yield. From time to time, each farmer gets into the position of a *turntaker* (TT) where he is authorized to take a certain quantity of water. During this time, the other farmer is in the position of a *turnwaiter* (TW).

At each turn, the strategic decision faced by the turntaker is whether to *steal* (S) additional water from the ditch or to take the authorized quantity of water. This is illustrated in Figure 1 by the two choices – S and $\neg S$ – that branch from the origin of the game tree where the turntaker makes the first move. The choices available to the turnwaiter are to *monitor* (M) what the turntaker is doing or to refrain from monitoring ($\neg M$). Technically, the turnwaiter has two *decision nodes* since monitoring has different consequences depending on whether the turntaker steals water or not. The turntaker does not, however, know in advance what the turntaker is doing. Accordingly, he is not able to distinguish between his two decision nodes. This state of his information is expressed by saying that the two nodes belong to the same *information set*. The information set of the turnwaiter is graphically indicated in the game tree by a dashed line encircling his decision nodes. (If the turnwaiter knew what the turntaker had done, each of his two decision nodes would form an information set of its own. In this case, each node would be enclosed by a dashed line).

We assume that monitoring is the only way that a stealing event can be detected. However, monitoring is not perfect. Even if the turntaker takes more water than he is authorized to do, a monitoring turnwaiter will only detect stealing with a certain probability, the ***detection probability*** α ($0 \leq \alpha \leq 1$). In the game tree, this is indicated by the *chance move* that follows on the path (S,M).

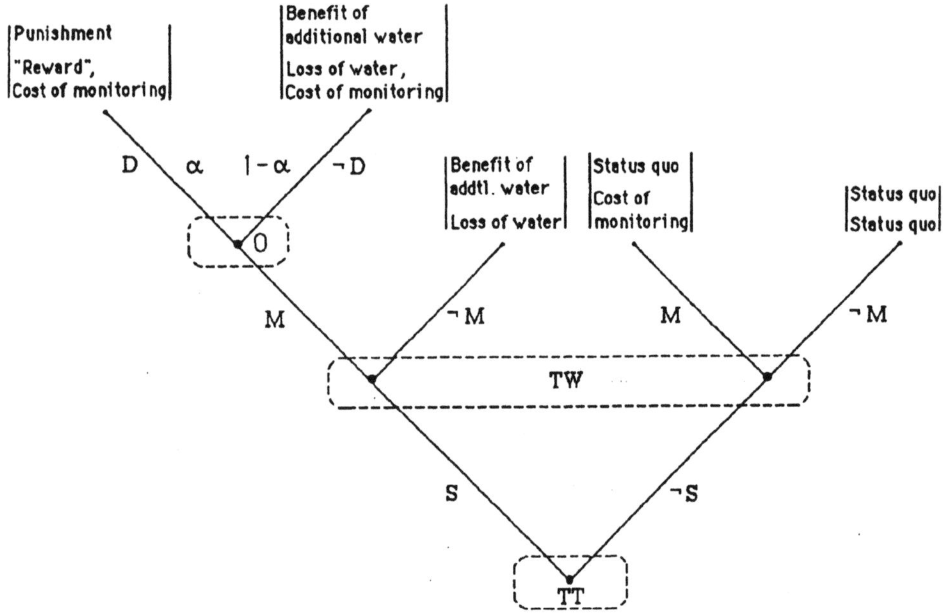

FIGURE 1: Game tree characterizing the strategic interaction between a turntaker and a turnwaiter.

A game tree characterizes all kinds of interactions which are possible within the framework of the model. Each specific interaction may be described by a path from the *origin* of the tree to one of its *end nodes*. In our simple model, there are five end nodes which are characterized by the pair of decisions of both players and the result of the chance move. Each end node of a game tree describes the end situation that is reached after a specific course of the interaction. Therefore, the possible *outcomes* of the interaction for all the players may be assigned to the end nodes.

In our example, the outcomes for player 1, the turntaker, are shown above the outcomes for player 2, the turnwaiter. Moving in the figure from right to left, the pairs of outcomes may be described as follows:

- If the turntaker does not steal and the turnwaiter does not monitor, the turntaker gets the same quantity of water as is authorized and without any costs being invested by the turnwaiter. This can be thought of as the *status quo* outcome for both players.
- If the turntaker does not steal and the turnwaiter monitors, the turntaker receives a status quo outcome and the turnwaiter has to pay the cost – whatever it is – of monitoring.

- If the turnwaiter decides not to monitor when the turntaker has actually stolen water, the turntaker gets the benefit of additional water while the turnwaiter loses the same quantity of water from what would be available to him.
- Even if the turnwaiter monitors, there is a chance that stealing remains undetected. In this case, the turntaker gets the benefit of additional water while the turnwaiter not only loses water but has to pay the cost of monitoring in addition.
- If, however, the turnwaiter monitors and detects that the turntaker has decided to steal, he can effectively prevent the stealing event. Accordingly, the turntaker does not get more water than authorized. The turnwaiter has to pay the cost of monitoring, but he does not lose any water. The detection of a stealing event has additional consequences which are not modeled explicitly but described by the terms 'punishment' for the turntaker and 'reward' for the turnwaiter (see Section 4 for details).

In the situation described above both players have two *pure strategies* (S and $\neg S$ for player 1, and M and $\neg M$ for player 2). We also consider the possibility that the players behave according to a *mixed strategy* which is a probability distribution over the set of their pure strategies. In our case, a mixed strategy for any player – player i – is represented by a real number, s_i, between zero and one ($0 \leq s_i \leq 1$). The number s_i represents the probability that the first strategy is taken by player i, whereas $1-s_i$ represents the probability that the second strategy is taken by him. Accordingly, the *strategy set* S_i of player i corresponds to the closed interval $[0,1]$.

The strategic decisions of the players are completely described by a *strategy combination* $\mathbf{s}$, which is an n-tuple of strategies containing one mixed strategy for each of the n players:

$$\mathbf{s} = (s_1, s_2, \ldots, s_n) \in S_1 x S_2 x \ldots x S_n. \tag{1}$$

In our example, a strategy combination $(s_1,s_2) = (\sigma,\mu)$ should be interpreted as a pair of strategies where $s_1 = \sigma$ characterizes the probability of stealing of the turntaker whereas $s_2 = \mu$ represents the probability of monitoring of the turnwaiter.

In this paper, we also use an alternative interpretation of mixed strategies as *behavioral tendencies*. Instead of interpreting the mixed strategy $s_2 = \mu$ as the probability of choosing the pure strategy M, it may also be viewed as a monitoring rate, or as a level of monitoring that is intermediate between the two extreme options of not monitoring at all or monitoring at the maximal rate. When we have this interpretation in mind, we shall talk about the *monitoring rate*, μ, of a turnwaiter and similarly of the *stealing rate*, σ, of the turntaker. Strictly speaking, a monitoring rate in the second interpretation is a pure strategy in a larger game with a continuum of pure strategies. In view of our assumptions about the payoff structure (see Section 3.3 below), however, both interpretations lead to the same evaluation of a strategy combination by the players. Since both interpretations lead to identical results, we shall henceforth use that interpretation which appears to be the most natural in a particular context.

3.3 Modeling the Players' Incentives

We assume that the players have preferences over the outcomes of the game which remain fixed over several *time units.* A time unit corresponds to that period of time where one specific farmer is in the position of a turntaker. As usual in game theory, we represent preferences over outcomes on a scale of real numbers. For example, the turntaker has three different outcomes which might be ranked as follows:

1. benefits of an additional amount of water,
2. status quo amount of water,
3. punishment received when detected.

If we were to represent the status quo point by zero, then we can represent the benefits by a positive number B, and the punishment by a negative number $-P$.

The turnwaiter has five different outcomes whose ranking depends on more specific assumptions about the system. Again, the status quo outcome is set to zero. Losing water due to stealing is worse than the status quo; therefore it is represented by a negative number $-L_I$. No water is lost if the stealing event is detected. Instead, there will be a reward for the turntaker, which is characterized by the number R_I. The costs of monitoring are represented by a negative number $-C_M$. In order to keep the number of parameters as small as possible, we assume that $-C_M$ corresponds to fixed costs which may be combined *additively* with the other payoff parameters 0, $-L_I$, and R_I. Figure 2 illustrates how these assumptions about the incentive structures of the players can be incorporated into the game tree by replacing the neutral description of the outcomes by the subjective evaluation of them by the players.

We assume that the preferences over outcomes are reflected in preferences over strategies which the players develop in the course of time. It is clear that the outcome of a strategic interaction not only depends on a player's own strategy, but also on the strategies chosen by the other players and on chance events which come about because of the imperfectness of monitoring. Accordingly, a strategy s_i of player i does not usually result in a single outcome. Instead, s_i induces a frequency distribution over the set of all possible outcomes. This frequency distribution is also a function of $(s_1,\dots,s_{i-1},s_{i+1},\dots,s_n)$, the strategy combination of the other $n-1$ players. We assume that the frequency distribution over outcomes which are induced by a strategy s_i – for fixed behavioral tendencies of the other $n-1$ players – is at least 'intuitively' evaluated by player i.

What is more, we assume that the payoffs of the basic outcomes combine in a way that the expected value of payoffs is a good representation of the individual evaluation of a frequency distribution over the basic outcomes. In essence, this implies that the payoffs of our models may be viewed in terms of von Neumann and Morgenstern's notion of *utility* (see e.g. von Neumann and Morgenstern, 1944; Luce and Raiffa, 1957), although we want to avoid the normative interpretation of this concept. We are aware of the fact that this concept often gives a poor

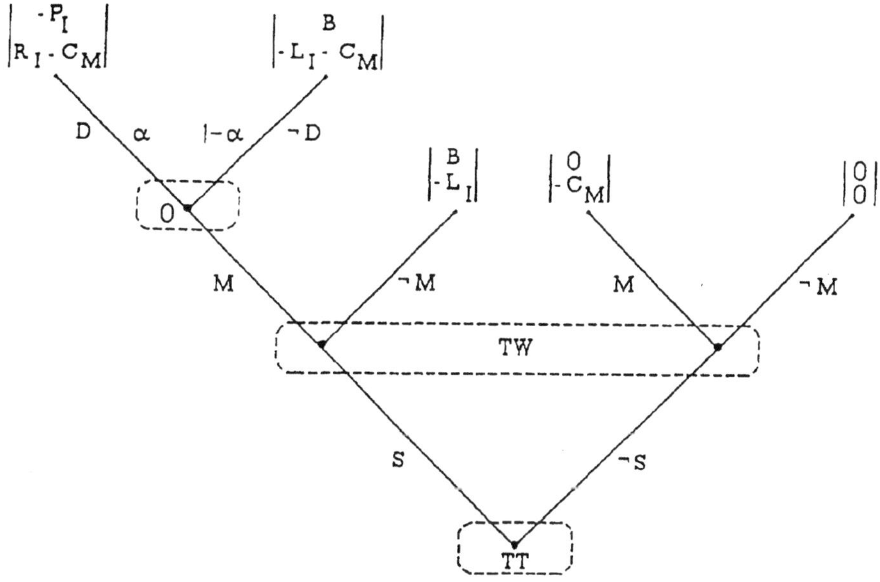

FIGURE 2: A two–person game in extensive form describing the strategic structure and the evaluation of outcomes in an interaction between a turntaker and a turnwaiter.

representation of the preference structure of human individuals: human beings do not usually have fixed preferences that are independent of the context, and they are often not risk neutral as is required by normative theory. Nevertheless, we assume that the von Neumann–Morgenstern interpretation of payoffs gives at least an approximate description of the motivation structure of the players in our irrigation games. Even if it is not a good representation of real–world situations, we take the view that this approach is an adequate starting point for the analytical investigation of a system, where our knowledge about the incentives of the participants is incomplete. In addition, we have the impression that the expected utility approach gives a reasonable description of real–world incentives in 'evolved systems' where the same individuals interact repeatedly with one another in the same basic situation over a long time horizon and where expectations can be based on a large body of experience.

Under the assumptions made above, it makes sense to assign an *expected payoff* $F_i(\mathbf{s})$ for each player i to each strategy combination $\mathbf{s} = (s_1,...,s_n)$. $F_i(\mathbf{s})$ is interpreted as player i's expected payoff for his strategy s_i, provided that the n–1 remaining players stick to the strategy combination $(s_1,...,s_{i-1},s_{i+1},...,s_n)$. The function

TT: \ TW:	$\neg M$	M
$\neg S$	0 ; 0	0 ; $-C_M$
S	B ; $-L_I$	$\alpha(-P_I)+(1-\alpha)B$; $\alpha(R_I-C_M)+(1-\alpha)(-L_I-C_M)$

FIGURE 3: The 2x2–bimatrix characterizing the payoff functions and thus the normal form game between a turntaker and a turnwaiter. The payoff of player 1 (the turntaker) is depicted in the upper left–hand corner of a cell, the payoff of player 2 (the turnwaiter) is depicted in the lower right–hand corner.

$$\mathbf{F} : S_1 x \ldots x S_n \longrightarrow \mathbb{R}^n, \quad \mathbf{F}(\mathbf{s}) := (F_1(\mathbf{s}),\ldots,F_n(\mathbf{s})), \tag{2}$$

that assigns the vector of expected payoffs to the strategy combination $\mathbf{s}$ may be interpreted as the payoff function of a *normal form game.*

In view of our assumption that the players are risk neutral, the payoff function $\mathbf{F}$ is fully specified by the values it takes on the set of *pure* strategy combinations. In our simple example, there are two players and four strategy combinations ((S,M), $(S,\neg M)$, $(\neg S,M)$, and $(\neg S,\neg M)$). Therefore, the payoff function may be characterized by four pairs of payoffs which may be arranged in the form of a 'bimatrix', i.e., a pair of matrices where each matrix corresponds to the payoffs of one of the two players. Figure 3 shows the *bimatrix game* that results from the extensive form game in Figure 2.

3.4 A Positive Interpretation of the Nash Equilibrium Concept

After having assigned a normal form game to our models, we now approach the question of how to analyse such a game. All *solution concepts* of normative non–cooperative game theory are based on the notion of an equilibrium point that was introduced by Nash (1950, 1951). A *Nash equilibrium point* is a strategy combination $\mathbf{s}^* = (s_1^*,\ldots,s_n^*)$ that is self–stabilizing in the sense that no player i has an incentive to deviate from his strategy s_i^* if he expects all other players to abide by $\mathbf{s}^*$ (see Harsanyi and Selten, 1988). Correspondingly, a Nash equilibrium point $\mathbf{s}$ is characterized by the fact that for each player i the strategy s_i^* which is prescribed to him by $\mathbf{s}^*$

is a *best reply* to the collection of strategies $(s_1^*,...,s_{i-1}^*,s_{i+1}^*,...,s_n^*)$ of the remaining $n-1$ players:

$$F_i(s^*) = \max_{s_i} F_i(s_1^*,...,s_{i-1}^*,s_i,s_{i+1}^*,...,s_n^*) \quad \text{for all } i. \qquad (3)$$

Although the notion of a Nash equilibrium point has been introduced in a normative context, predictions based on this concept are surprisingly successful in some experiments (e.g., O'Neill, 1987). Evolutionary game theory has demonstrated that Nash equilibrium points may arise as the outcomes of a dynamic process (e.g., Maynard Smith, 1982; Hofbauer and Sigmund, 1988). Inspired by these results, some efforts have been undertaken to provide a positive interpretation for the Nash equilibrium concept (see, e.g., Cross, 1983; Selten, 1991). Under quite general assumptions, it can be shown that only Nash equilibrium points are candidates for being the stable outcome of a process of learning or cultural evolution (Friedman, 1990; Nachbar, 1990).

Imagine, for example, a population of individuals which interact repeatedly with one another in a social setting where each 'player' has exactly two pure strategies which might be labeled I and J. Let us assume that the individuals are learning from their experience and that they try to adapt their behavior to their current situation. Even if we do not know the exact specifications of this learning process, it is reasonable to assume that it fulfills the following requirement: If pure strategy I yields a consistently higher payoff to player i than the alternative strategy J, player i should show an increased tendency to use strategy I in the future. A system which meets this requirement is called *locally adaptive* in the biological literature (Eshel, 1982). It is easy to see that all stable fixed points of locally adaptive dynamic processes correspond to Nash equilibrium points (Weissing, 1983; Friedman, 1990).

Although the Nash equilibrium property seems to be a necessary condition for stability in many reasonable learning models, it is usually far from being sufficient. In fact, it is easy to show that genuinely mixed Nash equilibrium points are almost always asymptotically unstable for a wide class of adaptive processes (Eshel and Akin, 1983; Crawford, 1985; Selten, 1991). The dynamic instability of Nash equilibrium points is usually caused by the strategic complexity of the situation and by asymmetries in the positions of the players.[18] Selten (1991), for example, has demonstrated that stable Nash equilibrium points do exist if the players have only two pure strategies available, and if the players are in positions that are similar enough to enable each player to imagine how to act reasonably in the role of another player.

The analysis of our models will be based on the Nash equilibrium concept (see Section 4.2 for an equilibrium point analysis of the game in Figure 3). We believe that this approach is justified if the strategic structure of the situation is not too complex, and if the social background of the players is similar enough to enable them to anticipate each others' behavior. For these reasons, we concentrate on models where the players may be treated symmetrically, and where all participants have only two pure strategies available. The rather simple strategic structure of our models is compensated, however, by the complex configuration of payoff parameters.

3.5 Symmetry Assumptions

A significant part of our analysis is based on the assumption that the farmers who constitute the players of our models can be treated symmetrically if they are in a symmetric position. For the interpretation of our solution concept, it is important to distinguish several aspects of this symmetry assumption.

First of all, there is *symmetry in payoffs* for the turnwaiters. This means that all turnwaiters are equally affected by a stealing event and that their motivations to monitor the turntaker's behavior are comparable to each other. In some of our models, symmetry in payoffs is reflected by assigning an identical set of payoff parameters to all the turnwaiters. Second, we often assume that the turnwaiters are *structurally equivalent* in that similar behavior leads to similar consequences for the farmer society. In the models of this chapter, two turnwaiters are structurally equivalent if they have the same detection probability, α. Third, we usually assume that all players come from the same farmer population with a prevailing *norm of behavior*. By this we mean that all farmers behave similarly whenever they find themselves in a similar situation. They also react similarly whenever they experience a similar change in their situation. Fourth, we usually assume that the irrigators of our models have the same, reliable *knowledge* about the system and, in particular, the other players' norm of behavior. What is more, we assume that all players take this knowledge for granted and that they take it at least intuitively into consideration when they try to anticipate their colleagues' behavior. Finally, we assume that all players have similar *starting conditions*. Otherwise, a player could get into a peculiar situation just by historical accident.

For our models, these assumptions imply that players in a symmetric situation may be treated symmetrically, even though each player makes his decision individually and independently from all the other players. As a consequence, we put special emphasis on *symmetric* Nash equilibrium points (Nash, 1951), i.e., on equilibrium points where players who are in the same position use the same strategies.[19] In the models of this chapter, an equilibrium point is symmetric if the turnwaiters do not differ in their tendency to monitor.

Focusing on symmetric Nash equilibrium points simplifies the analysis of our models considerably. In addition to the symmetric equilibrium points, there is often a large number of asymmetric equilibria, and it is usually difficult to get much insight into their general structure. Quite often, several equilibria coexist for the same parameter constellation. Such a situation makes it difficult if not impossible to arrive at a clearcut prediction from our models since no criteria are available on which to base a distinction between them.[20]

Assuming that the farmers can be treated symmetrically with respect to their incentives limits the empirical relevance of our models to relatively small, *homogeneous systems* where farmers own (or rent) approximately equal parcels of land and where they water a similar spectrum of crops. While there are many such systems in the world, it is important to stress the limitations of this assumption. Since water flows downhill, some asymmetries are always caused by whether one farmer is located uphill or downhill from another farmer. Small systems reduce the magnitude of this asymmetry. Further asymmetries result from the geographical proximity of one farmer to a neighbor as contrasted to a farmer located several fields away. The asymmetries of neighboring land and distance are frequently used in the design of irrigation systems to reduce the number of turntakers that any turnwaiter may need to monitor to those on either side of the turnwaiter himself.

Since the various symmetry assumptions are so crucial for our analysis, we also study the consequences of relaxing them to a limited extent. In Section 4.6, we analyse a model where one or several of the symmetry assumptions do not hold true for one 'distinguished' turnwaiter. We compare the asymmetric case with the symmetric case and perform some comparative statics.

Asymmetries in payoffs or structural parameters lead almost automatically to asymmetries in norms of behavior and thus represent deep failures of the symmetry assumption. Even if all turnwaiters were comparable to one another in terms of payoffs and detection probabilities, however, they might still differ with respect to their knowledge or their norm of behavior. Above all, such a situation is likely to occur if a new farmer enters an historically grown homogeneous system. Asymmetries of this kind can easily be analysed within the scope of our models. In fact, we shall show that the entry of a newcomer may destabilize the whole system, even if the newcomer merely replaces a resident that is identical to the other farmers in all relevant physical aspects. Thus, in most cases, we use the symmetry assumptions to increase analytical tractability, but in the course of our analysis, we address what happens if one or another of our symmetry assumptions is relaxed.

4. Game–Theoretical Analysis of Irrigation Models Without Guards

Having described our general approach, we will now lay out in a more precise manner the series of models that we plan to analyse starting with the simplest model.

4.1 A General Description of the Models

For all our models, we imagine a population of farmers who share a single irrigation system that is their main source of water supply for their crops. The authorization to take water is rotated among the farmers according to a predetermined formula. At each point of time, only one farmer is in the position of a *turntaker* (TT) who is authorized to take a certain quantity of water. The other farmers wait for their turn to take water and are called *turnwaiters* (TW). There is a potential conflict of interest between the turntaker and the turnwaiters, since the turntaker is in a position to extract more than the authorized amount of water from the system, whereas the turnwaiters are negatively affected by a stealing event. The models of this paper focus on this conflict. We shall analyse the question of how this conflict might be resolved in the absence of any 'official' inspection and sanctioning mechanism.

We have attempted to model the rotation scheme as generally as possible in order to make it applicable to a broad number of empirical cases (see Section 5.1). All rotation schemes divide the irrigation season into small time periods, where two consecutive periods are separated from another by a rotation event. All of our models focus on these time periods. In particular, our assumptions about the incentives of the players refer to single time periods.[21]

The strategic decision facing the turntaker is whether to take the authorized quantity of water or to steal additional water from the irrigation ditch. *Stealing* is defined as taking actions that are not allowed by the rotation system in use. The ways that a turntaker may steal include irrigating more land than is authorized, watering crops more than they need at the current time, taking a longer turn than authorized, storing water in a local pond, or any method devised by the farmer to obtain more water than intended given the particular rotation system involved. Let us assume that the turntaker always has a positive incentive to steal water. The only thing that deters a turntaker from stealing is the possibility that the stealing event might be detected by others and punishment imposed.

We assume that a stealing event, if detected, can be prevented effectively and that the turntaker cannot extract any benefit from a detected stealing event. In contrast, upon conviction of having stolen water, the turntaker incurs some impairment imposed by the community of farmers. Such a *punishment* may be composed of several factors including 'objective' components such as a fine denominated in money or in crops, a reduction in the water that the turntaker is authorized to take in the future, or a requirement to devote a certain amount of labor to canal maintenance. It may also be composed of more 'subjective' factors such as the scorn of fellow farmers and a loss of status in a community. In the present paper, we do not explicitly model how sanctions against a cheating turntaker are enforced by the farmer community.

Many field observations in self–organized irrigation institutions stress that farmers expend energy monitoring each other's behavior. It is an obvious consequence of monitoring the turntaker's behavior that stealing – once it occurs – is detected with a higher probability. Accordingly, we interpret *monitoring* as a strategic option of the turnwaiters. Monitoring involves observing the turntaker in an effort to ascertain if the turntaker cheats or not. A turnwaiter who does not monitor devotes his time to his own farming and other pursuits.

Monitoring involves a cost of allocating time and energy to go and observe the behavior of the turntaker. The *cost of monitoring* is affected by the physical design of the system, the clarity of the rules defining cheating and not cheating, and the ambiguity involved in monitoring the turntaker. Monitoring may be worth its costs if stealing is prevented by it. In fact, undetected stealing is costly for a turnwaiter since the water supply is reduced and becomes less predictable in the future when he himself will rotate into the turntaker position. In addition to this economic effect, the detection of a cheating event may have other socio–psychological consequences for the turnwaiters. For example, the detection of such an event might produce a feeling of 'satisfaction' in those farmers who have monitored the turntaker's behavior since it shows how important it is to keep one's eyes open. The detection of a stealing event may also have an impact on those turnwaiters who did not monitor, since it becomes obvious that at least some stealing is going on within the farmer community. All these socio–psychological factors may differ considerably from one irrigation system to the next. In fact, there are conceivable systems where monitoring is viewed as 'meddling in someone else's business' and is itself sanctioned rather than encouraged.

Let us now formalize these ideas by means of a (N+1)–person game. Player 1 of our game corresponds to the farmer who is in the turntaker position. The other N players recruit from the population of turnwaiters. However, only those turnwaiters will be considered as *players* who are directly affected by the turntaker's behavior and who have strategic influence on his decision. In an extremely small irrigation system, the players in our model would include all the farmers. In most systems, however, the players would include only those farmers who are located on the same field channel, who know who the turntaker is, who feel a direct loss when water is stolen, and who can positively or negatively sanction each other's behavior. In many cases, it will only be the farmer whose turn will come in the next period, who fulfills these requirements. In the latter case, only two farmers would be included among the players even if the irrigation system itself were quite large.

Player 1, the turntaker, is assumed to be the only player who has the opportunity to violate the rules of the system. He has two pure strategies which are called 'stealing' (S) and 'not stealing' ($\neg S$). We also consider the possibility that the turntaker behaves according to a mixed strategy σ ($0 \leq \sigma \leq 1$), where σ denotes the probability that S is chosen by player 1 rather than $\neg S$. According to the conventions introduced in Section 3.2, σ will also be interpreted as the *stealing tendency* of player 1.

Without knowing the turntaker's decision, the other N players, the turnwaiters, have to decide between 'monitoring' (M) and 'not monitoring' ($\neg M$). The mixed strategy of turnwaiter i ($1 \leq i \leq N$) is characterized by a real number μ_i ($0 \leq \mu_i \leq 1$), the probability of monitoring. μ_i will also be interpreted as the *monitoring tendency* of turnwaiter i. Stealing can only be detected by monitoring, but monitoring is not completely efficient. Even if a stealing event is in fact taking place, monitoring turnwaiter i will only detect it with a certain probability, the *detection probability* α_i ($0 \leq \alpha_i \leq 1$). If a turnwaiter monitors at a lower rate, his detection rate will decrease proportionally. Accordingly, turnwaiter i monitoring at rate μ_i will detect a stealing event with probability $\alpha_i\mu_i$, and he will not detect it with probability $1-\alpha_i\mu_i$.

The behavioral tendencies and the detection probabilities within the turnwaiter community are characterized by the vectors $\mu = (\mu_1, \mu_2, \dots, \mu_N)$ and $\alpha = (\alpha_1, \alpha_2, \dots, \alpha_N)$. We assume that the turnwaiters do not monitor in a coordinated manner but independently of one another and without knowing each others' monitoring tendency. This implies, for example, that a stealing event remains undetected with probability

$$\eta(\mu, \alpha) := (1-\alpha_1\mu_1)\cdot(1-\alpha_2\mu_2)\cdot \dots \cdot(1-\alpha_N\mu_N). \tag{4}$$

The complementary probability, $1-\eta(\mu,\alpha)$, corresponds to the probability that a stealing event will be detected by at least one of the turnwaiters. Accordingly, $\eta(\mu,\alpha)$ and $1-\eta(\mu,\alpha)$ will be interpreted as the *monitoring deficiency* and the *monitoring efficiency* of the turnwaiter community.

As described in Section 3.3, the players' incentives are represented as gains and losses relative to a *status quo* outcome on a scale of real numbers. Getting the authorized quantity of water will be considered the status quo outcome of player 1, whereas the status quo of a turnwaiter is given by the situation that he does not monitor and that he is not harmed by a stealing event.

For the turntaker, the payoff for stealing is either negative ($-P_I$) or positive (B) depending on whether the stealing event remains undetected or not. If a cheating event has gone undetected, the status quo payoff of a turnwaiter is lowered by a certain amount ($-L_I$) which corresponds to his individual loss of water and is equivalent to his share of the value of the water taken by the turntaker.

There will be no loss of water if a stealing event is detected by a farmer monitoring the turntaker's behavior. Monitoring involves a fixed cost ($-C_M$) which interacts additively with the other payoff components of a turnwaiter. The socio-psychological consequences of the detection of a stealing event will be subsumed under the payoff parameters R_I and R_0 . R_I denotes the 'reward' obtained by a turnwaiter who himself detected a stealing event, whereas R_0 represents the payoff consequences for the other turnwaiters after a stealing event has been detected by somebody else. We interpret R_0 as a kind of satisfaction that stealing has been prevented. The components of R_I are likely to be primarily composed of 'subjective' benefits obtained as a result of the social approval or disapproval that the detection of a cheating event brings to a turnwaiter from the other farmers. In systems where the farmers are aware of the importance of keeping cheating levels down, R_I is likely to be positive and substantial. However, R_I could also take a negative value where farmers do not share a commitment to the rotation system and consider monitoring to be a form of meddling in other farmers' affairs. In such systems, a farmer may fear a physical assault if he were to report even blatant stealing by other farmers (see, for example, Harriss, 1977, 1984).

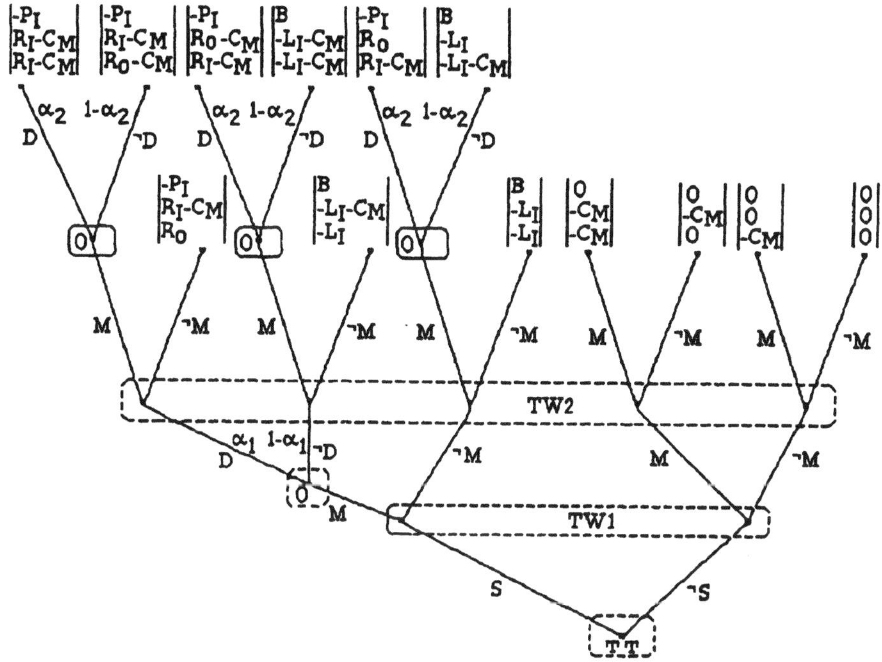

FIGURE 4: Strategic structure and payoff relationships for an irrigation game without guards, illustrated by the interaction between a turntaker and two turnwaiters (i.e., N = 2). For simplicity, it is assumed that the two turnwaiters have identical payoff parameters.

For convenience, all the 'absolute' payoff parameters of the model are summarized in Table 1 (see Section 4.5). Table 1 also contains several 'relative' payoff parameters which will be defined below.

Usually, we concentrate on the symmetric case where the payoff parameters of the N turnwaiters (i.e. C_M, L_I, R_I, and R_O) are identical. However, the asymmetric case will also be considered. In that case, superscripts to the payoff parameters (C_M^i, L_I^i, R_I^i, and R_O^i) will indicate that the payoff of turnwaiter i differs from that of other turnwaiters.

For the special cases N = 1 and N = 2, the strategic structure and the payoff structure of the interaction are illustrated by the game trees in Figure 2 and Figure 4. It should be obvious, how this description of the situation can be generalized to systems with a larger number of players. The resulting (N+1)–person game will be referred to as an *irrigation game without guards*.

TW: / TT:	$\neg M$	M
$\neg S$	0 ; 0	0 ; $-C_M$
S	B ; $-L_I$	$B-\alpha(B+P_I)$; $\alpha(R_I+L_I)-L_I-C_M$

FIGURE 5: Normal form of a two–person irrigation game without guards. The arrows indicate the fixed part of the best–reply structure of the game.

4.2 The Bilateral Interaction Between the Turntaker and One Turnwaiter

In this section, we shall perform an equilibrium point analysis of an irrigation game without guards for the special case $N = 1$, i.e., for the interaction of a turntaker with just one turnwaiter. It has been shown in Section 3.3 that in this case the payoff function of the game is fully specified by the 2x2–bimatrix in Figure 3.

This bimatrix is reproduced in Figure 5. However, two arrows have been added which indicate part of the *best reply structure* of the irrigation game. Consider first the horizontal arrow in the upper part of the diagram. This arrow corresponds to the payoff relationship $0 > -C_M$ for player 2, and it indicates the fact that the turnwaiter prefers $\neg M$ to M *provided that* the turntaker sticks to his strategy $\neg S$. We say that $\neg M$ is a *better reply* to $\neg S$ than M. Similarly, the vertical arrow in the left half of the diagram indicates that player 1's pure strategy S is a better reply to $\neg M$ than his pure strategy $\neg S$. In view of the payoff relationship $B > 0$, the turntaker prefers S to $\neg S$ if he has reasons to believe that the turnwaiter does not monitor.

In order to complete the best reply structure to pure strategies we have to know player 1's best reply to M and player 2's best reply to S. This, however, depends on the payoff relationship between $B-\alpha(B+P_I)$ and 0 on the one hand, and between $\alpha(R_I+L_I)-L_I-C_M$ and $-L_I$ on the other. Depending on the parameter constellation of the model, there are four possible best reply structures to pure strategies (see Figure 6).

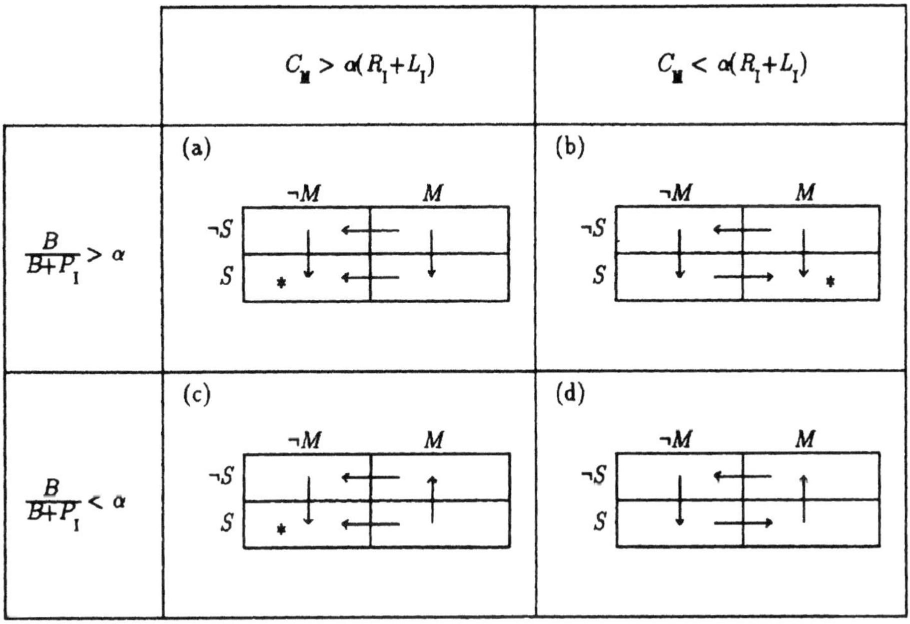

FIGURE 6: The four different best reply structures of a two–person irrigation game without guards. Nash equilibrium points in pure strategies are indicated by an asterisk.
(a) $(S,\neg M)$ is the unique Nash equilibrium point;
(b) (S, M) is the unique Nash equilibrium point;
(c) $(S,\neg M)$ is the unique Nash equilibrium point;
(d) There is no Nash equilibrium point in pure strategies.

A Nash equilibrium point of a two–person game is a strategy combination (s_1^*,s_2^*) where s_1^* is a best reply to s_2^*, and s_2^* is a best reply to s_1^*. Correspondingly, the equilibrium points in *pure* strategies can be derived directly from the best reply diagrams in Figure 6. In fact, (s_1^*,s_2^*) is a Nash equilibrium point in pure strategies, if in a best reply diagram both the horizontal and the vertical arrows point into the cell that corresponds to the strategy combination (s_1^*,s_2^*). In Figure 6, such cells are marked by an asterisk. A Nash equilibrium point in pure strategies exists in cases (a), (b), and (c), and it is easy to see that in these cases the equilibrium is unique. In contrast, no equilibrium in pure strategies exists in case (d). Instead, the best reply structure in Figure 6(d) implies the existence of a unique Nash equilibrium point in mixed strategies (see, for example, Harsanyi and Selten, 1988) which will be denoted by (σ^*,μ^*). The equilibrium rates of stealing, σ^*, and monitoring, μ^*, will be derived below.

Let us introduce the terms

$$B_{rel} := \frac{B}{B+P_I} \quad \text{and} \quad R_{marg} := \alpha\cdot(R_I+L_I)\,, \tag{5}$$

which may be interpreted as the *relative benefit of stealing water* and the *marginal benefit of monitoring* (see Section 4.3). The best reply diagrams in Figure 6(a) and (c) show that the turnwaiter's pure strategy $\neg M$ dominates M if the cost of monitoring, C_M, exceeds the marginal benefit of monitoring. Since S is the turntaker's best reply to $\neg M$, the pure strategy combination $(S,\neg M)$ is the unique Nash equilibrium in these cases. The diagrams in Figure 6(a) and (c) indicate that the turntaker's pure strategy S dominates $\neg S$ if the relative benefit of stealing water is larger than the detection probability α. The turnwaiter's best reply to S depends on the relation between the cost and the marginal benefit of monitoring. If C_M is smaller than R_{marg} (Figure 6(b)), the unique Nash equilibrium is given by the pure strategy combination (S,M): the turnwaiter should monitor even if it does not deter the turntaker from stealing since monitoring is the only way to reduce the efficiency of stealing.

The game has no dominated strategies only in case (d) where the cost of monitoring is smaller than the marginal benefit of monitoring, and where the relative benefit of stealing water is smaller than the detection probability. The cyclic best–reply structure in that case is characteristic for a large class of 'pursuit and evasion' games (see, e.g., Tsebelis, 1989; Selten, 1991). In order to derive σ^* and μ^*, the (mixed) tendencies to steal and to monitor at equilibrium, we also have to consider the best replies to mixed strategies. For the turntaker, the expected payoff for his pure strategies at equilibrium are given by

$$F_1(S,\mu^*) = \mu^*\cdot(B-\alpha(B+P_I)) + (1-\mu^*)\cdot B, \tag{6}$$

$$F_1(\neg S,\mu^*) = 0. \tag{7}$$

S is a better reply to μ^* than $\neg S$ – which will be symbolized by $S \succ \neg S$ – if and only if $F_1(S,\mu^*)$ is larger than $F_1(\neg S,\mu^*)$. Therefore, we have

$$S \succ \neg S \;\Leftrightarrow\; B > \mu^*\cdot\alpha(B+P_I) \;\Leftrightarrow\; \mu^* < B_{rel}/\alpha. \tag{8}$$

For the turnwaiter, the expected payoffs for M and $\neg M$ at equilibrium are given by

$$F_2(\sigma^*,M) = \sigma^*\cdot(-L_I-C_M+\alpha(R_I+L_I)) + (1-\sigma^*)\cdot(-C_M), \tag{9}$$

$$F_2(\sigma^*,\neg M) = \sigma^*\cdot(-L_I). \tag{10}$$

M is a better reply to σ^* than $\neg M$ ($M \succ \neg M$) if $F_2(\sigma^*,M)$ is larger than $F_2(\sigma^*,\neg M)$:

$$M \succ \neg M \;\Leftrightarrow\; \sigma^*\cdot\alpha(R_I+L_I) > C_M \;\Leftrightarrow\; \sigma^* > C_M/R_{marg}\,. \tag{11}$$

The strategy combination (σ^*,μ^*) is a Nash equilibrium point if σ^* is a best reply to μ^* and μ^* is a best reply to σ^*. In view of the linearity of the payoff functions, a genuinely mixed strategy σ^* is a best reply to μ^* if and only if both its pure strategy components (S and $\neg S$) are best replies to μ^* (see, for example, Harsanyi and Selten, 1988: Lemma 2.3.1). This means that S and $\neg S$ are payoff equivalent against μ^* which will be symbolized by $S \approx \neg S$. Similarly, μ^* is a best reply to σ^* if and only if M and $\neg M$ are payoff equivalent against σ^* ($M \approx \neg M$). In view of (8) and (11), these payoff equivalence relationships may be represented by

$$S \approx \neg S \quad \Leftrightarrow \quad \mu^* = B_{rel}/\alpha\,, \tag{12}$$

$$M \approx \neg M \quad \Leftrightarrow \quad \sigma^* = C_M/R_{marg}\,. \tag{13}$$

Taken together, (12) and (13) determine the mixed strategy equilibrium (σ^*,μ^*) for case (d) in Figure 6. Notice that at equilibrium the stealing tendency of the turntaker is determined by the cost–benefit ratio of monitoring for the turnwaiter whereas the monitoring tendency of the turnwaiter is determined by the relative benefit of stealing for the turntaker. The fact that the equilibrium strategy of a player is more directly related to the payoff parameters of *the other* players than to his own payoffs is rather typical for Nash equilibrium points in mixed strategies (at least in two–person games; see Tsebelis, 1989; Bianco, Ordeshook, and Tsebelis, 1990). Quite often, this inverse relationship leads to conclusions which appear counterintuitive at first sight. These conclusions, however, just reflect the interactive nature of the situation and the dependence of a player's payoff on the behavior of the other players. We come back to this point in Section 5.5. Let us summarize the equilibrium analysis given above:

<u>PROPOSITION 1:</u> (Equilibrium regimes for N = 1)[22]

A two–person irrigation game without guards has usually (see below) exactly one Nash equilibrium point. Depending on the parameter constellation of the game, the equilibrium is of one of the following three types:

(a) $\sigma^* = 1$, and $\mu^* = 0$, if the cost of monitoring, C_M, is larger than the marginal benefit of monitoring, R_{marg}.

(b) $\sigma^* = 1$, and $\mu^* = 1$, if C_M is smaller than R_{marg}, and if the relative benefit of stealing water, B_{rel}, is larger than the detection probability α.

(c) $\sigma^* = C_M/R_{marg}$, and $\mu^* = B_{rel}/\alpha$, if C_M is smaller than R_{marg}, and if B_{rel} is smaller than α.

It is easy to see that there is a continuum of Nash equilibrium points if the cost of monitoring is equal to the marginal benefit of monitoring, or if the relative benefit of stealing water is equal to the detection probability and C_M is smaller than R_{marg}. We shall henceforth neglect all such 'border cases' which are defined by equalities between external payoff parameters. Such specific

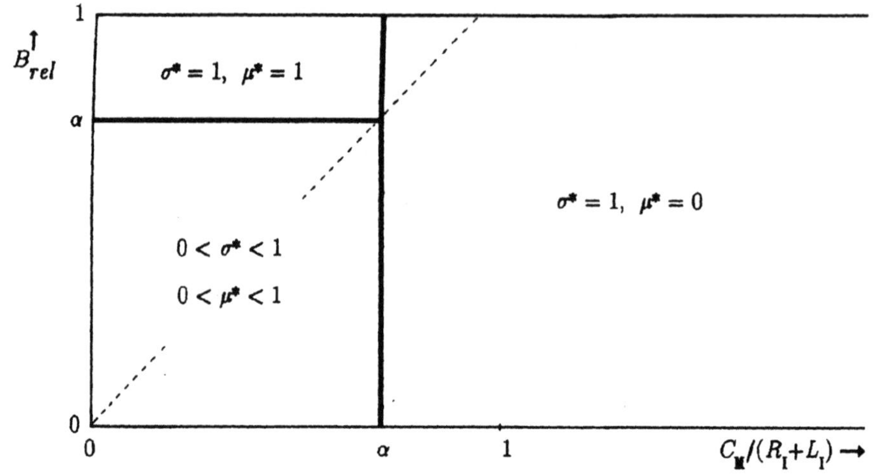

FIGURE 7: Equilibrium diagram illustrating the parameter regimes corresponding to the three different classes of Nash equilibrium points for two–person irrigation games without guards.

Graphical convention for this and the following figures:
Double lines separating equilibrium regimes from another indicate those 'border case' parameter constellations for which there is a continuum of equilibrium points.

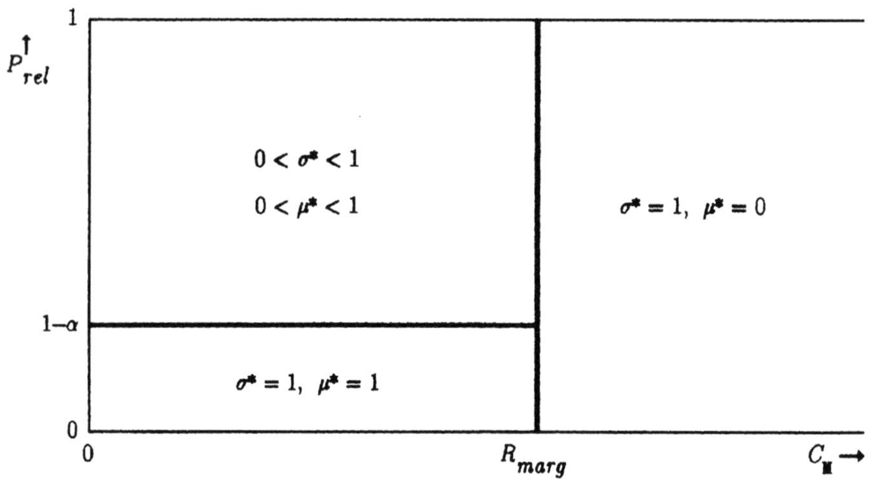

FIGURE 8: Alternative representation of the equilibrium regimes for two–person irrigation games without guards.

parameter constellations are not very interesting from an empirical point of view. They are structurally unstable since they are easily destroyed by the smallest parameter fluctuations. It is therefore very unlikely to encounter such a 'border case system' in a real world situation.

The most interesting consequences of Proposition 1 will be stated as a corollary:

COROLLARY 2: (Stealing tendency at equilibrium)

If the turntaker interacts with only one turnwaiter (i.e. $N = 1$), his stealing tendency is always positive at equilibrium. The equilibrium stealing tendency, σ^*, is smaller than one if and only if the detection probability α is larger than both, the relative benefit of stealing water, B_{rel}, and the cost–benefit ratio $C_M/(R_I+L_I)$.

The parameter regimes for the three different types of equilibria are illustrated by the *equilibrium diagram* in Figure 7 which summarizes the results of Proposition 1. This figure shows how the equilibrium regimes of a two–person irrigation game without guards depend on the relation between the detection probability, α, the relative benefit of stealing water, B_{rel}, and the cost–benefit ratio $C_M/(R_I+L_I)$. Figure 8 depicts the equilibrium diagram in a slightly different but equivalent way. On the abscissa, the cost of monitoring, C_M, is related to the marginal benefit of monitoring, R_{marg}. On the ordinate, the term

$$P_{rel} := \frac{P_I}{B+P_I} = 1-B_{rel}, \tag{14}$$

which may be interpreted as the *relative cost of a detected stealing event* for the turntaker, is related to $1-\alpha$, the probability of not being detected by a monitoring turnwaiter. The representation in Figure 8 has the advantage that it can easily be generalized to irrigation games with more than one turnwaiter (see Figure 10).

4.3 Basic Aspects of the Best Reply Structure

Let us now turn to the general case where a turntaker is facing N turnwaiters which all have a strategic influence on his decision. The structure of the resulting irrigation game is illustrated by Figure 4. Whether stealing is profitable for the turntaker or not depends on his payoff parameters B and P_I, and on the probability that a stealing event is detected by one of the turnwaiters. This probability is given by the monitoring efficiency $1-\eta(\mu,\alpha)$, where $\mu = (\mu_1,\mu_2,\dots,\mu_N)$ and $\alpha = (\alpha_1,\alpha_2,\dots,\alpha_N)$ characterize the monitoring tendencies and detection probabilities of the turnwaiters. Therefore, the expected payoffs for the two pure strategies of the turnwaiter are given by

$$F_{TT}(S,\mu) = \eta(\mu,\alpha)\cdot B + (1-\eta(\mu,\alpha))\cdot(-P_I), \tag{14}$$

$$F_{TT}(\neg S,\mu) = 0. \tag{15}$$

This implies that the best reply of the turntaker to the mixed strategy combination μ depends on the relation between the monitoring deficiency, $\eta(\mu,\alpha)$, and the relative cost of a detected stealing event, P_{rel}:

$$S \succ \neg S \iff \eta(\mu,\alpha)\cdot(B+P_I) > P_I \iff \eta(\mu,\alpha) > P_{rel}. \tag{16}$$

Let us now analyse the situation from the point of view of turnwaiter i. Let us assume that the payoff parameters of this turnwaiter are given by C^i_M, L^i_I, R^i_I, and R^i_O, and that he is faced with a turntaker stealing at a rate σ and N–1 colleagues whose monitoring tendencies and detection probabilities are characterized by the vectors

$$\mu_{-i} := (\mu_1,..,\mu_{i-1},\mu_{i+1},..,\mu_N) \text{ and } \alpha_{-i} := (\alpha_1,..,\alpha_{i-1},\alpha_{i+1},..,\alpha_N). \tag{17}$$

Whether monitoring is profitable or not for turnwaiter i depends crucially on the monitoring efficiency of the other N–1 turnwaiters. This is given by $1-\eta_{-i}$ where η_{-i} denotes the probability that a stealing event is not detected by any of the other turnwaiters:

$$\eta_{-i} := \eta(\mu_{-i},\alpha_{-i}) = \prod_{k\neq i} (1-\alpha_k\mu_k). \tag{18}$$

Notice that η_{-i} can be written in the form

$$\eta_{-i} = \eta(\mu,\alpha)/(1-\alpha_i\mu_i). \tag{19}$$

The different ***payoff configurations*** for turnwaiter i ('ego') and their probabilities are represented in Figure 9. From this, it is easy to derive the expected payoffs for M and $\neg M$:

$$F_{TW}(M,\sigma,\mu_{-i}) = \sigma\cdot\{\alpha_i R^i_I + (1-\alpha_i)(R^i_O - \eta_{-i}(R^i_O + L^i_I))\} - C^i_M, \tag{20}$$

$$F_{TW}(\neg M,\sigma,\mu_{-i}) = \sigma\cdot(R^i_O - \eta_{-i}(R^i_O + L^i_I)). \tag{21}$$

The payoff difference between monitoring and not monitoring is given by

$$F_{TW}(M,\sigma,\mu_{-i}) - F_{TW}(\neg M,\sigma,\mu_{-i}) = \sigma\cdot R_{marg}(\mu_{-i},\alpha) - C^i_M, \tag{22}$$

where

$$R_{marg}(\mu_{-i},\alpha) := \alpha_i\cdot\{ R^i_I - R^i_O + \eta_{-i}(R^i_O + L^i_I) \}. \tag{23}$$

Decis. of TW	Event	Probability of event	Decision of TT: S	Decision of TT: $\neg S$
			σ	$1-\sigma$
M	TW i ('ego') detects	α_i	$R_I^i - C_M^i$	$-C_M^i$
	Another TW detects	$(1-\alpha_i)(1-\eta_{-i})$	$R_O^i - C_M^i$	$-C_M^i$
	Nobody detects	$(1-\alpha_i)\eta_{-i}$	$-L_I^i - C_M^i$	$-C_M^i$
$\neg M$	Another TW detects	$1-\eta_{-i}$	R_O^i	0
	Nobody detects	η_{-i}	$-L_I^i$	0

FIGURE 9: Payoff configurations for a turnwaiter who has to decide between M and $\neg M$. The payoff of turnwaiter i ('ego') for each of his pure strategies depends on the strategic decision of the turntaker and on whether a stealing event is detected by him or another turnwaiter. The probabilities of these events depend on the stealing tendency of the turnwaiter, σ, on the detection probability of turnwaiter i, α_i, and on $1-\eta_{-i}$, the efficiency of the other N–1 turnwaiters.

$R_{marg}(\mu_{-i},\alpha)$ corresponds to the payoff difference between monitoring and not monitoring in case that stealing is going on and that the fixed costs of monitoring, C_M^i, are neglected. This term may therefore be interpreted as a measure for the *marginal benefit of monitoring*. Notice that R_{marg} is constant if there is only one turnwaiter. In that case, $\eta_{-i} \equiv 1$, and R_{marg} is given by (5).

(22) implies that the best reply of turnwaiter i to a stealing rate σ of the turntaker and monitoring tendencies $\mu_1,..,\mu_{i-1},\mu_{i+1},..,\mu_N$ of the other N–1 turnwaiters may be determined from:

$$M \succ \neg M \iff \sigma \cdot R_{marg}(\mu_{-i},\alpha) > C_M^i . \tag{24}$$

Taken together, (16) and (24) characterize the best reply structure of a N+1–person irrigation game without guards.

Notice in particular that $\neg M \succ M$ if $\sigma = 0$, and that $S \succ \neg S$ if $\mu_1 = \mu_2 = \ldots = \mu_N = 0$ (since $\eta(0,\alpha) = 1$). Accordingly, not monitoring is the unique best reply of a turnwaiter to a turntaker who does not steal. On the other hand, stealing is the unique best reply of a turntaker to a turnwaiter community which does not monitor at all. This implies that Corollary 2 holds true in general:

<u>PROPOSITION 3:</u> (Positive stealing tendency at equilibrium)

An irrigation game without guards does not admit Nash equilibrium points with $\sigma^* = 0$. The stealing tendency is always positive at equilibrium.

Let us now have a closer look at those Nash equilibrium points (σ^*,μ^*) where the stealing tendency is *not* maximal, i.e., where $0 < \sigma^* < 1$. Since σ^* is a completely mixed equilibrium strategy, both S and $\neg S$ are best replies to μ^*. In view of (16), this is equivalent to:

$$\eta(\mu^*,\alpha) = P_{rel}, \tag{25}$$

i.e., the monitoring deficiency of the turnwaiter community equilibrates at the relative cost of a detected stealing event.

As we have seen above $\mu^* \neq 0$ since $\eta(0,\alpha) = 1 > P_{rel}$. On the other hand, we will also neglect the non–generic border case $\eta(1,\alpha) = P_{rel}$ and assume

$$P_{rel} \neq (1-\alpha_1)\cdot(1-\alpha_2)\cdot\ldots\cdot(1-\alpha_N) \tag{26}$$

which implies $\mu^* \neq 1$ at an equilibrium with $\sigma^* < 1$.

Taken together, $\mu^* \neq 0$ and $\mu^* \neq 1$ imply that at least one turnwaiter i monitors at an intermediate rate $0 < \mu_i^* < 1$. (19) and (25) show that the monitoring deficiency of the remaining N–1 turnwaiters may be viewed as a function of α_i and μ_i^* which is given by

$$\eta_{-i} = P_{rel} / (1-\alpha_i\mu_i^*). \tag{27}$$

Notice that η_{-i} is positively correlated with μ_i^* : The monitoring tendency of turnwaiter i equilibrates at a higher value if the monitoring deficiency of the other N–1 turnwaiters is higher, and vice versa.

Treating α_i as a fixed parameter, the marginal benefit of monitoring for turnwaiter i can also be viewed as a function of μ_i^* :

$$R_{marg}(\mu_i^*) = \alpha_i\cdot\{ R_1^i - R_0^i + P_{rel}\cdot(R_0^i+L_1^i)/(1-\alpha_i\mu_i^*) \}. \tag{28}$$

Like η_{-i}, $R_{marg}(\mu_i^*)$ is *positively* correlated with μ_i^*: Since the overall deficiency of the turnwaiter population equilibrates at P_{rel}, the marginal benefit of monitoring increases with an increase in the monitoring tendency μ_i^*.

On the basis of (25), the best reply of turnwaiter i at equilbrium can be determined on the basis of his equilibrium strategy μ_i^*:

$$M \succ \neg M \iff \sigma^* > CB_{mon}(\mu_i^*) := C_M^i \,/\, R_{marg}(\mu_i^*). \tag{29}$$

$CB_{mon}(\mu_i^*)$, the ratio between the cost and the marginal benefit of monitoring at equilibrium, will be called the *cost–benefit ratio of monitoring.* If $0 < \mu_i^* < 1$ is a genuinely mixed equilibrium strategy, turnwaiter i is indifferent between M and $\neg M$. Accordingly, the stealing tendency σ^* equilibrates at the cost–benefit ratio of monitoring:

$$\sigma^* = CB_{mon}(\mu_i^*). \tag{30}$$

This equation shows that the stealing tendency at equilibrium, σ^*, is *negatively* correlated with the equilibrium rates of monitoring of those turnwaiters who monitor at an intermediate rate. Putting all this together, we get:

<u>PROPOSITION 4:</u> (Equilibria with intermediate stealing tendency)

Let (σ^*,μ^*) be a Nash equilibrium point of an irrigation game without guards for which the stealing tendency is not maximal (i.e. $0 < \sigma^* < 1$). If the game is non–degenerate (i.e., if (26) holds true), $\mu^* \neq 0$ and $\mu^* \neq 1$. Accordingly, at least one turnwaiter monitors, and not all turnwaiters monitor at maximal rate. If turnwaiter i is one of the turnwaiters who monitor at an intermediate rate ($0 < \mu_i^* < 1$), σ^* equalizes at $CB_{mon}(\mu_i^*)$, the cost–benefit ratio of monitoring for turnwaiter i.

For the special case of a symmetric irrigation game, we get the corollary:

<u>COROLLARY 5:</u> (Symmetric equilibria with intermediate stealing tendency)

Let (σ^*,μ^*) be a symmetric Nash equilibrium point (i.e. $\mu_1^* = \mu_2^* = \dots = \mu_N^* = \mu^*$) of a symmetric irrigation game without guards for which (26) holds true. If the stealing tendency is not maximal (i.e. $0 < \sigma^* < 1$), all turnwaiters monitor at an intermediate rate ($0 < \mu_i^* < 1$), and (σ^*,μ^*) is a completely mixed strategy combination. The equilibrium strategies σ^* and μ^* are given by the equations

$$(1-\alpha\mu^*)^N = P_{rel}, \text{ and } \sigma^* = CB_{mon}(\mu^*). \tag{31}$$

Let us now investigate some conditions which lead to a maximal rate of stealing at equilibrium. First, notice that the monitoring deficiency $\eta(\mu,\alpha)$ of a turnwaiter community is a decreasing function of the monitoring tendency of each turnwaiter. Thus $\eta(\mu,\alpha)$ attains its maximum η_{max} at $\mu = 0$ and its minimum η_{min} at $\mu = 1$:

$$\eta_{max} := \eta(0,\alpha) = 1, \tag{32}$$

$$\eta_{min} := \eta(1,\alpha) = (1-\alpha_1)\cdot(1-\alpha_2)\cdot\ldots\cdot(1-\alpha_N). \tag{33}$$

In view of (16), stealing is always the unique best reply of the turntaker if η_{min} is larger than the relative cost of stealing:

$$S \succ \neg S \quad \text{if} \quad \eta_{min} > P_{rel}. \tag{34}$$

Notice that $\eta_{min} > P_{rel}$ is equivalent to $B_{rel} > 1-\eta_{min}$: stealing is the unique best reply of the turntaker if the relative benefit of stealing water is larger than the maximal efficiency of monitoring.

(23) shows that the marginal benefit of monitoring for turnwaiter i is a linear function of η_{-i}. We have not specified the sign of the payoff parameters R_0^i. A negative sign is admissible, but in that case we assume that R_0^i is smaller in absolute value than L_I^i, the loss of water felt by turnwaiter i. For the rest of this paper, we shall therefore concentrate on the case

$$R_0^i + L_I^i > 0 \quad \text{for all } i. \tag{35}$$

If (35) holds true, the marginal benefits of the turnwaiters are increasing functions of η_{-i} and they attain their minimal and maximal values at the minima and maxima of η_{-i}. Treating α as a constant, let us define the maximal and minimal marginal benefits of monitoring by

$$R_{max} := \max\{\, R_{marg}(\mu_{-i},\alpha) \mid \mu \in [0,1]^N,\ 1 \le i \le N \,\}, \tag{36}$$

$$R_{min} := \min\{\, R_{marg}(\mu_{-i},\alpha) \mid \mu \in [0,1]^N,\ 1 \le i \le N \,\}. \tag{37}$$

(24) yields that $\neg M$ is always the unique best reply for a turnwaiter if the cost of monitoring, C_M^i, exceeds the maximal marginal benefit of monitoring:

$$\neg M \succ M \quad \text{if} \quad C_M^i > R_{max}. \tag{38}$$

Putting all this together, we get:

PROPOSITION 6: (Conditions for a maximal stealing tendency at equilibrium)

Let (σ^*, μ^*) be a Nash equilibrium point of an irrigation game without guards.

1. If $\eta_{min} > P_{rel}$, the stealing tendency is maximal at equilibrium (i.e. $\sigma^* = 1$).
2. The equilibrium stealing tendency is also maximal if $C^i_M > R_{max}$ for all i.
   In that case, the equilibrium is given by $\sigma^* = 1$ and $\mu^* = 0$.

4.4 The Interaction Between the Turntaker and N Symmetric Turnwaiters

Let us now restrict our attention to a system where all N turnwaiters are in a symmetric position. For an irrigation model without guards this means that the payoff parameters C_M, L_I, R_I, and R_0 do not differ from one turnwaiter to the other, that the detection probabilities of the turnwaiters are the same, and that we may concentrate on symmetric strategy combinations (σ,μ) where μ is of the form $\mu = (\mu,\mu,...,\mu) = \mu\cdot 1$. Treating the detection probability α of the N turnwaiters as a fixed parameter we may consider the detection deficiency and the marginal benefit of monitoring as functions of the monitoring tendency μ:

$$\eta(\mu) := (1-\alpha\mu)^N, \tag{39}$$

$$R_{marg}(\mu) := \alpha\cdot\{ R_I - R_0 + (1-\alpha\mu)^{N-1}(R_0+L_I) \}. \tag{40}$$

Notice that η_{min}, R_{max}, and R_{min} are given by

$$\eta_{min} = (1-\alpha)^N, \tag{41}$$

$$R_{max} = \alpha\cdot(R_I + L_I), \tag{42}$$

$$R_{min} = \alpha\cdot\{ R_I - R_0 + (1-\alpha)^{N-1}(R_0+L_I) \}. \tag{43}$$

A comparison of (39) and (40) shows that the marginal benefit of monitoring may be viewed as a function of the monitoring deficiency of the turnwaiters: $R_{marg}(\mu) = f(\eta(\mu))$ where the function f is given by

$$f(\eta) := \alpha\cdot\{ R_I - R_0 + \eta^{(N-1)/N}\cdot(R_0+L_I) \}. \tag{44}$$

The function f is concave and monotonically increasing in η. Hence, it is invertible, and its inverse function g is monotonically increasing and convex. If $f(\eta) = R$, η is related to R via

$$\eta = g(R) := \left\{ \frac{R/\alpha + R_0 - R_I}{R_0 + L_I} \right\}^{\frac{N}{N-1}}. \tag{45}$$

Notice that the function η (which is given by (39)) is also invertible. If $\eta(\mu) = P$, μ is of the form:

$$\mu = \eta^{-1}(P) = [1 - P^{1/N}]/\alpha. \tag{46}$$

It is now easy to characterize the symmetric equilibrium points of a symmetric irrigation game without guards:

<u>PROPOSITION 7:</u> (A characterization of the symmetric equilibrium points)

For generic parameter constellations a symmetric irrigation game without guard has a unique Nash equilibrium point that will be denoted by (σ^*,μ^*), $\mu^* = \mu^*\cdot 1$. Each equilibrium point belongs to one of the following four categories:

1. Equilibrium points with maximal stealing tendency ($\sigma^* = 1$):
   (a) For $C_M \geq R_{max}$, σ^* and μ^* are given by $\sigma^* = 1$ and $\mu^* = 0$.
   (b) For $R_{max} > C_M \geq R_{min}$ and $P_{rel} \leq g(C_M)$, $\sigma^* = 1$ and $0 < \mu^* < 1$. In this case, μ^* is given by:
   $$\mu^* = \eta^{-1}(g(C_M)). \tag{47}$$
   (c) For $R_{min} > C_M$ and $P_{rel} < \eta_{min}$, $\sigma^* = 1$ and $\mu^* = 1$.
2. Equilibrium points with intermediate stealing tendency ($0 < \sigma^* < 1$):
   If $P_{rel} > \max\{\eta_{min}, g(C_M)\}$, the unique equilibrium point is completely mixed, and it is given by:
   $$\sigma^* = C_M/f(P_{rel}), \quad \text{and} \quad \mu^* = \eta^{-1}(P_{rel}). \tag{48}$$

The four different equilibrium regimes of a symmetric irrigation game are illustrated by Figure 10. Regions I, II, and III correspond to parameter constellations that induce a maximal stealing rate at equilibrium. In these regions, the equilibrium rate of monitoring is unaffected by a change in P_{rel}, the relative cost of detected stealing, and it is decreasing with an increase in C_M, the cost of monitoring. Region IV corresponds to the mixed strategy equilibrium characterized by (48). Here, σ^* and μ^* both increase with a decrease in P_{rel}. Whereas μ^* remains unaffected by a change in C_M, the stealing rate increases with an increase of the cost of monitoring. Only the border case $P_{rel} = \eta_{min}$, $R_{min} > C_M$ admits a multitude of symmetric Nash equilibria.

In contrast to the two-person case, the marginal benefit of monitoring, R_{marg}, can no longer be considered as constant if more than one turnwaiter is present. Outside the interval from R_{min} to R_{max}, Figure 10 corresponds perfectly to the two-person equilibrium diagram in Figure 8.

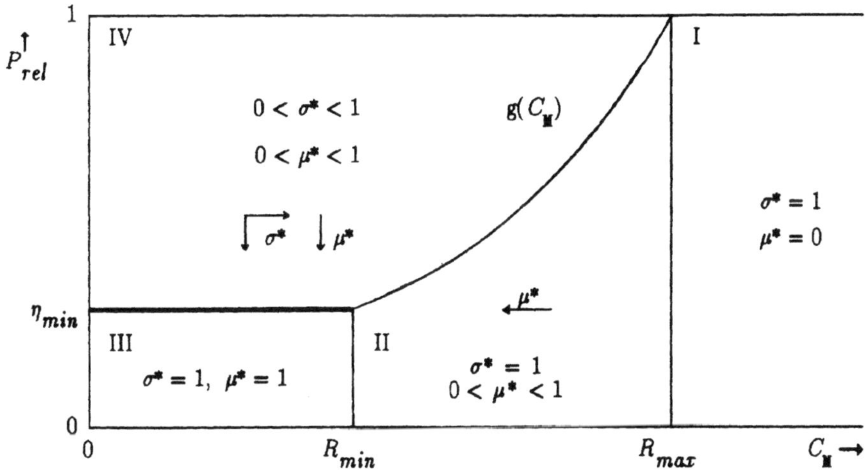

FIGURE 10: Equilibrium diagram illustrating the parameter regimes (I to IV) corresponding to the four different classes of symmetric Nash equilibrium points in (N+1)–person irrigation games without guards. Regions I, II, and III correspond to cases 1.(a), (b), and (c) in Proposition 7, respectively; region IV corresponds to the mixed–strategy equilibrium described in 2.
The arrows in regions II and IV indicate how a mixed equilibrium strategy is affected by changes in the parameters of the system.

PROOF OF PROPOSITION 7:

1. Suppose that (σ^*,μ^*) is a symmetric equilibrium with $\sigma^* = 1$. In view of (24) the best reply structure of the turnwaiters is given by

$$M \succ \neg M \iff R_{marg}(\mu^*) > C_M . \tag{49}$$

Accordingly, $\mu^* = 0$ is a best reply to $\sigma^* = 1$ if and only if $C_M \geq R_{marg}(0) = R_{max}$. Proposition 5 shows that for $C_M \geq R_{max}$, $\sigma^* = 1$ and $\mu^* = 0$ are in fact in equilibrium.
$\mu^* = 1$ is a best reply to $\sigma^* = 1$ if and only if $R_{marg}(1) = R_{min} \geq C_M$. On the other hand $\sigma^* = 1$ is a best reply to $\mu^* = 1$ if and only if $\eta(1) = \eta_{min} \geq P_{rel}$. Correspondingly, $\sigma^* = 1$ and $\mu^* = 1$ are in equilibrium if and only if $R_{min} \geq C_M$ and $\eta_{min} \geq P_{rel}$.

A mixed strategy $0 < \mu^* < 1$ is a best reply to $\sigma^* = 1$ if and only if $C_M = R_{marg}(\mu^*) = f(\eta(\mu^*))$. The equation $f(\eta(\mu^*)) = C_M$ is equivalent to $g(C_M) = \eta(\mu^*)$, and it implies that μ^* is given by (47). The correspondence between (47) and $f(\eta(\mu^*)) = C_M$ also implies that (47) defines a mixed monitoring tendency if and only if $R_{max} > C_M > R_{min}$. In this case, (16) shows that $\sigma^* = 1$ and μ^* are in equilibrium if and only if $P_{rel} \leq \eta(\mu^*) = g(C_M)$.

Param:	Interpretation:	Constraints:	Def:
0	Status quo payoff for turntaker	0	Sec. 4.1
B	Benefit of additional water	$B > 0$	
$-P_I$	Punishment after detected stealing	$P_I > 0$	
B_{rel}	Relative benefit of stolen water	$0 < B_{rel} < 1$	(5)
P_{rel}	Relative cost of detected stealing	$0 < P_{rel} < 1$	(14)
0	Status quo payoff for turnwaiter	0	Sec. 4.1
$-C_M$	Fixed cost of monitoring	$C_M > 0$	
$-L_I$	Loss of water due to undetected stealing	$L_I > 0$	
R_I	'Reward' for detecting a stealing event	$R_I + L_I > 0$	
R_0	Payoff if another TW has detected stealing	$R_0 + L_I > 0$	
R_{marg}	Marginal benefit of monitoring	$R_{marg} > 0$	(23)
R_{max}	Maximal marginal benefit of monitoring	$R_{max} \geq R_{marg}$	(36)
R_{min}	Minimal marginal benefit of monitoring	$R_{min} \leq R_{marg}$	(37)
CB_{mon}	Cost–benefit ratio of monitoring	$CB_{mon} > 0$	(29)
N	Number of turnwaiters	$N \geq 1$	4.1
α	Detection probability for individual TW	$0 < \alpha < 1$	4.1
η	Monitoring deficiency of TW community	$0 < \eta < 1$	(4)
$1-\eta$	Monitoring efficiency of TW community	$0 < 1-\eta < 1$	(4)
η_{-i}	Monitoring deficiency of other TW's	$0 < \eta_{-i} < 1$	(18)
η_{min}	Minimal monitoring deficiency	$0 < \eta_{min} < 1$	(33)

<u>TABLE 1:</u> Basic and derived parameters of an irrigation game without guards. Superscripts distinguishing the payoff parameters of turnwaiter i from those of the other turnwaiters are omitted.

2. Let us now suppose that (σ^*,μ^*) is a symmetric equilibrium with $\sigma^* \neq 1$. Proposition 6 implies that we may assume $P_{rel} > \eta_{min}$. According to Corollary 5, (σ^*,μ^*) is a completely mixed strategy combination which is given by (31). It is easy to see that (31) is equivalent to (48) since $R_{marg}(\mu^*) = f(\eta(\mu^*)) = f(P_{rel})$.
On the other hand, (48) defines a mixed strategy equilibrium if $P_{rel} > \eta_{min}$ and $C_M < f(P_{rel})$. In view of the monotonicity of the function f, the last inequality is equivalent to $P_{rel} > g(C_M)$.

* * *

4.5 Comparative Statics for the Symmetric Model

In this section we shall investigate how the symmetric Nash equilibrium point of a symmetric irrigation game without guards is affected by a change in the parameters of the game. An irrigation game without guards is characterized by a list of basic parameters which may be put into three categories: (1) the payoff parameters of the turntaker, (2) the payoff parameters of the turnwaiters, and (3) structural parameters describing the detection efficiencies of the turnwaiters. All the basic parameters and the more important derived parameters of our models are summarized in Table 1.

The equilibrium diagram in Figure 10 shows that a change in a parameter may induce a change in the equilibrium behavior of the players in two different ways: On the one hand, the boundaries of the equilibrium regions I, II, III, and IV might change and induce a switch from one class of equilibrium to another. On the other hand, the mixed tendencies to steal and to monitor might change within Region IV, and the mixed tendency to monitor might change within Region II.

Table 2 shows how a change in the parameters of the game (P_{rel}, C_M , L_I , R_I , R_O , α, or N) is reflected in a change in the stealing and monitoring tendencies in equilibrium region IV (σ^*_{IV} and μ^*_{IV}), the monitoring tendency in equilibrium region II (μ^*_{II}), and the borders of the equilibrium regions which are determined by η_{min}, R_{min}, R_{max}, and $g(C_M)$. The stealing rate in Region IV, for example, is given by (48):

$$\sigma^*_{IV} = \frac{C_M}{\alpha\cdot\{ R_I - R_O + P_{rel}^{(N-1)/N}\cdot(R_O + L_I) \}} . \qquad (50)$$

Correspondingly, σ^*_{IV} is positively correlated with C_M , R_O , and N, and it is negatively correlated with P_{rel} , L_I , R_I , and α. This means that the stealing rate will *increase* if the situation changes in a way which is associated with

- a decrease in P_{rel}, the relative cost of detected stealing;
- a decrease in α, the detection probability for an individual turnwaiter;
- an increase in C_M , the cost of monitoring;
- a decrease in L_I , the loss of water due to undetected stealing;
- a decrease in R_I , the (psychological) 'reward' a turnwaiter gets after detecting stealing;
- an increase in R_O , the 'satisfaction' that a turnwaiter gets after seeing that a stealing event was detected by one of his colleagues;
- an increase in N, the size of the system.

Param:	σ^*_{IV}	μ^*_{IV}	μ^*_{II}	η_{min}	R_{min}	R_{max}	$g(C_M)$
P_{rel}	–	–	0	0	0	0	0
C_M	+	0	–	0	0	0	+
L_I	–	0	+	0	+	+	–
R_I	–	0	+	0	+	+	–
R_0	+	0	–	0	–	0	+
α	–	–	+/–	–	+/–	+	–
N	+	–	–	–	–	–	+

TABLE 2: Comparative statics for symmetric irrigation games without guards. A '+' indicates a positive correlation, a '–' indicates a negative correlation, '0' indicates no correlation, and '+/–' indicates that the sign of the correlation depends on the constellation of the other parameters. See text for explanation.

An increase in σ^*_{IV} is expected to occur if at least one of the parameters changes in the described way, and if none of the other parameters changes in the opposite direction. The stealing rate will *decrease* if all changing parameters have a tendency to shift into the opposite direction. The direction of change in σ^*_{IV} cannot be predicted on basis of the directions of parameter shifts alone if one of the parameters changes in a way indicated above whereas another parameter changes in the opposite direction. If, for example, C_M and α both increase, σ^*_{IV} might increase or decrease depending on whether the ratio C_M/α increases or decreases.

Notice that the stealing rate at equilibrium, σ^*_{IV}, is influenced by *all* parameters of the system. The monitoring rate, μ^*_{IV}, on the other hand, is only influenced by the payoff of the turntaker, P_{rel}, and by the structural parameters α and N (see Section 5.5 for a discussion). μ^*_{II}, in contrast is not influenced by P_{rel}, and there is no clearcut relation between μ^*_{II} and α. In fact, it depends on the other parameters how a change in α is reflected by a change in μ^*_{II}.

Table 2 focuses on the influence of single parameters on the behavior of the players at equilibrium. One of the most important conclusions is that, *all other things being equal*, an increase in the size of the system leads to an increase in the stealing rate at equilibrium. As has been stressed above, this conclusion can be extended to comprise the simultaneous change in several parameters of the game provided that all changes occur in a specific direction. This point is made more explicit in Table 3.

Table 3 is based on the assumption that a change in N, the number of turnwaiters in a system, induces (or: is correlated with) a change in the other parameters of the system. Accordingly, these other parameters are viewed as being functions of N. For example, water

		$\frac{\partial}{\partial N}\alpha$	$\frac{\partial}{\partial N}C_M$	$\frac{\partial}{\partial N}L_I$	$\frac{\partial}{\partial N}R_I$	$\frac{\partial}{\partial N}R_O$	$\frac{\partial}{\partial N}P_{rel}$
$\frac{\partial}{\partial N}\sigma^*_{IV} > 0$	if	≤ 0	≥ 0	≤ 0	≤ 0	≥ 0	≤ 0
$\frac{\partial}{\partial N}\mu^*_{IV} < 0$	if	≥ 0	–	–	–	–	≥ 0
$\frac{\partial}{\partial N}\mu^*_{II} < 0$	if	0	≥ 0	≤ 0	≤ 0	≥ 0	–
$\frac{\partial}{\partial N}\eta_{min} < 0$	if	≥ 0	–	–	–	–	–
$\frac{\partial}{\partial N}R_{min} < 0$	if	0	–	≤ 0	≤ 0	≥ 0	–
$\frac{\partial}{\partial N}R_{max} < 0$	if	≤ 0	–	≤ 0	≤ 0	–	–
$\frac{\partial}{\partial N}g(C_M) > 0$	if	≤ 0	≥ 0	≤ 0	≤ 0	≥ 0	–

TABLE 3: Dependence of equilibrium behavior on N, the number of turnwaiters. In contrast to Table 2 the parameters of the game are not considered to be independent from another. Instead, the payoff parameters and the detection probability are viewed as functions of N. See text for explanation.

might become more precious for an individual farmer if the farmer community served by an irrigation channel grows larger. Consequently, the benefit of stolen water might increase with an increase in the size of the system. Mathematically, this will be indicated by $\partial B/\partial N \geq 0$ or, equivalently, $\partial P_{rel}/\partial N \leq 0$, where, for simplicity, N is regarded as a continuous parameter. Similarly, the loss of water due to undetected stealing might be more severe in a larger system. On the other hand, the loss felt by an individual turnwaiter might be smaller since the overall loss due to a single stealing event is spread over a larger community. The inequality $\partial L_I/\partial N \leq 0$ indicates that the individual loss is negatively correlated with the size of the system.

Table 2 shows that, all other things being equal, an increase in N induces an increase in σ^*_{IV} and $g(C_M)$ and a decrease in μ^*_{IV}, μ^*_{II}, η_{min}, R_{min}, and R_{max}. Table 3 indicates how robust these conclusions are if a simultaneous change in the other parameters of the system is taken into account. In particular, the first row of Table 3 shows that the stealing tendency at equilibrium, σ^*_{IV} can be expected to *increase* with the size of the system if:

- α, L_I, R_I, and P_{rel} are *negatively* (or at least not positively) correlated with N, and
- C_M and R_O are *positively* (or at least not negatively) correlated with N.

Nothing definite can be said about the direction of change in σ^*_{IV} if the parameters of the system are related to N in a different way. The stealing rate might, for example, well decrease with the size of the system if the cost of monitoring is smaller in a larger system.

Table 3 also shows that α and P_{rel} should be positively (more precisely: not negatively) correlated with N in order to ensure that μ^*_{IV} will in fact decrease with an increase in N. The interdependence between N and the other parameters has no influence on the change in μ^*_{IV}. In contrast to μ^*_{IV}, μ^*_{II} is not affected by P_{rel}. Notice that a decrease in μ^*_{II} can only be inferred from an increase in N if the detection probability α is neither positively nor negatively influenced by the size of the system.

Most of the relations which are summarized by Tables 2 and 3 follow directly from the definitions of σ^*_{IV}, μ^*_{IV}, μ^*_{II}, η_{min}, R_{min}, R_{max}, and $g(C_M)$. The sign of $\partial\sigma^*_{IV}/\partial N$, for example can easily be derived from the relation $\sigma^*_{IV} = C_M/f(P_{rel})$. $\partial\sigma^*_{IV}/\partial N$ is positive if $\partial C_M/\partial N \geq 0$ and $\partial f(P_{rel})/\partial N < 0$. In view of (43), the latter inequality holds true if $\partial\alpha/\partial N \leq 0$, $\partial L_1/\partial N \leq 0$, $\partial R_1/\partial N \leq 0$, $\partial R_0/\partial N \geq 0$, and $\partial/\partial N\, [P_{rel}^{(N-1)/N}] < 0$.

In order to analyse the last inequality we have to investigate how an expression of the form $F(N)^{\varphi(N)}$ changes with N if F and φ are continuously differentiable, scalar functions of N which fulfill the additional requirements that $0 < F(N) < 1$ and $\varphi(N) > 0$ for all N. Since $F(N)^{\varphi(N)}$ may be written in the form $F(N)^{\varphi(N)} = \exp(\varphi(N)\cdot\ln F(N))$, its derivative is given by

$$\frac{d}{dN}\left[F(N)^{\varphi(N)}\right] = F(N)^{\varphi(N)}\cdot\frac{d}{dN}[\varphi(N)\cdot\ln F(N)]. \tag{51}$$

Accordingly, the left–hand side of (51) has the same sign as

$$\frac{d}{dN}[\varphi(N)\cdot\ln F(N)] = \ln F(N)\cdot\frac{d}{dN}\varphi(N) + \frac{\varphi(N)}{F(N)}\cdot\frac{d}{dN}F(N). \tag{52}$$

In view of $\varphi(N)/F(N) > 0$ and $\ln F(N) < 0$ for $0 < F(N) 1$ we get:

$$\frac{d}{dN}\left[F(N)^{\varphi(N)}\right] < 0 \text{ if } \frac{d}{dN}\varphi(N) > 0 \text{ and } \frac{d}{dN}F(N) \leq 0. \tag{53}$$

(Nothing definite can be said if $d\varphi/dN$ and dF/dN are of the same sign.) Applying (53) to the functions $F(N) := P_{rel}(N)$ and $\varphi(N) := (N-1)/N$ yields

$$\frac{\partial}{\partial N}\left[P_{rel}^{(N-1)/N}\right] < 0 \text{ if } \frac{\partial}{\partial N}P_{rel} \leq 0. \tag{54}$$

In particular, we get that $\partial\sigma^*_{IV}/\partial N > 0$ if $\partial P_{rel}/\partial N \leq 0$. All other relations in Table 3 are derived in a similar way.

4.6 Relaxing the Symmetry Assumptions: Models With a Distinguished Turnwaiter

In this section we relax the symmetry assumptions on which the analysis of the preceding two sections was based. For simplicity, we will concentrate on a system with just one 'distinguished' turnwaiter: whereas turnwaiters 1,...,N–1 are still in a symmetric position, turnwaiter N differs from his colleagues either in his payoffs, or in his detection probability, or just in his norm of behavior. Examples for distinguished turnwaiters are newcomers to an established system, farmers who irrigate a different amount of land or a different spectrum of crops, or 'elders' of a community who might have different payoffs or different norms of reaction due to the expectation of the other players.

As before, let P_{rel} describe the incentive of the turntaker, whereas C_M, L_I, R_I, R_0, and α denote the payoff and structural parameters of the N–1 'symmetric' turnwaiters. In contrast, the distinguished turnwaiter N has payoff parameters $C_M^{\prime}$, $L_I^{\prime}$, $R_I^{\prime}$, $R_0^{\prime}$, and a detection probability $\alpha^{\prime}$. As a special case, all the parameters which characterize turnwaiter N might be the same as those of his colleagues, and turnwaiter N might be distinguished from the other turnwaiters just by a different norm of behavior.

Due to the symmetry of the turnwaiters 1,...,N–1, we will concentrate on Nash equilibrium points which are characterized by a stealing tendency σ^* of the turntaker, a common monitoring tendency μ^* of the first N–1 turnwaiters, and a (different) monitoring tendency μ^*_{dist} of the distinguished turnwaiter. We will contrast such an equilibrium point with the symmetric equilibrium $(\sigma^*_{sym}, \mu^*_{sym})$, $\mu^*_{sym} = \mu^*_{sym} \cdot 1$ of the 'corresponding' symmetric game, i.e., the game where all N turnwaiters are characterized by the parameters C_M, L_I, R_I, R_0, and α.[23] In particular, we are interested in the question how the equilibrium stealing frequency, σ^*_{sym}, is influenced by the fact that one of the turnwaiters gets into a distinguished position.

For simplicity, we will concentrate on those parameter constellations for which the stealing rate is not maximal at equilibrium. In the symmetric case, this assumption implies that σ^*_{sym} and μ^*_{sym} are given by Corollary 5:

$$\pi(\mu^*_{sym}) = (1-\alpha\mu^*_{sym})^N = P_{rel} \quad \text{and} \quad \sigma^*_{sym} = CB_{mon}(\mu^*_{sym}). \tag{55}$$

It is plausible to assume that an intermediate stealing rate in the symmetric case (i.e. $0 < \sigma^*_{sym} < 1$) corresponds to an intermediate stealing rate, $0 < \sigma^* < 1$, in the model with a distinguished turnwaiter – at least if the payoff and structural parameters of the distinguished turnwaiter differ only slightly from those of his colleagues. In fact, $\sigma^* = 1$ implies $\sigma^* > CB_{mon}(\mu^*_{sym})$. According to the dynamic interpretation of Nash equilibria, all N turnwaiters are expected to increase their monitoring tendency (at least if they start close to μ^*_{sym}). This process will not stop before the stealing tendency is reduced to an intermediate level.

Therefore, we will assume that $0 < \sigma^* < 1$, which implies $S \approx \neg S$ or, equivalently,

$$(1-\alpha\mu^*)^{N-1} \cdot (1-\alpha'\mu^*_{dist}) = (1-\alpha\mu^*_{sym})^N = P_{rel}. \tag{56}$$

Accordingly, the monitoring deficiency at the 'asymmetric equilibrium' is exactly the same as the monitoring deficiency at the 'symmetric equilibrium'.[24] Notice that $\mu^* < \mu^*_{sym}$ implies that $\alpha'\mu^*_{dist} > \alpha\mu^*_{sym}$ whereas $\mu^* > \mu^*_{sym}$ implies $\alpha'\mu^*_{dist} < \alpha\mu^*_{sym}$. Therefore, we have either

$$\mu^* > \mu^*_{sym} > (\alpha'/\alpha)\cdot\mu^*_{dist}, \tag{57}$$

or

$$\mu^* < \mu^*_{sym} < (\alpha'/\alpha)\cdot\mu^*_{dist}. \tag{58}$$

We will neglect the possibility that all N turnwaiters use a pure strategy, since, in view of (56), this corresponds to a border case in parameter space. Border cases such like this left aside, an irrigation game with a distinguished turnwaiter has up to five coexisting Nash equilibrium points. In addition to the completely mixed equilibrium corresponding to the symmetric case, there are four other classes of Nash equilibrium points which are characterized by the fact that either the distinguished turnwaiter or the symmetric turnwaiters specialize on one of their two pure strategies.

In the rest of this section, we will focus on two scenarios. Let us first consider a scenario, where the distinguished turnwaiter is 'inherently' in a better position with respect to monitoring than his colleagues. By this we mean that turnwaiter N has better detection abilities ($\alpha' \geq \alpha$), and/or that he is *inherently higher motivated for monitoring* than the other N–1 turnwaiters in that he has a lower cost of monitoring, a higher incentive to prevent losses due to stealing, a higher satisfaction after having detected a stealing event, and/or a smaller tendency to rely on the monitoring activity of his colleagues:

$$C'_M \leq C_M, \quad L'_I \geq L_I, \quad R'_I \geq R_I, \quad R'_O \leq R_O. \tag{59}$$

It this case, the inequality $\alpha'\mu^*_{dist} > \alpha\mu^*$ implies

$$CB^N_{mon}(\mu^*_{dist}) < CB_{mon}(\mu^*), \tag{60}$$

i.e., the cost–benefit ratio of monitoring is lower for the distinguished turnwaiter than that of the other N–1 turnwaiters. (60) implies that μ^*_{dist} and μ^* cannot both be genuinely mixed: either the distinguished turnwaiter monitors at an intermeditate rate and his colleagues do not monitor at all, or the N–1 symmetric turnwaiters monitor at an intermediate rate (which is lower than μ^*_{sym}) whereas the distinguished turnwaiter specializes on monitoring ($\mu^*_{dist} = 1$).

The second scenario is based on the assumption that the 'symmetric' turnwaiters have better prerequisites for monitoring than the distinguished turnwaiter. This means that they have a higher detection probability ($\alpha \geq \alpha'$), and/or that they are inherently higher motivated for monitoring in the sense that

$$C_M \leq C_M' , \quad L_I \geq L_I' , \quad R_I \geq R_I' , \quad R_O \leq R_O' . \tag{61}$$

Similar to the case considered above, $\alpha\mu^* > \alpha'\mu^*_{dist}$ implies that μ^*_{dist} and μ^* cannot both be genuinely mixed: either the 'symmetric' turnwaiters monitor at an intermeditate rate (which is higher than μ^*_{sym}) and the distinguished turnwaiter does not monitor at all, or the distinguished turnwaiter monitors at a small, but intermediate rate whereas his N–1 colleagues have to monitor at maximal rate.

Let us now consider the case that the symmetric turnwaiters monitor at an intermediate rate ($0 < \mu^* < 1$). In view of $M \approx \neg M$, this implies

$$\sigma^* = CB_{mon}(\mu^*). \tag{62}$$

Note that the *cost–benefit function* CB_{mon} for the first N–1 turnwaiters does not differ from the corresponding cost–benefit function in the symmetric case. It was shown in Section 4.3, that CB_{mon} is a decreasing function of μ^*, i.e. σ^* is negatively correlated with μ^*. Applying this result to (55) and (62), we get:

<u>PROPOSITION 8:</u> (Intermediate monitoring rates of symmetric turnwaiters)

Let σ^*, μ^*, and μ^*_{dist} characterize the Nash equilibrium point of an irrigation game with a distinguished turnwaiter and let the monitoring tendency of turnwaiters 1,...,N–1, μ^*, be at an intermediate level ($0 < \mu^* < 1$). Then the equilibrium stealing tendency, σ^*, is smaller than σ^*_{sym} , the stealing rate in the symmetric case, if and only if the symmetric turnwaiters monitor at a higher rate than in the symmetric case:

$$\sigma^* < \sigma^*_{sym} \quad \Leftrightarrow \quad \mu^* > \mu^*_{sym} . \tag{63}$$

Proposition 8 has some interesting implications when applied to the two scenarios considered above. If, for example, the distinguished turnwaiter is inherently higher motivated for monitoring, he will either specialize on monitoring (which induces an increase in the stealing rate), or the inherently higher motivated player will monitor at a *lower* rate inducing his fellow turnwaiter to monitor at such a high rate that the stealing rate decreases nevertheless.

PROPOSITION 9: (The influence of inherent motivation on equilibrium behavior)

Under the assumptions of Proposition 8 (in particular $0 < \mu^* < 1$) the following holds true:

(1) There are two possible equilibrium scenarios if the distinguished turnwaiter has inherently better prerequisites for monitoring:
First, the distinguished turnwaiter may specialize on monitoring ($\mu^*_{dist} = 1$). In this case, the remaining N–1 turnwaiters monitor at such a low rate that the stealing rate increases when compared to the symmetric case ($\sigma^* > \sigma^*_{sym}$).
Second, the distinguished turnwaiter may monitor at a lower rate than his colleagues ($0 \leq \mu^*_{dist} < \mu^*$). The N–1 symmetric turnwaiters compensate for that, leading to a decrease in the stealing rate ($\sigma^* < \sigma^*_{sym}$).

(2) There are also two possible equilibrium scenarios if the N–1 symmetric turnwaiters have inherently better prerequisites for monitoring:
First, the distinguished turnwaiter may not monitor at all ($\mu^*_{dist} = 0$). The other turnwaiters compensate for this, and the stealing rate decreases ($\sigma^* < \sigma^*_{sym}$).
Second, the distinguished turnwaiter may monitor at a higher rate than his colleagues ($\mu^* < \mu^*_{dist} \leq 1$). The other turnwaiters do not compensate for this, and the stealing rate increases ($\sigma^* > \sigma^*_{sym}$).

Notice in particular that two of the four equilibrium scenarios described in Proposition 9 are rather counterintuitive in that just those turnwaiters who are inherently in a *better* position for monitoring monitor at a *lower* rate. The interpretation and relevance of these *paradoxical equilibria* will be discussed in Sections 5.4 and 5.5.

Let us now consider those equilibrium points where the N–1 symmetric turnwaiters use one of their two pure strategies. $\mu^* = 0$, for example, implies that σ^* and μ^*_{dist} are given by

$$\mu^*_{dist} = (1 - P_{rel}) \,/\, \alpha' , \quad \text{and} \quad \sigma^* = C'_{M} \,/\, [\alpha' \cdot (R'_{I} + L'_{I})] \,. \tag{64}$$

Of course, such an equilibrium only exists if (64) yields values for σ^* and μ^*_{dist} which lie between 0 and 1. This is always true if the parameters of the distinguished turnwaiter do not differ too much from those of his colleagues. The strategy combination defined by (64) and $\mu^* = 0$ is, however, only in equilibrium if the symmetric turnwaiters prefer $\neg M$ to M, i.e., if

$$CB_{mon}(0) \geq \sigma^*, \tag{65}$$

or, equivalently,

$$\frac{C_{M}}{\alpha \cdot [\, R_{I} - R_{0} + P_{rel}(R_{0} + L_{I}) \,]} \geq \frac{C'_{M}}{\alpha' \cdot (\, R'_{I} + L'_{I} \,)} \,. \tag{66}$$

(66) is always true if the distinguished turnwaiter is inherently in a better position for monitoring than the other turnwaiters. In that case, a comparison between σ^* and σ^*_{sym} (given by (55) and (64)) yields that σ^* is smaller than σ^*_{sym}.

Since inequality (66) only depends on R_0 and not on R_0', it might also be true in some cases where the symmetric turnwaiters have inherently better prerequisites for monitoring. In such a case, we get another 'paradoxical' equilibrium where just those turnwaiters monitor at the lowest rate who are in the best position for monitoring. The parameter constellations which yield a paradoxical equilibrium with $\mu^* = 0$ are rather specific, but they do not form border cases in parameter space. Nevertheless, we consider such parameter combinations as unrealistic enough to neglect these equilibria in the sequel.

Finally, we have to consider those equilibrium points where $\mu^* = 1$, i.e., where the N–1 symmetric turnwaiters specialize on monitoring. By similar arguments as those given above, one can show that an equilibrium of this kind usually does exist if the symmetric turnwaiters are inherently in a better position for monitoring than the distinguished turnwaiter. In this case, the stealing tendency, σ^*, is larger than the stealing tendency in the symmetric case. For specific parameter constellations, one gets again a paradoxical equilibrium, where the symmetric turnwaiters monitor at a maximal rate even if the distinguished turnwaiter has inherently better prerequisites for monitoring. Again, we will henceforth neglect this type of equilibrium.

Without considering the two types of paradoxical equilibria mentioned above, Table 4 summarizes the results of this section. We would like to stress that the analysis comprises the special case

$$\alpha' = \alpha, \; C_M' = C_M, \; L_I' = L_I, \; R_I' = R_I, \; R_0' = R_0, \tag{67}$$

where the distinguished turnwaiter only differs from his colleagues by a different norm of behavior. In this case, the completely mixed equilibrium corresponds to the symmetric equilibrium $(\sigma^*_{sym}, \mu^*_{sym})$, and it can no longer be considered paradoxical. In addition, however, there are four asymmetric equilibria which have one pure-strategy component. Notice that – rather counterintuitively – the stealing rate σ^* is larger just in those cases where either the distinguished turnwaiter or his colleagues specialize on monitoring. σ^* is smaller than σ^*_{sym} in those equilibria where nobody specializes on monitoring, and where either the distinguished turnwaiter or the symmetric turnwaiters do not monitor at all. In our opinion, this result sheds some light on the question whether stealing can be reduced by introducing the position of a 'monitoring' turnwaiter who is supposed to do most of the monitoring of the farmer community.

Type of difference	Type of equilibrium	Monitoring tendencies	Stealing rate
Distinguished turnwaiter inherently in better position	Paradoxical	$0 \leq \mu^*_{dist} \leq \mu^*$	$\sigma^* \leq \sigma^*_{sym}$
	$\mu^*_{dist} = 1$	$\mu^* < \mu^*_{sym}$	$\sigma^* > \sigma^*_{sym}$
	$\mu^* = 0$	$\alpha \cdot \mu^*_{sym} < \alpha' \cdot \mu^*_{dist}$	$\sigma^* < \sigma^*_{sym}$
Symmetric turnwaiters inherently in better position	Paradoxical	$\mu^* \leq \mu^*_{dist} \leq 1$	$\sigma^* \geq \sigma^*_{sym}$
	$\mu^* = 1$	$\alpha' \cdot \mu^*_{dist} < \alpha \cdot \mu^*_{sym}$	$\sigma^* > \sigma^*_{sym}$
	$\mu^*_{dist} = 0$	$\mu^*_{sym} < \mu^*$	$\sigma^* < \sigma^*_{sym}$

TABLE 4: Coexisting Nash equilibrium points for irrigation games with a distinguished turnwaiter.

The distinguished turnwaiter is inherently in a better position for monitoring if his detection chances are higher ($\alpha' \geq \alpha$) and/or if he is inherently higher motivated for monitoring (i.e., if (59) holds true). The symmetric turnwaiters are inherently in a better position for monitoring if $\alpha \geq \alpha'$ and/or if (61) holds true.

In both cases, one additional 'paradoxical' equilibrium is not shown since it only exists for specific parameter constellations.

Notice that both cases comprise (67), the case where the distinguished turnwaiter only differs in his norm of behavior.

See text for further explanation.

5. Conclusions and Implications

5.1 The Applicability of the Models to Natural Settings

The models examined in this paper are relatively simple when contrasted to the complexity of naturally occurring irrigation systems. The models capture, however, major aspects of the underlying strategic structure facing farmers deciding whether to follow allocation rules and to monitor each others' water diversion activities on irrigation systems characterized by rotation systems without guards. The initial two–person game provides a stark outline of the choices available to a turntaker and a turnwaiter and of the types of payoff structures to be found in such situations. The (N+1)–person game is more general. Several conclusions can be derived from it that have important policy implications.

Irrigation systems can be broadly divided into three general classes according to how water is allocated: (1) farmers take whatever amounts of water they are able or wish to take whenever there is water in the canal, (2) a predefined share of the quantity of available water is allocated to each farmer through a carefully designed intake device or weir, and (3) the authorization to take water is rotated among the farmers according to a predetermined formula. The models discussed in this chapter all relate to the third type of irrigation systems that are used extensively throughout the world. We modelled a rotation scheme in as general a manner as we could so as to apply to the broadest set of empirical cases.

At least two types of rotation systems, frequently used in practice, are included in the empirical reference of the models in this chapter.

1. The allocation of a fixed time slot during a fixed time period to each farmer.[25]

   An example of this type of rotation system would be that each farmer is authorized to open his farm headgate for X hours on Y days of the week and at no other time. Every farmer knows exactly when in a week he will be allowed to irrigate, but the amount of water available at that time may not be the optimal amount needed by his crops.

2. The allocation of a position in a list with authority to irrigate a set amount of land when his 'turn' comes.[26]

   An example would be a system that defines the order in which each farmer can irrigate and the amount of land to be irrigated, but does not restrict the farmer to a particular amount of time for irrigation. Every farmer is allowed to take the optimal amount of water needed by the crops planted in a defined area, but not to 'waste' water by applying more than this amount of water or to apply water to other land.

In addition to our focus on irrigation systems with rotation rules, we also assume the absence of a formal position of guard. It almost seems inconceivable that as scarce and valuable a commodity as irrigation water would be allocated on a regular basis without the assignment of special responsibilities to an agent to punish those who steal. Many relatively small, indigenous systems have existed for long periods of time without utilizing a specialized guard. Netting (1974) describes a long-established Swiss system using fixed time-slots where no irrigator knew the full rotation scheme but all knew and agreed perfectly on the order of rotation for the two or three time-slots just prior and after their own. In essence, what happened on the other days of the week were irrelevant to an irrigator; he or she became a turntaker or turnwaiter only during a particular window of time during the overall rotations scheme. The effective number of participants at any one point in time, therefore, was very small. In this system, water has been distributed regularly and peacefully for centuries without any formal guards.

Spooner (1974) provides a good description of the system used in Deh Salm, an Iranian village, where the water share system is based on a 12 1/2 day cycle that is further divided into 25 periods of sunrise to sunset (and the obverse) and further divided into another 25 shares for a total of 625 shares in the rotation cycle. As described by Spooner (1974: 51):

> The system has no overseer or referee. When a man considers that it is time for his share, he proceeds to the divisor (sar-i-ju) and either diverts the flow from one channel to the other, or if that is not necessary, follows the flow to the point where the change has to be made. He walks along the channel in front of the water, clearing it out to improve the flow at points where animals have trodden or the sides have collapsed, until he arrives at the set of segments (kal) that he wishes to irrigate.

Spooner reports that while "disputes are a regular part of daily life in Deh Salm, they were never (while I was there) occasioned by differences over the distribution of water" (ibid.). Further, Spooner reports that he "did not encounter any case of one individual infringing upon the rights of another by keeping the water flow beyond the time limit of his share" (ibid.: 44).[27] Like the Swiss system that Netting described, no one in Deh Salm knew the entire sequence of share ownership. Because blocks of families own land adjacent to one another along a channel, everyone within a block knows and agrees upon the exact order within that block.

Although it is easy to document the existence of irrigation systems without guards, almost all of these are farmer-owned and organized (see Section 2.2) rather than government-owned and organized systems (see Section 2.1). The exact proportion of farmer- or government-organized irrigation systems without guards is not known. However, data in the CPR data base about the types of guards utilized on 44 irrigation systems located in 13 countries provides an initial approximation.[28] Of the 15 government-owned irrigation systems included in the CPR data base, only 3 (20%) do not have formal guards. Of the 29 farmer-owned systems, 12 (41%) do not employ formal guards.

Romana P. de los Reyes (1980) has also conducted a survey of 51 'communal' irrigation systems in the Philippines that contains information about formal guards. Of these 51 self-organized systems, 33 (65%) do not employ guards, while 18 (35%) have created a formal position as a guard (de los Reyes, 1980: 24). Whether a formal position of guard has been established is related to the size of the communal irrigation system as shown in Table 5.

	Size of Communal Irrigation System		
	Small (less than 50 hectares)	Medium (50 to 100 hectares)	Large (more than 100 hectares)
Number of Systems	30	11	10
Number of Systems without Guards	24	6	3
Percent Without Guards	80	55	30

<u>TABLE 5:</u> Correspondence between the size of an irrigation system and the existence of a formal position of a guard (after de los Reyes, 1980: 32, Table XX)

From these two sources, it would appear that a substantial proportion of smaller, farmer–owned systems operate without formal guards but that few government–owned systems operate without guards. Because all of the models in this chapter analyze irrigation games without guards, the models are most likely to apply to farmer–owned systems than to government–owned systems and to smaller, rather than larger, systems.

It is important to note that on 5 out of the 11 (45%) farmer–managed systems without guards for which data are available in the CPR data base, the farmers monitor each other sufficiently that the irrigators follow their own rules in a regular manner. This, of course, means that on 6 of these 11 systems, farmers were not following rules routinely. Thus, it is empirically possible for irrigators on small, self–organized systems to develop norms of behavior that involve low rates of stealing. Only half of the farmer–owned systems are successful in this regard, however. Thus, our analytical findings concerning how changes in the number of irrigators, costs and efficiencies of monitoring and other parameters increasing stealing rates provide an initial theoretical explanation for the differences in rule conformance among farmer–organized systems.

It is also interesting to note that on the three government–owned systems without guards, farmers did not routinely follow rules and stealing was endemic. Most government–owned systems tend to be larger than farmer–owned systems. From our analysis, we would expect that the monitoring rate of turnwaiters would be lower and the stealing rate of turntakers would be higher on the larger, government–owned systems.

A significant part of this chapter is focused on irrigation systems without guards where all participants have identical payoff paramters, norms of behavior, starting conditions, and knowledge about the system and thus the same detection probabilities. Findings from models that assume symmetry are most applicable to irrigation systems that serve farmers irrigating similar sized parcels where the difference in outcomes for those located close to the head of the system as contrasted to those located near to the tail of the system are minimal. Systems that were initially constructed by a small set of farmers who have remained relatively stable over time come closer to meeting the assumption of identical knowledge and starting conditions. The assumption of identical norms of behavior will be more closely approximated on socially homogeneous systems than on systems that are characterized by ethnic, racial, religious, political, or economic heterogeneity.

5.2 Equilibrium Scenarios for Irrigation Games Without a Guard Position

The most interesting general finding from our analysis of systems without guard positions is that no equilibria exist where the rate of stealing is equal to zero. In all of the games analysed in Section 4, some stealing occurs at equilibrium. This strong finding – which holds true for the asymmetric games as well as for the symmetric ones – derives from two key assumptions of our models. On the one hand, monitoring is *only* profitable if there is a positive chance to prevent a stealing event. If no stealing occurs, the costs of monitoring outweigh its benefits. On the other hand, stealing is *always* profitable if no monitoring occurs. Since we assume that stealing can only be prevented after being detected by a monitoring turnwaiter, the benefits of stealing always give the turntaker a positive incentive to steal if no monitoring takes place. A strategy combination without stealing cannot be in equilibrium since a zero stealing rate induces the turnwaiters not to monitor, and a zero monitoring rate in turn gives the turntaker a positive incentive to steal.

The different equilibrium regimes for irrigation games without guards are most clearly illustrated by the two–person interaction between the turntaker and one turnwaiter. The best–reply structure diagram for the corresponding two–person game (Figure 6) shows that there are four best reply structures dependant upon the relationship between the detection probability, α, the relative benefit of stealing, $B_{rel} = B/(B+P_I)$, and the ratio $C_M/(R_I+L_I)$, which can be interpreted as a cost–benefit ratio for monitoring. In two of the four possible cases ((a) and (b)), the turntaker faces a dominant strategy to steal (where the relative benefit of stealing is higher than the detection probability). In two of the cases ((a) and (c)), the turnwaiter faces a dominant strategy not to monitor. Thus in cases (a), (b), and (c), the best response strategy of the turntaker is to steal.

In the two–person game, the only situation where the turntaker changes from the pure strategy of stealing at every opportunity to a mixed strategy occurs when the detection probability is larger than both, the turnwaiter's cost–benefit ratio for monitoring and the turntaker's relative benefit of stealing (Corollary 2). Even in this case, however, the change of strategy is from stealing at every opportunity to stealing at a rate above zero.

The equilibrium diagram in Figure 8 illustrates the three different equilibrium scenarios for a two–person irrigation game without guards. The equilibrium monitoring rate is zero if and only if the cost of monitoring, C_M, is larger than a threshold value, $R_{marg} = \alpha\cdot(R_I+L_I)$, which can be interpreted as the marginal benefit of monitoring. Of course, a zero monitoring rate implies a maximal rate of stealing. The rate of stealing is also maximal if the relative benefit of stealing is larger than the detection probability, or, equivalently, if the relative cost of stealing, P_{rel}, is smaller than $1-\alpha$, the minimal probability of becoming detected by a monitoring turnwaiter. If the cost of monitoring is smaller than the marginal benefit, the turnwaiter should monitor at maximal rate. Even if monitoring does not deter the turntaker from stealing, the turnwaiter should monitor nevertheless since that is his only means to reduce the efficacy of stealing.

The upper left–hand region of the equilibrium diagram corresponds to the only parameter constellations (small cost of monitoring, high cost of detected stealing) which do not lead to a Nash equilibrium in pure strategies. In fact, both players have to play a genuinely mixed strategy at equilibrium. Formulae (12) and (13) show that the monitoring tendency at equilibrium is directly proportional to the relative benefit of stealing water whereas the stealing rate corresponds to the ratio between the cost and the marginal benefit of monitoring. Thus the equilibrium strategy of each player is a function of the payoff parameters of *the other player*, a fact that will be discussed in more detail in Section 5.5.

The equilibrium diagram for the interaction of a turntaker with N symmetric turnwaiters (Figure 10) closely resembles that of the two–person game. Again, stealing is never totally eliminated, and it is only reduced to intermediate levels if the cost of monitoring is low when compared to its benefits, and if the cost of detected stealing is high when compared to the detection efficiency of the turnwaiter community (Region IV in the equilibrium diagram). In Region IV, the stealing rate and the rate of monitoring are larger than zero, but they are both at their lowest level in the northwest (the upper left–hand) corner of the region. Monitoring rates are at their lowest levels, in fact, along the entire northern edge of Region IV and increase with movements in a southern direction which correspond to a reduction of the difference between the cost of detected stealing and the monitoring efficiency of the turnwaiters. Stealing rates increase from the northwest corner in both an eastern and a southern direction.

As in the two–person case, there are two regions in parameter space (I and III) which induce a Nash equilibrium in pure strategies. In Region I, the cost of monitoring exceeds the maximal marginal benefit of monitoring. Monitoring is therefore a dominated strategy, and, not being controlled by the turnwaiters, the turntaker steals at a maximal rate. In Region III, the relative cost of detected stealing is smaller than the minimal detection efficiency of the turnwaiter community. This makes stealing a dominating strategy. Since the cost of monitoring is very low, the turnwaiters should monitor nevertheless in order to reduce the efficacy of stealing.

The main difference between the two– and the (N+1)–person case is related to the fact that the marginal benefit of monitoring of a turnwaiter is no longer constant if he has colleagues who might also detect a stealing event. If the cost of monitoring is between the minimal and the maximal marginal benefit of monitoring, the turnwaiters should always monitor at an intermediate rate. It depends on the relationship between the cost of monitoring and the cost of detected stealing whether the stealing rate at equilibrium is intermediate (Region IV) or maximal (Region II). Note that in the latter case the turntaker plays a pure strategy whereas the turnwaiters behave according to a mixed strategy.

5.3 The Dependence of Equilibrium Behavior on Model Parameters

Table 2 summarizes the comparative statics for an irrigation system without guards where the N turnwaiters are in a symmetric position. It shows how the unique symmetric Nash equilibrium of such a game is influenced by the parameters of the game: (1) the relative costs for the turntaker if it is discovered that he tries to steal water, (2) the costs of monitoring, (3) the loss suffered by a turnwaiter from a stealing event, (4) the rewards to a turnwaiter from a discovery of a stealing event, (5) the payoff that 'other' turnwaiters receive when someone else detects a stealing event, (6) the probability that a stealing event will be detected by a monitoring turnwaiter, and (7) the size of the system, i.e., the number of farmers whose behavior affects one another. Using these comparative statics, we can address such questions as:

- What happens to the rates of stealing and monitoring if farmers discover a new mode of monitoring that increases the probability of detecting a stealing event?
- What happens to the rates of stealing and monitoring as the number of farmers whose behavior is interdependent increases?

The first two columns of Table 2 provide information on the effect of a parameter change on the stealing and monitoring rates within that parameter region of the irrigation game which induces an equilibrium with an intermediate stealing rate (Region IV in Figure 10). The third column of Table 2 shows how the monitoring rate is affected within Region II, where a mixed monitoring tendency of the turnwaiters is associated with a maximal stealing rate of the turntaker. The last four columns report on how the parameters of an irrigation game affect the shape and extent of Regions I, II, III, and IV.

In this section, we are mainly concerned with the question how the stealing rate might be reduced by a change in the parameters of the system. Looking down the first column of Table 2, all of the variables affect the rate of stealing within Region IV. An increase in any of four variables will lead to a decrease in the stealing rate: the costs of detected stealing, the loss due to stolen water, the reward for discovering stealing, and the detection probability. Stealing will be less at equilibrium if three additional variables are decreased rather than increased. These are: the costs of monitoring, the reward if someone else detects stealing, and the size of the system. We will discuss what each of these relationships mean in natural settings.

The Relative Costs of Punishment for Stealing

In contrast to the two–person case, the stealing tendency of the turntaker is not only affected by the payoff parameters of the turnwaiters but also by the payoff parameters of the turntaker himself. Contrary to some claims in the literature (see Section 5.5), an increase in the costs of detected stealing may efficiently decrease the stealing rate. This holds true despite of the fact that the turnwaiters have a lower tendency to monitor the turntaker's behavior. Notice that the cost of detected stealing, P_{rel}, is increased when the amount of punishment imposed on a stealing event is increased or when the benefit to be obtained from stealing is lowered.

In field settings, the variety of punishments used by self–organized irrigation systems is immense. One class of formal sanctions reallocates water from the offender to those offended. Other types of formal sanctions include monetary or commodity fines or additional work responsibilities. If irrigators do not willingly pay their fine, the other irrigators on many systems are authorized to send a large crew to harvest part of an offender's crop or to withhold water from a non–conforming irrigator.[29] Aspects of the punishments used on farmer–organized systems are thus subject to self–conscious design and could be changed by the irrigators themselves in an effort to change stealing behavior.

Other aspects of the punishments used on self–organized systems may happen more as a result of 'righteous self–indignation' than as a result of prior design. Even on very small, self–organized systems, finding the right combination of formal sanctions to keep stealing levels well below maximum is not an easy task. Formal sanctions are frequently backed up on these systems by the willingness of the irrigators to impose informal punishments in addition to the formal sanctioning mechanisms.

The Sunduwari system in Nepal, for example, is a very old system comprised of 24 households who all own about the same amount of land. The irrigators have thus had plenty of time to work out formal punishments in light of extensive information about how their system works and the propensities of participants to steal. Despite the age and small size of the system, the "major source of conflict is water stealing. Water must be guarded for the duration of one's turn. The fine for stealing water is to give the offended a longer turn and skip the offender's turn. There were also reports of beatings when water was stolen" (Nepal Irrigation Research Project, 1983: 4). Obviously, the reallocation of water from the offender to the offended was not perceived by participants as a sufficient punishment in all cases. Otherwise, one would not hear of reports of beatings.

Not all of the informal sanctions that back up the more formally imposed punishments involve physical violence. Modest social sanctions may quickly communicate to an offender that others are displeased with his behavior. The economic losses that someone accrues who becomes known as non–trustworthy in a small, closed society may also be substantial.

Detection Probability and the Costs of Monitoring

Qualitatively, an increase in the probability of detecting a stealing event has the same consequences as an increase in the relative costs of punishment: all other things being equal, an increase in α leads to a decrease in both the rate of stealing and the rate of monitoring. The level of stealing is also reduced by a reduction of the costs of monitoring. Surprisingly, a change in the costs of monitoring has no effect on the monitoring tendencies of the turnwaiters. In fact, the costs of monitoring affect the level of stealing due to a 'secondary effect' (see Section 5.5). The initial effect of a reduction in the costs of monitoring is to increase the level of monitoring. The increased level of monitoring leads to a decrease in the level of stealing which, in turn, leads to a return to the initial rate of monitoring. Thus, a lowering of the cost of monitoring affects the stealing rate at equilibrium but not the monitoring rate.

In field settings, many factors affect both the costs of monitoring and the efficacy of monitoring (the detection probability). While, theoretically, one can carefully separate these two parameters, variables that are likely to affect one of these parameters in a field setting are also likely to affect the other. Consequently, we will discuss both of these parameters together.

Michael Hechter is one of the few analysts of self–organized systems to have pondered very much about the costs or efficacy of monitoring. In particular, he points to several strategies that groups can adopt to increase the efficacy of monitoring (Hechter, 1987: 150–157). Among the variables Hechter identifies as increasing the efficacy of monitoring or reducing the costs are (1) increasing visibility through architecture and the creation of public rituals and (2) minimizing errors of interpretation by establishing clear–cut rules and recruiting participants who share similar understanding of the world.

The design of the physical layout of an irrigation system and of the devices and rules used by farmers in distributing water can, in some systems, have a major impact on monitoring costs and detection probabilities. Systems that are constructed so that the actions of the farmer taking water are visible at low cost to others who are waiting should have high detection probabilities. Similarly, rules requiring a sequential rotation system along any one canal greatly reduce the ambiguities of who is supposed to be taking water and who is next in line. Further, such rules bring those who are most directly affected by stealing activities to a similar physical location at about the same time.[30]

Self–organized irrigation systems are frequently constructed so that they are divided into many discrete physical units within a larger system. At times they are "arranged so that each unit is served directly from the main canal or a lateral and is not dependent on a water supply that passes over the territory of another mini–unit" (Coward, 1980: 207). This type of physical design has two consequences. First, the number of farmers whose stealing and monitoring actions directly affect one another is kept quite small even when the number of farmers served by the entire system is quite large. Thus, by physically subdividing the system into relatively separable units, the actual number of farmers whose actions directly affect one another is kept quite small.[31] Second, the efficacy with which each farmer can monitor other farmers is also relatively high. These systems also tend to organize themselves around these mini-units, and when formally organized, to employ a much higher level of personnel responsible for distributing and monitoring activities than government-organized systems of about the same overall size.

Hechter stresses the importance of homogeneous participants in minimizing the errors of interpretation as to what constitutes a legal strategy. Detection probabilities are, of course, lowered if an observed action by a turntaker is not clearly considered stealing (or not stealing) by all participants. The assumption of symmetric players leads to maximal homogeneity. The importance of this assumption has already been demonstrated in our analyses of three types of asymmetry.

The Marginal Value of Water and the Loss Suffered from Stealing

Another parameter to have a negative impact on the rate of stealing is the loss imposed on the turnwaiter, L_I, which corresponds to the value of water stolen by the turntaker. As all the other payoff parameters of the turnwaiters, L_I has no impact on the monitoring tendency at equilibrium, and the stealing rate is only affected due to secondary effects. If the water supply to an irrigation system were to be decreased somewhat making the marginal value of water higher and thus increasing the loss of water from a stealing event, the first response would be an increase in the monitoring rate, then a decrease in the stealing rate, and then a readjustment of the monitoring rate back to that of the original equilibrium.

From this finding, one would expect that stealing levels would be less on systems where the marginal value of water was higher, holding other variables constant. It should, however, be noted that in most natural settings L_I, the value of stolen water for a turnwaiter, will be positively correlated with B, the benefit of stolen water for the turntaker. Correspondingly, L_I will be negatively correlated with P_{rel}, the relative cost of punishment. If the value of water becomes higher for the turntaker as well as for the turnwaiters, the increase in L_I is associated with a decrease in P_{rel}. Accordingly, the decrease in stealing tendency mediated by the increase in L_I might be compensated or even overcompensated by the increase in stealing tendency which is mediated by a decrease in P_{rel}.

In contrast to parameters like the punishment after detected stealing and the cost and efficacy of monitoring, L_I and B cannot easily be used as a policy instrument in order to decrease the stealing rate. Rather, they are variables that describe both the level of physical scarcity of water and the importance of water for the crop spectrum planted. Accordingly, they might be considered as a constraint imposed by the physical structure of the irrigation system. It would be interesting to compare different systems in order to see whether – other things being similar – there is a correlation between stealing levels and the marginal value of water.

The Rewards to Turnwaiters

The reward that a turnwaiter receives upon discovery of a stealing event can take many forms. It reflects the psychological satisfaction (or dispair) that one farmer receives upon discovering that his 'neighbor' is stealing water. It may also reflect an increase (or decrease) in the social esteem that one farmer receives from the other farmers for successfully discovering and stopping a stealing event. On a self-organized system without guards, this payoff is not subject to much self-concious change as it is derived more from psychological or cultural origins than from a decision by a group to create a reward. The positive payoff for detected stealing will probably be higher on those systems where all farmers are strongly pre-committed to following the rules of their system and are also cognizant of the destabilizing nature that allowing rule infractions to be unpunished represents.

An increase in the reward to a turnwaiter for discovery of stealing affects rates of stealing through its secondary effect since the initial effect is to increase the level of monitoring. The increased level of monitoring leads to a decrease in the level of stealing which, in turn, leads to a decrease back to the initial rate of monitoring. Thus, a change in the reward structure for turnwaiters affects the stealing rate at equilibrium but not the monitoring rate. Interaction effects of this type are discussed more thoroughly in Section 5.5 below.

The payoff that 'other' turnwaiters receive when someone else detects a stealing event, R_0, has an intriguing relationship to the level of stealing. An increase in the payoff that the other turnwaiters obtain when one of them discovers stealing leads to an increase in the rate of stealing. One might think of this relationship in the following way. R_0 is somewhat like the satisfaction that someone receives when that person gains a benefit by the dint of someone else's work. To the extent that this satisfaction rises, the turnwaiters would be less likely to monitor themselves. As monitoring rates decrease, stealing rates increase, which in turn leads the turnwaiters to increase their monitoring rate to its prior level. Now, however, the rate of stealing is higher than it was. So, if the turnwaiters obtain too much satisfaction from each other's success, they will monitor less than otherwise and suffer a higher level of stealing. If this payoff were to fall, hold other parameters constant, the series of impacts would have the opposite direction.

The Number of Irrigators in the Rotation System

One of the strong findings from our analysis is that an increase in the number of irrigators is associated with an increase in stealing and a decrease in monitoring, holding other parameters constant. Obviously, in field settings, thoughtful individuals trying to offset the adverse consequences of increases in size can adjust the other parameters. Farmers in many self-organized systems can attempt to offset increases in the temptation to steal by increasing their rate of monitoring on larger systems. A survey of over 600 farmers served by the 51 communal irrigation systems in the Philippines described above provides strong evidence that farmers patrol

Percent of Farmers that Patrol Canals	Size of Communal Irrigation System: Small	Medium	Large
Yes	61	55	80
No	39	45	20
(N)	(288)	(110)	(219)

TABLE 6: Correspondence between the monitoring rate of the farmers and the size of an irrigation system (after de los Reyes, 1980: Table A26).

their own canals even when some of them are also patrolled by guards accountable to the farmers for distributing water. Further, the proportion of farmers who report patrolling the canals serving their farms increases on the larger self–organized systems as shown in Table 6.

This empirical evidence, that monitoring rates are positively associated with the size of a communal irrigation system, helps point out the danger of presuming that all other variables are held constant in nature settings. Consequently, all of the parameters we have discussed above – costs of punishment, detection probabilities and monitoring costs, losses, and rewards – can be viewed as potential 'levers' for policy intervention by those who wish to change the rates of monitoring and thus of stealing in any particular setting. To offset directly the adverse effects of larger size on rates of stealing, irrigators need to increase the relative costs of punishment, the rewards that a turnwaiter receives for detection, and/or the effectiveness of monitoring activities. The evidence from de los Reyes' study of communal organizations in the Philippines appears to confirm the feasibility of changing these other parameters to lead more irrigators to patrol their own canals.

5.4 Elders and Newcomers: On the Significance of Symmetry

The discussion in Sections 5.2 and 5.3 centers around the equilibrium behavor in an irrigation game without guards where the N turnwaiters are in a symmetric position. In Section 4.6, the symmetry assumption is relaxed by introducing the possibility of some kinds of asymmetry among the irrigators. In particular, we examine what happens if one of the turnwaiters differs in some manner from the other turnwaiters.

We have attempted to model this in a very general way. A turnwaiter might be 'distinguished' from his colleagues because he differs in regard to his skills (monitoring is less expensive or the rate of detection is higher), his assets (the costs associated with undetected stealing imposed on this player have less adverse consequences than when imposed on others), his knowledge of the system (including his knowledge of the norms of behavior of the other irrigators), and in his norm of behavior. The distinguished turnwaiter may simply be a neighbor of the turntaker who has better detection capabilities, or it may be that turnwaiter whose turn is next to come and who is thereby particularly affected by a stealing event. Alternatively, the individual may be a 'newcomer' who is distinguished by having different norms of behavior or knowledge levels. Another interpretation is that the distinguished player has been given the formal position of authority – like that of a guard – and thus has different payoffs, detection probabilities, and norms of behavior.

In our analysis we concentrate on those parameter constellations which in the symmetric case lead to a moderate amount of stealing. We consider three different scenarios. In the first scenario, the distinguished turnwaiter is 'inherently' in a better position for monitoring than the other turnwaiters. By this we mean that the distinguished turnwaiter has better detection abilities and/or inherently a higher motivation for monitoring (see (59)). In the second scenario, the distinguished turnwaiter is inherently in a worse position for monitoring, because he has a smaller detection probability and/or his incentives for monitoring are smaller (see (61)). The third scenario, finally, assumes that there are no 'physical' differences between the turnwaiters at all (see (67)) and that the distinguished turnwaiter only differs from his colleagues in regard to his norm of behavior.

A first consequence of asymmetry is that there is no longer a unique Nash equilibrium. Some special cases left aside, there are three coexisting Nash equilibrium points for a parameter constellation that corresponds to one of the first two scenarios. Even five different equilibria coexist in the last scenario, which is inherently symmetric (see Table 4). This observation exemplifies how strongly the uniqueness result in Proposition 7 depends on the assumption of identical norms of behavior.

We have the impression that the positive interpretation of the Nash equilibrium concept (see Section 3.4) loses some of its appeal when applied to a situation where several Nash equilibria coexist. On the one hand, our current knowledge of learning and cultural adaptation processes is much too scarce to derive criteria which might distinguish between coexisting equilibria (see Section 5.5 for some ideas in this direction). Therefore, there is no way to single out one equilibrium which – under the specific circumstances – appears to be the most plausible one. On the other hand, permanent non–equilibrium behavior might result in a situation where each participant continuously has to change his strategy since the behavior of the other players is governed by equilibrium predictions which are different from his own.

One of the coexisting Nash equilibrium points of an irrigation game with a distinguished turnwaiter is an equilibrium in genuinely mixed strategies. This equilibrium corresponds in a natural way to the unique (symmetric) Nash equilibrium of the associated symmetric game. In a non–symmetric context, however, the mixed–strategy equilibrium is rather 'paradoxical' in prescribing a counterintuitive behavior to the turnwaiters. In fact, just those turnwaiters monitor less who are inherently in a better position to monitor.

The other equilibrium points described in Table 4 are non–paradoxical in that a turnwaiter's monitoring tendency is positively associated with his capacity for monitoring. All non–paradoxical equilibria lead to specialization of one or more of the turnwaiters: either those turnwaiters, who are in a better position for monitoring, monitor at a maximal rate, or those turnwaiters, who are in a worse position for monitoring, do not monitor at all. Even if they are non–paradoxical, these equilibria, too, have a counterintuitive aspect. In fact, the stealing tendency of the turntaker is *higher* (when compared to the symmetric case) if there are turnwaiters who specialize on monitoring, and it is *lower* if some of the turnwaiters do not monitor at all.

We will now try to explain these counterintuitive results. Our explanation is based on four basic observations: (1) Since the stealing rate is always positive at equilibrium, all that the turnwaiters can hope for is to keep it at an intermediate level. This implies that the overall monitoring activity of the turnwaiter community will not increase above a level at which the monitoring deficiency is equal to the relative cost of undetected stealing (see (56)). Notice that this restriction corresponds to a negative feedback between the monitoring rates of the turnwaiters in a better position for monitoring and those who are in a worse position for monitoring. (2) Monitoring is only favored if the cost–benefit ratio of monitoring is smaller than the stealing rate (see (29)). At equilibrium, the turntaker will adjust his stealing rate in a way that the cost–benefit ratio of monitoring of those turnwaiters is matched, who monitor at an intermediate rate. (3) Those turnwaiters, who are in a better position for monitoring, have 'inherently' a lower cost–benefit ratio of monitoring. Other things being equal, the stealing rate will go down if the turntaker has to adjust his stealing rate to the cost–benefit ratio of those turnwaiters, who are in a better position for monitoring, and the stealing rate will go up if it is adjusted to the cost–benefit ratio of the turnwaiters, who are in a worse position. (4) At equilibrium, the cost–benefit ratio of monitoring of a turnwaiter is negatively correlated with his rate of monitoring. Any increase in the rate of monitoring of a turnwaiter induces the other turnwaiters to monitor less. As a consequence, the marginal benefit of monitoring gets larger (see (28)) and the cost–benefit ratio of monitoring is reduced.

Let us now consider a Nash equilibrium in genuinely mixed strategies. Such an equilibrium can only be achieved if all turnwaiters adjust their behavior in such a way that their cost–benefit ratio of monitoring becomes equal to the stealing rate. Accordingly, those turnwaiters, who are in a better position for monitoring, have to increase their cost–benefit ratio (which is inherently lower), whereas the turnwaiters, whose position is worse, have to decrease their inherently higher cost– benefit ratio of monitoring. In view of the negative correlation between the monitoring rate of a turnwaiter and his cost–benefit ratio of monitoring, this means that just those turnwaiters have to monitor more, whose cost–benefit ratio is inherently higher, whereas those, who are in a better positiong for monitoring, have to monitor less.

Notice that the argument given is quite general, and that it applies to more complex situations of asymmetry (e.g., to systems with several distinguished turnwaiters): *Every* Nash equilibrium point in genuinely mixed strategies has the property that a turnwaiter who is inherently in a better position to monitor has a lower tendency to monitor than a colleague who is inherently in a worse position for monitoring. In the next section we shall argue, however, that it is rather unlikely that a learning process leads to such a 'paradoxical' equilibrium.

The results above imply that *all* non–paradoxical equilibria have the property that one class of turnwaiters sticks to a pure strategy. Accordingly, the turntaker will adjust his stealing rate to the cost–benefit ratio of the other class of turnwaiters, i.e., to those turnwaiters who monitor at an intermediate rate. Let us first consider the case that those turnwaiters, who are in a better position for monitoring, monitor at a maximal rate. This implies that the turntaker will adjust his behavior to the cost–benefit ratios of those turnwaiters who are in a worse position for monitoring. Being in a worse position, the cost–benefit ratio of these players is relatively high. Therefore, it is plausible (and can be shown analytically) that the stealing rate is also relatively high. On the contrary, the stealing rate will be low, if it is adjusted to the low cost–benefit ratio of those turnwaiters, who are in a better position for monitoring. This will occur, if the turnwaiters, who are in a worse position, do not monitor at all, whereas the turnwaiters in a better position monitor at a high, but intermediate rate.

Again, these arguments are quite general, and they can to a certain extent be extended to irrigation games in which more than one turnwaiter is in an asymmetric position with respect to the others. Quite generally, one should expect the stealing rate to *increase* in all those cases where the agents, who are in the best position for monitoring, specialize on this task, whereas the other agents monitor at a small, but positive rate.

In particular, this result holds true for a symmetric situation where the turnwaiters only differ in their norm of behavior (corresponding to the third scenario). This is of particular relevance for all those cases where a turnwaiter is distinguished from his colleagues by being in a formal position of a guard who is supposed to monitor at a maximal rate. Thus, some of our results concerning irrigation games without guards provide the beginning of a bridge to a formal analysis of irrigation games with guards. In this chapter, we will not pursue this topic as it is the focus of a future effort.

5.5 Primary and Secondary Interaction Effects

Several of the results derived in this chapter are counterintuitive at first sight. Consider, for example, the completely mixed Nash equilibrium point of a symmetric irrigation game without guards. In the two–person case (Proposition 1), the equilibrium strategy of a player only depends on the detection probability and the payoff parameters of the *other* player. As a consequence, a change in the payoff parameters of the turntaker has no effect on the turntaker's behavior, whereas it strongly affects the equilibrium behavior of the turnwaiter. If N symmetric turnwaiters interact with the turntaker, their equilibrium tendency of monitoring does also not depend on their own payoff parameters but rather on structural parameters and the relative cost of stealing for the turntaker (see (48) and Table 2). Similarly, in an irrigation game with a distinguished turnwaiter, all those equilibria have counterintuitive aspects which lead to an intermediate stealing rate of the turntaker.

Results like these arise quite naturally in a game–theoretic context. In a sense, they reflect the interactive nature of the situation and the interdependence of the players. Consider, for example, a two–person irrigation game without guards which is characterized by a parameter constellation leading to a Nash equilibrium in mixed strategies. Suppose that a change in the parameters of the system leads to a slight increase in B_{rel}, the relative benefit of stealing water. All other things being equal, such a parameter change would make it more profitable for the turntaker to steal water. This 'primary effect' of an increase in B_{rel}, however, is not decisive, since all other things will not remain equal. In fact, even a slight increase in the stealing rate increases the turnwaiter's incentive to go out monitoring (see (11)). Accordingly, the increase in B_{rel} has the 'secondary effect' of increasing the monitoring rate. The game–theoretic analysis reveals that – as long as the parameter regime of the mixed–strategy equilibrium is not left – the secondary effect is large enough to overcome the primary effect: The increase in the monitoring rate will continue until stealing is no longer profitable for the turntaker. Finally, the stealing rate will be reduced to its previous level, whereas the monitoring rate remains high enough to ensure that the stealing rate will not increase.

It is difficult to give a positive interpretation for the fact that – in the mixed–strategy context – the primary effects of a player's payoff parameters on his own behavior are often dominated by the secondary effects of other players' behavior on his strategy choice. Notice, however, that in the above example the net increase in the monitoring rate is not costly for the turnwaiter. In fact, he remains *indifferent* between monitoring and not monitoring since *his* payoff parameters have not changed (see (13)). This is perhaps the main reason for the counterintuitive effect that an increase in the relative benefit of stealing does not result in an increased stealing rate. Since he is indifferent between monitoring and not monitoring, the turnwaiter has no difficulty in increasing his monitoring rate to an extent which just balances the parameter change in favor of the turntaker.

The counterintuitive aspects of mixed–strategy equilibria has repeatedly been noticed in the literature (see, e.g., Wittman, 1985; Holler, 1990). The policy importance of secondary effects has been stressed by George Tsebelis in a series of recent papers (1989, 1990a, 1990b). Tsebelis argues that – in games that have one, mixed–strategy equilibrium – policy analysts are frequently misled by an analysis based only on primary effects. "We expect a change in the payoffs of one player to influence that player's behavior, when in reality such a modification gets completely absorbed by a change in the strategy of the opponent" (Tsebelis, 1989: 89).

As we have demonstrated above, Tsebelis' assertion is supported by many of the results derived in this chapter. But Tsebelis' analysis is not quite as general as he wishes to assert. An inspection of the (paradoxical) mixed–strategy equilibrium of an asymmetric irrigation game with a distinguished turnwaiter reveals that the equilibrium strategies of the turntaker, the distinguished turnwaiter, and the 'symmetric' turnwaiters all depend on their own payoffs as well on the payoffs of all other types of players. In this example, the primary effects are not dominated by the secondary effects. Both types of effects together shape the equilibrium behavior of the players. This example shows that Tsebelis' argument loses much of its impact outside the class of two–person games, a fact that has also been stressed by Peter Ordeshoek: "Only for two–player games are we guaranteed that the algebra of solutions reduces to the simple form upon which Tsebelis bases much of his analysis" (Bianco, Ordeshoek and Tsebelis, 1990).

Interestingly, Tsebelis' assertion has to be relativized already in the context of *symmetric* irrigation games without guards. Table 2 shows that the relative cost of stealing water – which is entirely composed of payoff parameters of the turntaker – has both an immediate, primary impact on the stealing tendency of the turntaker and a secondary impact on the monitoring tendency of the turnwaiters. Thus, the splitting of one unified agent into two or more *identical* players is sufficient to destroy Tsebelis' main result that the (mixed–strategy) equilibrium behavior of a player *only* depends on the payoff parameters of his opponents.[32] In particular, our analysis shows how inadequate it may be to model the interaction between criminals and the police as a two–person game, where 'the criminals' and 'the police' are both considered as unified agents.

On the other hand, we would like to stress that – in the symmetric context – part of Tsebelis' argument survives the three– and more–person setting. In fact, Table 2 shows that the (mixed) equilibrium strategy of the turnwaiters are only affected by structural parameters and the payoff parameters of the turntaker, not by the payoff parameters of the turnwaiters themselves. This result is not an artifact, and it holds true for all mixed–strategy equilibria in games between a unified agent and N symmetric opponents, where 'symmetric' means that all opponents are characterized by the same structural and payoff parameters and that they all stick to the same equilibrium strategy.[33]

In order to see this, notice first that the unified player will only play a mixed strategy at equilibrium if all the pure strategies used give him the same payoff against the strategy combination of his opponents (see, for example, Harsanyi and Selten, 1988: Lemma 2.3.1). Now, the expected payoff for a pure strategy is a function of the unified player's own payoff parameters and the strategy being used by all his N opponents. As a consequence, the payoff equality for the unified player's pure strategies corresponds to an equation involving two components: the payoff parameters of the unified player and the strategy used by the other players. Solving this equation yields the equilibrium strategy of the N symmetric players as a function of the payoff parameters of the unified player. This general, algebraic argument explains why the turnwaiter strategy at equilibrium only depends on the turntaker's payoff parameters.

Consider now the equilibrium condition for the N symmetric players. As before, each of these players will only play a mixed strategy if all the pure strategies used yield the same payoff. Now, however, a pure strategy's payoff does not only depend on own-payoff parameters and the strategy of the unified player, but also on the strategy used by the other N–1 symmetric players. Therefore, the equilibrium condition corresponds to an equality involving the own payoff parameters and both types of strategy. As we have shown above, the strategy of the other N–1 symmetric players is a function of the payoff parameters of the unified players. Accordingly, the equilibrium condition for a symmetric player yields an equality involving the strategy of the unified player and payoff parameters of both types of player. As a consequence, the equilibrium strategy of the unified player is usually a function of the payoff parameters of *both* types of player.

A similar argument shows that Tsebelis' assertion breaks down completely if there is no unified player at all, and if two groups of identical agents interact with another. In that case, the (mixed-strategy) equilibrium behavior of each opponent is usually a function of the payoff parameters of both types of agent.

We would like to add one further aspect to the discussion on the importance of secondary effects. Tsebelis applies his result to the class of games which have only one, mixed-strategy equilibrium. If the Nash equilibrium concepts provides an adequate description of the behavior of the players, secondary effects are really decisive in these games (at least in the two-person context). What happens, however, if a mixed-strategy equilibrium governed by secondary effects coexists with one or more alternative equilibria which are more easily explained by immediate, primary effects?

As an example, take the 'paradoxical' mixed-strategy equilibrium of an irrigation game with a distinguished turnwaiter and compare it with the coexisting equilibrium point where those turnwaiters, who are in a better position for monitoring, monitor at a maximal rate. It is obvious that the paradoxical equilibrium is mainly governed by secondary effects, whereas the equilibrium leading to specialization on monitoring is dominated by primary effects. What kind of equilibrium should we expect to find in a natural setting? One might argue that the mixed-strategy equilibrium – even if the behavior of the turnwaiters appears to be paradoxical – is the best candidate. In fact, this equilibrium corresponds to the unique symmetric Nash equilibrium point of a symmetric irrigation game without guards. If the asymmetric game results from the symmetric game by a slight disturbance (i.e., if the distinguished turnwaiter differs only slightly from his colleagues), its paradoxical equilibrium is a slightly disturbed version of the symmetric equilibrium of the symmetric game. If small perturbations in a game usually have small effects on the equilibrium behavior, the paradoxical equilibrium should result in the asymmetric game.

We are not convinced by this line of argument. In fact, the paradoxical equilibrium is stabilized by the negative feedback between the monitoring rate of the distinguished turnwaiter and the monitoring rates of his colleagues (see Sections 4.6 and 5.4). If those turnwaiters, who are in a better position for monitoring, *suppose* that their colleagues will monitor more than they themselves do, it is profitable for them to monitor less. Otherwise, they should monitor more, and, in fact, they should increase their monitoring rate to a maximum. According to the positive interpretation of the Nash equilibrium concept, such an equilibrium is viewed as the outcome of a dynamical process of learning and cultural evolution. Different, coexisting equilibria might correspond to different, coexisting attractors of the dynamical process. This implies that it depends on the starting conditions where a specific system ends up in the long run. We can well imagine that an irrigation system with a distinguished turnwaiter might end up at a paradoxical equilibrium, *provided* that there is an initial asymmetry leading the turnwaiters in a better position to monitor less and those in a worse position to monitor more. If however, the turnwaiters, who are in a better position to monitor, initially have a higher tendency to monitor than their colleagues, we expect the non-paradoxical equilibrium to result.

As long as we have only a rudimentary understanding of learning and cultural evolution in natural settings, we can, of course, only speculate about the likely outcome of such processes. Nevertheless, we have the impression that primary effects govern the *immediate* reactions to a change of a situation whereas secondary effects only gain importance in a longer perspective.[34] If the final outcome of a learning process strongly depends on the initial conditions, immediate effects should be of utmost importance. Due to positive feedback, these primary effects will almost invariably push the system on a track which finally leads to an equilibrium that is also dominated by effects of this kind.

Notes:

1. Although these are important questions for both policy analysts and game theorists, the task of communicating about these problems across disciplinary boundaries is difficult given the difference in how key terms are defined. For a game theorist, the 'rules of a game' are the underlying physical and behavioral 'laws' that affect the entire structure of a game tree, the payoffs assigned to end points, and the information available at nodes in the tree. For a policy analyst, the 'rules of the game' are the normative prescriptions stating which action must, must not, or may be taken at each node of the tree as well as the legal payoff and information conditions. These prescriptions are followed to a greater or lesser extent in different situations.

Strictly speaking, one cannot explore rule–breaking behavior using the game theorist's definition of the rules of a game. According to the definition used in game theory, no actions included in a formal game are against the rules of that game. When we later model the choice of a player 'to steal' or 'not to steal', for example, the rules of the formal game we have constructed 'allow' stealing. Otherwise, stealing would not be an alternative action available to a player. Technically, to a game theorist, we do not explore rule–breaking behavior. Instead, we explore a game where 'the rules' allow stealing to occur. We then analyse what combination of parameters affect the level of stealing at equilibrium. For the policy analyst, however, rules are a major instrument for changing the structure of the games played in natural settings. If a 'rule' were not being broken in our models, there would be nothing like a 'punishment' as a payoff imposed on a player who is discovered to be stealing, nor would the person who discovered stealing be rewarded in any way.

Since this paper is written by a game theorist and a policy analyst, we need to clarify for the reader which meaning of the term 'rules' is being used at different junctures of this paper. In Section 2, which addresses the policy questions, we use the policy analyst's concept of rules as normative prescriptions that may or may not be followed. Consequently, in Section 2, we refer to rule–breaking, rule–conformance, and rule–enforcement. When we do so, the reader should use the policy analyst's interpretation of this concept and not the game theorist's concept. In Sections 3 and 4, we construct a series of games where stealing and monitoring are embodied in a formal game and a player does not 'break the rules' when choosing to steal. In Section 5, we return to the policy analyst's concept of rules.

2. Harsanyi (1977: 278) concludes that "the really important policy problem" is transforming undesirable non–cooperative game situations into situations of mutual advantage to all parties. This can be achieved "only by establishing effective law–enforcing agencies and/or by inculcating on both (or on all) sides attitudes more favorable to spontaneous law observance; but it does not tell us how such agencies can be best established and how such attitudes can be best imparted."

3. Mehra (1981) reports that the variability of crop yields after the construction and operation of major irrigation systems in India increased rather than decreased. Levine (1980: 55) reports that Iranian irrigators using a traditional system with minimal facilities had been able to achieve water–use efficiencies of approximately 25 percent prior to the construction of the Dez Pilot Irrigation Project as "a comprehensive system, with a full range of controls, measuring structures, organizational structure, and all the other accouterments of a large modern system". After six years of operation, the average water–use efficiency in the project area had fallen to between 11 and 15 percent. Bromley (1982) reports similar reductions in productivity after major projects were completed and put in operation throughout Asia.

4. These general conclusions are documented in many sources. See IBRD, 1985; Ascher and Healey, 1990; Cernea, 1985; Uphoff, 1985.

5. For a description of some of the allocation rules used see Tang, 1989.

6. Where farmers are served by more than one irrigation system, the difference in their strategies is readily apparent. In some parts of South India, for example, farmers own land served by large–scale, government–managed irrigation systems as well as by small, farmer–managed systems. The same farmers follow elaborate rotation systems for allocating water and invest substantial amounts of their own labor input to maintain their own systems while refusing to invest any effort in following the rules for water or work allocation of the government–managed system (Wade, 1990).

7. The relationship between the unreliability of a water supply and decreases in agricultural yield is a key to understanding the importance of organization at the field channel level. Chris Panabokke, a Project Leader on Kalankuttiya block of the Mahaweli system discussed above, stresses the importance of reliability of water in the decisions made by farmers to diversify their cropping patterns which could potentially increase their yield substantially.

> A major factor constraining the farmers' ability to diversify their cropping pattern ... is the unreliability of supply at the farm level. By that I mean the ability to feel certain that water is coming at a certain time and in a certain amount regularly. There is the further problem of how to share the water below the turnout as the data shows great variability in the amount of water going to field channels. When water is not delivered to the turnout as scheduled, there tends to be a free–for–all when it arrives (quoted in Colmey, 1988: 4).

8. For a recent analysis stressing the importance of farmer organization see USAID, 1983: 90 and discussion throughout Cernea, 1985.

9. See the section entitled "Ignoring Local Irrigation Organizations" in Coward, 1985: 33–36.

10. David Groenfeld describes two such systems where there are 'farmer leaders' but no 'farmer organizations'.

> In Kalankuttiya, there is a farmer representative who is elected every three years; however, many farmers don't communicate with him, in Dewahuwa, a farmer representative is selected by farmers to coordinate the farmers within a turnout group. However, a turnout group can have as many as 50 farmers who may or may not be located in the turnout, may or may not be owners of the land they cultivate, and may or may not know each other on a personal level. Farmer representatives for each turnout meet periodically with irrigation officials, but it would be inaccurate to say that they represent a group consensus among turnout farmers (quoted in Colmey, 1988: 4).

11. The frequency with which farmers are blamed for irrigation project failures is the stimulus for the following satirical characterization of the six phases of irrigation project development:

> The first phase is the designers' high enthusiasm and publicized expectations. Second comes disillusionment, when the implementors discover that the designs are sorrow–fully inadequate. The third phase is one of panic, when the operational staff discovers that the system will not operate as designed. Fourth comes the search for the guilty, characterized by a round robin of blame among designers, implementors, operators, and extension workers. Naturally, the fifth phase consists of blaming the innocent – that is, the farmer who had nothing to do with designing, implementing, operating, or extending the system. Thus, reports sadly conclude that ignorant and stubborn farmers remain set on destroying structures, stealing water, and creating all kinds of other problems and in general will not cooperate with well–meaning project authorities. Phase six is the time for praise; if a system works at 40 to 50 percent of the design efficiency the praise and honor for the success go not to the planners, engineers, technicians, or the farmers, but the politicians (Freeman and Lowdermilk, 1985: 91–92).

12. See Cernea, 1985; Uphoff, 1985; Ascher and Healy, 1990.

13. This has become a rather critical problem in countries such as India, the Phillippines, and Nepal where government policy is to invest in already operating, local systems. All too often, efforts to improve these systems have failed. The May, 1988 issue of the newsletter of the International Irrigation Management Institute, FMIS, describes one such 'improved' system in Uttar Pradesh, India where farmers have had to revert to reliance to rain–fed agriculture after they were 'helped' by the Irrigation Department rebuilding their system.

14. Elster himself is not completely sure that the dilemma of mutual monitoring is always 'decisive'. He points to the possibility that tasks may be organized so that monitoring can be accomplished without additional effort.

15. Non-cooperative game theory has been used to model the individual decisions of irrigators or other groundwater producers independently using an underground aquifer as a source of water (Negri, 1989). Cooperative game theory has been used in a number of prior studies to analyse the problem of fair allocation of joint costs of multiple purpose water resource projects (see Straffin and Heaney, 1981; Heaney and Dickinson, 1982; Young, Okada, and Hashimoto, 1982), various solutions to problems associated with international river basins (Rogers, 1969), and regional water quality management and other oligopoly type situations (Bogardi and Szidarovsky, 1976).

16. The existence of offsetting primary and secondary effects in many types of social situations where rational 'opponents' interact with one another is the major subject of Tsebelis (1989). See Section 5.5 for further discussion.

17. The initial inspiration for our work was the in-depth descriptions of Glick (1970) and Maass and Anderson (1986) of the rules and strategies of farmers in Valencia and other huertas of Eastern Spain that have operated their own irrigation systems for many centuries.

18. Even in strategically simple, symmetric situations Nash equilibrium points are not necessarily dynamically stable with respect to an adaptive learning or evolution process. In such situations, the instability of a Nash equilibrium is often associated with more complicated dynamic behavior often involving some form of cycling. Intuitively, it is quite plausible that a game with a best-reply structure as the the irrigation games in Figure 6(d) has a tendency to produce cycling. However, even if the Nash equilibrium is not stable in these cases, it can often be given a dynamical meaning (e.g., as a long-term time average, see Hofbauer and Sigmund, 1988).

The possible instability of Nash equilibria has led biologists to introduce a second stability condition, the 'ESS condition', in addition to condition (3) which characterizes a Nash equilibrium point (see Maynard Smith, 1982; Selten, 1983; Weissing, 1991). In a symmetric game where the players have only two pure strategies available, the notions of an evolutionarily stable strategy coincides with that of a symmetric Nash strategy, provided that the game admits only one symmetric Nash equilibrium point. In such games, the Nash equilibrium (resp. ESS) is always dynamically stable with respect to the 'replicator dynamics', the most prominent selection dynamics (see Hofbauer and Sigmund, 1988; Weissing, 1991).

19. Normative game theory also concentrates on symmetric equilibrium points. It is a basic demand on every normative solution concept that it must reflect the symmetries of a game. In particular, a solution concept would be regarded as unfair, if symmetric players are not treated symmetrically (see Harsanyi and Selten, 1988). Symmetry is also crucial for evolutionary game theory. Every evolutionarily stable strategy (ESS) corresponds to a symmetric Nash equilibrium point, and asymmetric equilibria are not suitable for describing behavior in intra-specific conflicts (see Selten, 1983). In our models, we try to give a different justification of the symmetry assumptions. We presume that players come to similar conclusions if they have similar preferences, assets, and social background.

20. See Section 4.6 for the coexistence of several equilibrium points and Section 5.5 for a discussion of some factors which might lead to a distinction between coexisting equilibria. Equilibrium selection in a normative context is the main theme of van Damme (1983, 1987) and Harsanyi and Selten (1988).

21. In a future paper, we will attack the question of how the incentives of the players might change in response to the equilibrium behavior of the farmer community in the past.

22. Since a two-person irrigation game without guards either has an equilibrium in pure strategies or an equilibrium point in completely mixed strategies, the equilibrium is always dynamically stable with respect to Selten's anticipatory learning process (see Selten, 1991).

23. Here and in the rest of this section, we slightly abuse terminology. Strictly speaking, none of our games is really symmetric since the turntaker is always in a distinguished position. In the context of irrigation games without guards, a game is considered to be *symmetric* if all the turnwaiters are in a symmetric position, i.e. if all turnwaiters have the same payoff parameters and the same detection probability.

24. Again, this is an abuse of terminology. The symmetric Nash equilibrium point of a 'symmetric' irrigation game without guards (in the sense of Note 25) is *symmetric* since it respects all the symmetries which are inherent in the game (see e.g. Harsanyi and Selten, 1988). In this sense, however, the Nash equilibria described in Section 4.6 are also symmetric since they respect the symmetry between the turnwaiters 1,...,N–1. Nevertheless, we will sometimes call them 'asymmetric' in order to distinguish them from the symmetric equilibrium of the corresponding symmetric game.

25. This type of rotation system is used very frequently in practice. Examples of it exist in huertas of Murcia and Orihuela (Maass and Anderson, 1986). The warabandi system used extensively in India is a rotation system of this type (see Walker, 1983). The waqt system in Iraq used preset time periods from sunrise to sunset and sunset to sunrise as one rotation period or event (see Fernea, 1970).

26. An example of this type of system has existed in Valencia, Spain for centuries. It is described by Maass and Anderson (1986: 28):

> On days when water is running in a lateral ... those farmers who want to irrigate will take it in turn (pro turno), generally in order from the head to the tail of the channel. Once a farmer opens his headgate, he takes all the water he needs, without any restriction of time; and he defines his own needs, principally in terms of the water requirements of the crops he has chosen to plant. The only limitation is that he may not waste water. If a farmer fails to open his headgate when the water arrives there, he misses his turn and must wait for the water to return to the farm on the next rotation.

Each farmer in this system is authorized to irrigate only on that land designed by title as land that can be irrigated (regadiu).

27. Spooner spent three months in the village. Thus, his report should be interpreted as providing evidence that the probability of cheating was relatively low rather than evidence that cheating never occured.

28. These data are located at the Workshop in Political Theory and Policy Analysis at Indiana University. See Tang, 1989 for a description of the systems and the data base.

29. On the irrigation system in Nepal visited by E. Ostrom in 1989, the irrigators had constructed a fenced pen where they could place an animal – usually a cow – that was taken from the recalcitrant farmer until he paid the fine owed to the water users' group.

30. Many sequential rotation systems that are used in farmer–managed systems are criticized by irrigation specialists as being too rigid and potentially inefficient in their consequences. If a farmer has a higher valued use for available water, but the farmer is not next in line, it is very difficult to adjust these sequential water distribution systems to get the water to the farmer who will receive the highest value from the marginal delivery. The analysis here raises the important question that there may be other factors to take into account in evaluating the allocation rules of an irrigation system than the short–run efficiency of water use. If an allocation scheme is used that farmers cannot effectively monitor at relatively low costs, short–term efficiencies can rapidly be lost as monitoring rates fall and stealing rates rise.

31. Of course for a subdivision of a large system into relatively separable subunits to be effective, there have to be clear and unambiguous rules for allocating water among these subunits that are themselves monitored effectively. Most large, self-organized irrigation systems that have successfully survived for long periods of time are organized at three or four different tiers.

32. It should be stressed, however, that Tsebelis' argument remains to be true if these identical players are not affected by the behavior of the other players of their own type. In our models, a turnwaiter's payoff depends to a certain extent on the strategic decision of the other N–1 turnwaiters (see (18), (20), and (22)). The fundamental difference between the two-person and the (N+1)-person context would disappear if that were not the case: The turntaker's equilibrium behavior would only depend on the payoff parameters of the turnwaiters if each turnwaiter were not affected by the other turnwaiters' behavior. Similarly, the interaction between M criminals and N policemen would, in Tsebelis' example, be governed by secondary effects *provided* that the criminals are not affected by the behavior of the other criminals and the policemen are not affected by the behavior of the other policemen.

33. In the following arguments, we tacitly assume that all player have only two pure strategies at their disposal.

34. Experimental evidence on learning in interactive situations seems to be in line with this general prediction (Reinhard Selten, personal communication).

References:

Ascher, W. and R. Healy (1990): Natural Resource Policy Making in Developing Countries. Durham, North Carolina: Duke Univ. Press.

Asian Development Bank (1973): Regional Workshop on Irrigation Water Management. Manila, Phillippines: Asian Development Bank.

Bianco, W.T., P.T. Ordeshook, and G. Tsebelis (1990): Crime and Punishment: Are One-Shot, Two-Person Games Enough? American Political Science Review **84**: 569–586.

Bogardi, I. and F. Szidarovszky (1976): Application of Game Theory in Water Management. Applied Mathematical Modelling **1**: 16–20.

Bromley, D.W. (1982): Improving Irrigated Agriculture: Institutional Reform and the Small Farmer. Washington, D.C.: World Bank Staff Working Paper, No. 531.

Cernea, M.M (1985): Putting People First. Sociological Variables in Rural Development. New York: Oxford Univ. Press.

Chambers, R. (1980): Basic Concepts in the Organization of Irrigation. In: E.W. Coward Jr. (ed.): Irrigation and Agricultural Developement in Asia. Ithaca, N.Y.: Cornell Univ. Press, pp. 28–50.

Chambers, R. and I.D. Carruthers (1986): Rapid Appraisal to Improve Canal Irrigation Performance: Experience and Options. IIMI Research Paper No. 3. Digan Village, Sri Lanka: International Irrigation Managment Institute.

Colmey, J.L. (1988): Irrigated Non-Rice Crops: Asia's Untapped Resource. IIMI Review **2**: 3–7.

Corey, A.T. (1986): Control of Water Within Farm Turnouts in Sri Lanka. Proceedings of a Workshop on Water Managment in Sri Lanka. Documentation Series No. 10. Colombo: Agrarian Research and Training Institute.

Coward, E.W. Jr. (1980): Management Themes in Community Irrigation Systems. In: E.W. Coward Jr.(ed.): Irrigation and Agricultural Development in Asia. Ithaca, N.Y.: Cornell Univ. Press, pp. 15–27.

Coward, E.W. Jr. (1985): Technical and Social Change in Currently Irrigated Regions: Rules, Roles, and Rehabilitation. In: M.M. Cernea (ed.): Putting People First. Sociological Variables in Rural Development. New York: Oxford Univ. Press. pp. 27–51.

Crawford, V.P. (1985): Learning Behavior and Mixed Strategy Nash Equilibria. J. Economic Behav. Organis. 6: 69–78.

Cross, J.G. (1983): A Theory of Adaptive Economic Behavior. Cambridge, England: Cambridge Univ. Press.

de los Reyes, R.P. (1980): Managing Communal Gravity Systems. Farmers' Approaches and Implications for Program Planning. Quezon City, Phillippines: Institute of Phillippine Culture.

Elster, J. (1989): The Cement of Society. A Study of Social Order. Cambridge, England: Univ. of Cambridge Press.

Eshel, I. (1982): Evolutionarily Stable Strategies and Viability Selection in Mendelian Populations. Theoretical Population Biology 22: 204–217.

Eshel, I., and E. Akin (1983): Coevolutionary Instability of Mixed Nash Solutions. J. Mathem. Biology 18: 123–133.

Fernea, R.A. (1970): Shakh and Effendi: Changing Patterns of Authority Among the El Shabana of Southern Iraq. Cambridge, MA: Harvard Univ. Press.

Freeman, D.M., and M.L. Lowdermilk (1985): Middle–level Organizational Linkages in Irrigation Projects. In: M.M. Cernea (ed.): Putting People First. Sociological Variables in Rural Development. New York: Oxford Univ. Press. pp. 91–118.

Friedman, D. (1989): Evolutionary Economic Games. Working Paper.

Friedman, J.W., and R.W. Rosenthal (1986): A Positive Approach to Non–Cooperative Games. J. Econom. Behav. Organiz. 7: 235–251.

Glick, T.F. (1970): Irrigation and Society in Medieval Valencia. Cambridge, Ms.: The Belknap Press of Harvard Univ. Press.

Hansen, R.G., and W.F. Samuelson (1988): Evolution in Economic Games. J. Econom. Behav. Organiz. 10: 315–338.

Harriss, J.C. (1977): Problems of Water Management in Hambantota District. In: B.H. Farmer (ed.): Green Revolution? Technology and Change in Rice Growing Areas of Tamil Nadu and Sri Lanka. New York: MacMillan, pp. 364–376.

Harriss, J.C. (1984): Social Organisation and Irrigation: Ideology, Planning, and Practice in Sri Lanka's Settlement Schemes. In: T.P. Bayliss–Smith and S. Wanmali (eds.): Under–standing Green Revolutions. Cambridge, England: Cambridge Univ. Press, pp. 315–338.

Harsanyi, J.C. (1977): Rational Behavior and Bargaining Equilibrium in Games and Social Situations. New York: Cambridge University Press.

Harsanyi, J.C. and R. Selten (1988): A General Theory of Equilibrium Selection in Games. Cambridge, MA: The MIT Press.

Hart, H.C. (1978): Anarchy, Paternalism, or Collective Responsibility Under the Canals? Economic and Political Weekly 13 (52): A125–A134.

Heaney, J.P. and R.E. Dickinson (1982): Methods for Apportioning the Cost of a Water Resource Project. Water Resources Research 18: 476–482.

Hechter, M. (1987): Principles of Group Solidarity. Berkeley, CA: Univ. of California Press.

Hofbauer, J., and K. Sigmund (1988): The Theory of Evolution and Dynamical Systems. Cambridge: Cambridge Univ. Press.

Holler, M.J. (1990): The Unprofitability of Mixed–Strategy Equilibria in Two–Person Games. Economics Letters 32: 319–323.

International Bank for Reconstruction and Development (IBRD) (1985): Tenth Annual Review of Project Performance Audit Results. Washington, D.C.: World Bank, Operations Evaluation Department.

Levine, G. (1980): The Relationship of Design, Operation, and Management. In: E.W. Coward, Jr. (ed.): Irrigation and Agricultural Development in Asia. Ithaca: Cornell Univ. Press. pp. 51–64.

Luce, R.D., and H. Raiffa (1957): Games and Decisions: Introduction and Critical Survey. New York: Wiley.

Maass, A. and R.L. Anderson (1986): ... and the Desert Shall Rejoice. Conflict, Growth, and Justice in Arid Environments. Malabar, Fl.: Krieger Publ. Co.

Maynard Smith, J. (1982): Evolution and the Theory of Games. Cambridge, England: Cambridge Univ. Press.

Mehra, S. (1981): Instability in Indian Agriculture in the Context of the New Technology. Washington, D.C.: Intern. Food Policy Research Institution. Research Report No. 25.

Nachbar, J. (1990): 'Evolutionary' Selection Dynamics in Games: Convergence and Limit Properties. to appear in: Intern. J. Game Theory.

Nash, J.F. (1950): Equilibrium Points in n–Person Games. Proc. Nat. Acad. Sci. USA 36: 48–49.

Nash, J.F. (1951): Non–Cooperative Games. Annals of Mathematics 54: 286–295.

Negri, D.H. (1989): The Common Property Aquifer as a Differential Game. Water Resources Research 25: 9–15.

Nepal Irrigation Research Project (1983): Sunduwari Irrigation System (Badichaur, Surkhet). Kathmandu, Nepal: Nepal Irrigation Research Project, mimeo.

Netting, R. (1974): In: T.E. Downing and M. Gibson (eds): Irrigation's Impact on Society. Tucson, Arizona: The University of Arizona Press.

O'Neill, B. (1987): Nonmetric Test of the Minimax Theory of Two–Person Zerosum Games. Proc. Nat. Acad. Sci. USA 84: 2106–2109.

Pressman, J.L. and A. Wildavsky (1973): Implementation. Berkeley, California: Univ. of California Press.

Rasmusen, E. (1989): Games and Information. An Introduction to Game Theory. Oxford, England: Basil Blackwell.

Rogers, P. (1969): A Game Theory Approach to the Problems of International River Basins. Water Resources Research 5: 749–760.

Selten, R. (1975): Reexamination of the Perfectness Concept for Equilibrium Points in Extensive Games. Intern. J. Game Theory 4: 25–55.

Selten, R. (1983): Evolutionary Stability in Extensive Two–Person Games. Mathem. Social Sciences 5: 269–363.

Selten, R. (1991): Anticipatory Learning in Two–Person Games. To appear in: R. Selten (ed.): Game Equilibrium Models. Vol. 1: Evolution and Game Dynamics. Berlin, Heidelberg, New York: Springer–Verlag.

Spooner, B. (1974): Irrigation and Society. The Iranian Plateau. In: T.E. Downing and M. Gibson (eds): Irrigation's Impact on Society. Tucson, Arizona: The University of Arizona Press.

Straffin, P.D. and J.P. Heaney (1981): Game Theory and the Tennessee Valley Authority. Intern. J. Game Theory 10: 35–43

Tang, S.Y. (1989): Institutions and Collective Action in Irrigation Systems. Bloomington, Indiana: Indiana University, Department of Political Science. Ph.D. dissertation.

Tsebelis, G. (1989): The Abuse of Probability in Political Analysis: The Robinson Crusoe Fallacy. American Political Science Review 83: 77–91.

Tsebelis, G. (1990a): Penalty Has No Impact on Crime: A Game–Theoretical Analysis. Rationality and Society 2: 255–286.

Tsebelis, G. (1990b): The Effect of Fines on Regulated Industries: Game Theory vs. Decision Theory. Journal of Theoretical Politics, forthcoming.

Uphoff, N. (1985): Fitting Projects to People. In: M.M Cernea (ed.): Putting People First. Sociological Variables in Rural Development. New York: Oxford University Press, pp. 359–398.

USAID (1983): Irrigation Assistance to Developing Countries Should Require Stronger Commitments to Operation and Maintenance. Washington, D.C.: General Accounting Office.

van Damme, E. (1983): Refinements of the Nash Equilibrium Concept. Berlin: Springer–Verlag.

van Damme, E. (1987): Stability and Perfection of Nash Equilibrium. Berlin: Springer–Verlag.

von Neumann, J. and O. Morgenstern (1944): The Theory of Games and Economic Behavior. New York: Wiley.

Wade, R. (1990): On the 'Technical' Causes of Irrigation Hoarding Behavior, or Why Irrigators Keep Interfering in the Main System. In: R.K. Sampath (ed.): Institutional Aspects of Canal Irrigation. Fort Collins, Colorado: Colorado State University, forthcoming.

Walker, H.H. (1983): Determinants for the Organisation of Irrigation Projects. In: Man and Technology in Irrigated Agriculture. Irrigation Symposium, 1982. Hamburg: Verlag Paul Parey, pp. 19–36.

Weissing, F. (1983): Populationsgenetische Grundlagen der Evolutionären Spieltheorie. Bielefeld: Materialien zur Mathematisierung der Einzelwissenschaften, No. 41 and No 42.

Weissing, F. (1991): Evolutionary Stability and Dynamic Stability in a Class of Evolutionary Normal Form Games. To appear in: R. Selten (ed.): Game Equilibrium Models. Vol 1: Evolution and Game Dynamics. Berlin: Springer–Verlag.

Wittman, D. (1985): Counter–intuitive Results in Game Theory. European Journal of Political Economy 1: 77–89.

Young, H.P., N. Okada, and T. Hashimoto (1982): Cost Allocation in Water Resources Development. Water Resources Research 18: 463–475.

EQUILIBRIUM SELECTION IN THE SPENCE SIGNALING GAME [1)]

by

Eric van Damme and Werner Güth

Abstract: The paper studies the most simple version of the Spence job market signaling model in which there are just two types of workers while education is not productivity increasing. To eliminate the multiplicity of equilibria, the general equilibrium selection theory of John Harsanyi and Reinhard Selten is applied. It is shown that without invoking Pareto comparisons, the Harsanyi/Selten theory selects Wilson's E_2-equilibrium as the solution. The main elements in the analysis are the study of primitive equilibria and of the tracing procedure. The analysis sheds light on the "evolutive" and the "eductive" aspects of Harsanyi and Selten's theory and it also allows a better understanding of the older, non game theoretic literature on signaling.

1. Introduction

Recently, signaling games have been extensively studied in economics and game theory. In such games, there is an informed party who sends a message to which one or several uninformed players react; the payoffs of the participants depend on the private information of the informed party, the signal that this player sends and the responses that the uninformed players take. Many economic models with informational asymmetries contain a signaling game as an essential ingredient. For a (very) partial overview of this huge literature, see Van Damme [1987, Sect. 10.4].

A signaling game typically admits a great multiplicity of equilibria. This is caused by the fact that the signal space is usually larger than the type space of the informed player so that there exist unused signals. The Nash equilibrium concept does not tie down the uninformed agents' beliefs and actions at such unreached information sets, and this arbitrariness of off the equilibrium path responses in turn allows many outcomes to be sustained in equilibrium. As a consequence, the class of signaling games has provided fertile playground for game theorists working on equilibrium refinements. Indeed many refinements of the Nash equilibrium concept have been defined initially only for such games although in most cases the concepts can be extended to general extensive form games. Again, see Van Damme [1987, Sect. 10.5] for a partial survey.

To reduce the multiplicity of equilibria in signaling games arising in economic contexts, variations on the theme of iterative elimination of dominated strategies have been most popular. Some of such (sometimes ad hoc) procedures can be justified by refering to the general notion of stable equilibria introduced in Kohlberg and Mertens [1986]. Indeed, various authors claim additional virtue for the result they obtain by stating that the outcome obtained is the only one satisfying the Kohlberg/Mertens stability criterion. However, the latter stability notion has its share of counterintuitive examples (see, for instance, Van Damme [1989]), and even the seemingly unobjectionable notion of iterative elimination of ordinary weakly dominated strategies is not

[1] Supported in part by the Sonderforschungsbereich 303 (DFG), Institut für Operations Research, Universität Bonn, W. Germany. The research was carried out while the authors participated in the research group 'Game Equilibrium Models' at the Center for interdisciplinary research of the University of Bielefeld.

without pitfalls as Binmore [1987] has argued. As a consequence, the present authors are not convinced that the theory that is currently accepted by the majority of researchers in the field will survive in the long run; there is scope for alternative theories.

Our aim in this paper is to illustrate such an alternative theory, viz. the equilibrium selection theory of John Harsanyi and Reinhard Selten [1988](the HS theory), by means of applying it to a simple signaling game arising in an economic context. Specifically, we will study the most basic version of the job market signaling model introduced in the seminal work of Spence [1973,1974]. We take this well-known model rather than an abstract signaling game for didactical reasons. Since many economic signaling models have a mathematical structure similar to Spence's model, arguments similar to the ones we will use will come up in their analyses and the reader can get a good idea for how the HS theory works by reading the present paper. In the literature, the reader can already find several applications of the HS theory, however, all of these are to games that admit multiple strict equilibria, and in these games stability *a la* Kohlberg/Mertens is not very powerful. (Recall that Kohlberg and Mertens do not claim that their theory is a selection theory, cf. Fn. 2 of their paper.) To our knowledge, the present paper is the first to apply the HS theory to a game in which the multiplicity of equilibria is caused solely by the existence of unreached information sets, i.e. by the perfection problem.

There is no doubt that there are fundamental differences between the HS (general) theory and alternative (partial) theories such as Kohlberg/Mertens stability. We refer the reader to the postscript of Harsanyi and Selten [1988] in which some of these are described. Perhaps the most profound difference is that HS work with the standard form of the game (this is basically the agent normal form), whereas stability is a normal form concept. Accordingly, HS reject ideas of 'forward induction' since they conflict with their 'subgame and truncation consistency'. Since in our game no player has two strategic moves along the same path[2)], this difference is not relevant for our model, however. (Also, cf. Mertens [1988]).

On the other hand, there are common elements in the various theories: HS iteratively eliminate inferior strategies of the perturbed standard form, Kohlberg/Mertens stability allows iterative elimination of dominated strategies in the normal form, and dominated strategies are inferior. Apart from the difference in game form, the main difference here is in the order of operations: HS perturb first and thereafter eliminate, whereas Kohlberg/Mertens follow the opposite order. A simple example in Appendix A illustrates that the order in which the operations are carried out may make a difference in general. However, in this paper we show that this is not so in the present case, i.e. if the unperturbed Spence game is dominance solvable (which holds if the proportion of low quality workers is relatively high), the Harsanyi/ Selten solution is the equilibrium that remains after all dominated strategies have been iteratively eliminated, i.e. the HS solution is the Pareto best separating equilibrium. Actually, the proof of this proposition constitutes the heart of the present paper, hence, the parameter constellation that is trivially solved according to conventional methods poses the greatest difficulty for the HS theory. The case where the proportion of low ability workers is small (so that the game is not dominance solvable) is relatively easy to solve by means of the HS theory. In this case, we show that only the Pareto best pooling equilibrium spans a primitive formation, therefore, the HS theory determines this equilibrium as the solution. To summarize, in our simple model, the solution obtained by the HS theory coincides with the E_2-equilibrium notion proposed by Wilson [1977]. It should be stressed that to obtain this result we do not make use of Pareto comparisons. Although such comparisons play a role in the HS theory, in our analysis we will never be in the position that the HS theory allows us to make such a comparison. Finally it should be remarked that it remains to be investigated for which class of games the Wilson and HS solutions coincide.

The remainder of the paper is organised as follows. In Sect. 2 we introduce the model and

---

[2] The informed player moves twice, but the second move (the choice of employer) is not really strategic, subgame perfectness determines this choice uniquely.

derive the Nash equilibria of the unperturbed game. (The model is chosen so that every Nash equilibrium is sequential). The HS theory cannot be applied to this game directly, rather one has to apply the theory to a sequence of uniformly perturbed games and then let the perturbances go to zero. In the Sects. 3 and 4 it is investigated which equilibria of the original game are uniformly perfect, i.e. which equilibria can be approximated by equilibria of such uniformly perturbed games. HS consider as the initial set of solution candidates of a perturbed game the so called primitive equilibria, i.e. those equilibria that span primitive formations. A formation is a subset of strategy combinations that is closed under taking best replies. A primitive formation is a minimal one, this concept generalizes the notion of a strict equilibrium point. In Sect. 5 we investigate the formation structure of the uniformly perturbed games. We show that, if the proportion of able workers is sufficiently high, only the Pareto best pooling equilibrium spans a primitive formation, hence, this equilibrium is the HS solution if there are many able workers. If there are only few able workers, then there are multiple primitive formations, in particular, the Pareto best pooling equilibrium as well as all separating equilibria in which the able worker does not invest too much in education span primitive formations. In this case, the HS solution is determined by applying the risk dominance criterion (in which use is made of the tracing procedure). In Sect. 6 it is shown that the Pareto best separating equilibrium risk dominates all other equilibria, hence, this equilibrium is the solution if there are few able workers. Sect. 7 offers a brief conclusion.

2. Model and Equilibria

Let Y be a finite set (of possible education choices) and let W be the finite set of possible wages that firms can offer. The entire paper is devoted to the analysis of the signaling game $\Gamma(Y, W, \lambda)$ described by the following rules:

(2.1) A chance move determines the type, respectively the productivity, t of player 1 (the worker); with probability λ the type is 1, with probability $1-\lambda$ the type is 0; only player 1 gets to hear his type.

(2.2) Player 1 chooses $y \in Y$.

(2.3) Two firms (the players 2 and 3) observe the $y \in Y$ chosen and they then simultaneously offer wages $w_2, w_3 \in W$.

(2.4) The worker observes the wages offered and chooses a firm.

(2.5) The payoff (von Neumann-Morgenstern utility) is 0 for a firm that does not attract the worker, it is $t - w$ for a firm that pays the wage w to a worker of type t, and it is $w - y$ (resp. $w - y/2$) for a worker of type 0 (resp. type 1) that receives the wage w after having chosen the education level y.

Several comments are in order concerning the above specification

(i) The main difference between our game and the basic model of Spence [1973] is that in our case the least able worker is not productive. This assumption simplifies the analysis somewhat since it ensures that every Nash equilibrium is a sequential equilibrium. With minor modifications our arguments also apply to Spence's specification, however.

(ii) To reduce the number of parameters, we have normalised the education cost such that these are twice as high for the type 0 worker as for the type 1 worker. Given our other assumptions, this normalisation is without loss of generality; the reader may verify that the HS theory yields the Wilson E_2-equilibrium as long as the least able worker has higher education cost.

(iii) Competition between the firms is modeled as a Bertrand game between two firms. It may be checked that the same results would be obtained with n firms, $n \geq 2$. Of course, if there is just one firm the solution is completely different (the firm offers a wage of zero and the workers do not invest in education).

(iv) Since the HS theory aplies only to finite games, we are forced to work with finite sets Y and W, hence, we discretize the continuous specification of Spence. Throughout, we will assume that this discretization is sufficiently fine, and at the end we will let the diameter of the grid go to zero.

Let $g > 0$ denote the smallest money unit. All wages have to be quoted in integer multiples of g, hence

$$W = \{kg;\ k = 0, 1, 2, ..., K\}. \tag{2.6}$$

To simplify the analysis somewhat, it will be assumed that

$$Y \subset W \text{ and } 0, \lambda, 1, 2 \in Y\ , \tag{2.7}$$

an assumption that is, however, not essential for our results to hold (Details are available from the authors upon request). To avoid some further technical uninteresting difficulties, let us assume that the grid of W is finer than that of Y, specifically

$$\text{if} \quad y, y' \in Y \quad \text{and} \quad y \neq y', \quad \text{then} \quad |y - y'| > 4g\ . \tag{2.8}$$

Finally, we introduce the following convenient notation:

$$\text{If} \quad x \in \mathbb{R}, \text{then} \quad x^- = max\{w \in W | w < x\}\ . \tag{2.9}$$

Next, let us start analysing the game $\Gamma(Y, W, \lambda)$. Note that this game admits several (trivial) subgames in stage (2.4): Subgame perfection requires that at this stage the worker chooses the firm offering the highest wage, and symmetry implies that the worker should randomize evenly over the firms in case they offer the same wage. Hence, it is natural to analyse the truncated game obtained by constraining the worker to behave in this way in stage (2.4). The HS theory allows this procedure to be followed.[3] Therefore, from now on, attention will be restricted to the truncated game. This game will be denoted $G(Y, W, \lambda)$

Denote by $s_t(y)$ the probability with which the type t worker takes the education choice y. Hence, (s_0, s_1) is a behavioral strategy of player 1 in $G(Y, W, \lambda)$. Let $s(y)$ be the probability that education level y is chosen

$$s(y) = \lambda s_1(y) + (1 - \lambda) s_0(y), \tag{2.10}$$

and let $\mu(y)$ be the expected productivity if y is chosen with positive probability. By Bayes' rule

$$\mu(y) = \lambda s_1(y)/s(y) \qquad \text{if} \quad s(y) > 0. \tag{2.11}$$

Assume $s(y) > 0$. In any Nash equilibrium of $G(Y, W, \lambda)$, firms, in response to y, play a Bertrand equilibrium of the game in which it is known that the surplus (i.e. the worker's productivity) is $\mu(y)$. The following Lemma specifies the equilibria of this game.

Lemma 2.1. *The game in which two firms compete for a surplus μ by means of wage offers $w_i \in W$ has the following equilibria:*

[3] Actually, the HS theory requires to first perturb the game before one starts replacing subgames (which are cells in the perturbed game) by their solutions. To simplify the exposition, we have interchanged the operations of perturbing and decomposition, what, for the case at hand, does not influence the final result.

(a) *If $\mu \in W$, both firms either offer μ, or μ^-, or μ^{--}.*

(b) *If $\mu \notin W$, both firms either offer μ^-, or μ^{--} or they randomize between μ^- and μ^{--} choosing μ^- with probability $(g - \mu + \mu^-)/g$, where g is as in (2.6) (i.e. $g = \mu^- - \mu^{--}$).*

Proof. A straightforward argument establishes that, if $\mu \notin W$, only μ^- and μ^{--} can be in the support of an equilibrium. By investigating the 2x2 game in which each firm chooses between μ^- and μ^{--}, conclusion b) follows easily from (2.6). If $\mu \in W$, the equilibrium support now possibly also includes μ and conclusion a) follows easily by investigating the 3x3 game in which each firm chooses between μ, μ^- and μ^{--}.

□

Remark 2.2. Note that in case (a) of Lemma 2.1, the equilibrium in which both firms offer μ is not perfect if $\mu > 0$ (bidding μ is dominated by bidding μ^-), while the other two equilibria are perfect (even uniformly perfect) if $\mu > 3g$. At a certain stage of the solution process, the HS theory eliminates the equilibrium μ^{--} since it is not primitive (when my opponent offers μ^{--}, I am indifferent between μ^{--} and μ^-, also see Sect. 5). The equilibrium (μ^-, μ^-), being strict, is definitely primitive. Consequently, in case (a), we will restrict attention to this equilibrium right from the beginning. (Also note that μ^{--} cannot occur in equilibrium if there are more than two firms). In case (b) all three equilibria are (uniformly) perfect. The mixed equilibrium is not primitive. To simplify the statements of results to follow somewhat, and without biasing the final result, we will restrict attention to the equilibrium (μ^-, μ^-) also in this case.

Let $s = (s_0, s_1)$ be an equilibrium strategy of player 1 in $G(Y, W, \lambda)$ and write $Y(s) = \{y;\ s(y) > 0\}$. The above lemma and remark, with $\mu = \mu(y)$ as in (2.11) determines the equilibrium response of the firms at y. Assume $y, y' \in Y(s)$ with $y < y'$. From optimizing behavior of the worker, one may conclude that $\mu(y) < \mu(y')$. Since it cannot be the case that both types of worker are indifferent between y and y' we must have $\mu(y) = 0$ or $\mu(y') = 1$, hence, in any equilibrium at most three education choices can occur with positive probability. The Nash concept does not specify the equilibrium responses of firms at education levels $y \in Y/Y(s)$. To generate the set of all equilibrium outcomes we may put $w_i(y) = 0$ at such y, since this is the best possible threat of the firms to avoid that y will be chosen. Using this observation it is not difficult to prove that the set of Nash equilibrium paths of $G(Y, W, \lambda)$ is as described in the following Proposition.

Proposition 2.3. *(s, w) with $s = (s_0, s_1)$ and $w = (w_2, w_3)$ with $w_i : Y(s) \to W$ is a Nash equilibrium path of $G(Y, W, \lambda)$ if and only if*

$$\text{(2.12)} \qquad \text{if} \quad s_t(y) > 0, \text{ then} \begin{cases} y \in \arg\max\limits_{Y(s)} \; w(y) - y/(t+1) \\ \text{and} \quad w(y) - y/(t+1) \geq 0 \end{cases}$$

$$\text{(2.13)} \qquad w(y) = w_1(y) = w_2(y) = \mu(y)^- \text{ for all } y \in Y(s)$$

In the following sections, emphasis will be on those equilibria in which player 1 does not randomize. (The other equilibria are not primitive, hence, the HS theory eliminates them, see Sect.6). These equilibria fall into two classes:

(a) pooling equilibria, in which the two types of worker take the same education choice, hence $Y(s) = \{\overline{y}\}$ for some $\overline{y} \in Y$. Then $\mu(\overline{y}) = \lambda$ and $w(\overline{y}) = \lambda^-$.Therefore, (2.12) shows that $\overline{y} \in Y$ can occur in a pooling equilibrium if and only if $\overline{y} < \lambda$.

(b) separating equilibria, in which the education choice y_0 of type 0 is different from the choice y_1 of type 1. Proposition (2.3) shows that the pair (y_0, y_1) can occur in a separating equilibrium if and only if $y_0 = 0$ and

$$1^- - y_1 \leq 0 \leq 1^- - y_1/2 \tag{2.14}$$

Hence, we have

Corollary 2.3. *For any $\overline{y} \in Y$ with $\overline{y} < \lambda$, there exists a pooling equilibrium of $G(Y, W, \lambda)$ in which both players choose $\overline{y}$. There exists a separating equilibrium in which type t chooses y_t if and only if $y_0 = 0$ and $1 \leq y_1 < 2$.*

3. The Uniformly Perturbed Game

$G(Y, W, \lambda)$ is a game in extensive form. The HS theory is based on a game form that is intermediate between the extensive form and the normal form, the so called standard form. In our case, the standard form coincides with the agent normal form (Selten [1975]) since no player moves twice along the same path. The agent normal form has $2|Y| + 2$ active players, viz. the two types of player 1, and for each $y \in Y$ and each firm i an agent iy that is responsible for the firm's wage offer at y. This agent normal form will again be denoted by $G(Y, W, \lambda)$. Note that, if (s, w) is a pure strategy combination ($s = (s_0, s_1)$ with $s_t(y_t) = 1$ for some $y_t \in Y$, $w = (w_2, w_3)$ with $w_i : Y \to W$), then the payoffs to agent t (the type t worker, $t = 0, 1$) are

$$H_t(s, w) \;=\; max_{i=1,2}\, w_i(y_t) - y_t/(t+1), \tag{3.1}$$

while the payoffs to agent iy are

$$H_{iy}(s, w) \;=\; \begin{cases} 0 & \text{if } s(y) = 0 \text{ or } w_i(y) < w_j(y) \\ \mu(y) - w_i(y) & \text{if } s(y) > 0 \text{ and } w_i(y) > w_j(y) \\ (\mu(y) - w_i(y))/2 & \text{if } s(y) > 0 \text{ and } w_i(y) = w_j(y) \end{cases} \tag{3.2}$$

To ensure perfectness of the final solution, the HS theory should not be directly applied to $G(Y, W, \lambda)$, but rather to a sequence of uniformly perturbed games $G_\epsilon(Y, W, \lambda)$ with $\epsilon > 0$, ϵ small, ϵ tending to zero. The latter games differ from the original one in that each agent cannot completely control his actions. Specifically in $G_\epsilon(Y, W, \lambda)$, if agent t intends to choose y_t, then he will by mistake also choose each $y \in Y$, $y \neq y_t$ with a probability ϵ and he will actually play the completely mixed strategy s_t^ϵ given by

$$s_t^\epsilon(y) \;=\; \begin{cases} \epsilon & \text{if } y \neq y_t \\ 1 - (|Y| - 1)\epsilon & \text{if } y = y_t, \end{cases} \tag{3.3}$$

More generally, if the type t worker intends to choose the mixed strategy s_t then he will actually play the completely mixed strategy s_t^ϵ given by

$$s_t^\epsilon(y) = s_t(y)(1 - |Y|\epsilon) + \epsilon \tag{3.4}$$

(Of course, ϵ should be chosen so small that $|Y|\epsilon < 1$). Similarly, if agent iy intends to choose $w_i(y)$, he will actually also choose all different wages with the same positive probability ϵ.

Formally, the uniformly perturbed game $G_\epsilon(Y, W, \lambda)$ has the same agents as players, these have the same strategies available (which are now interpreted as intended choices), but the payoff function H^ϵ is slightly different from that in (3.1) – (3.2) and takes into account the above described mistake technology. If ϵ is small, the payoffs H^ϵ of those agents moving on the equilibrium path are close to the payoffs as described by H, however, for agents iy that cannot be reached intentionally there is a discontinuity. Namely, since both types of player 1 choose y with probability ϵ in the perturbed game, such agents play a Bertrand game with surplus λ. It will be convenient to assume that actually firms never make mistakes. The reader may verify that this assumption does not influence our results, it just simplifies notation. In this case, the perturbed game payoff to agent iy is given by

$$(3.5) \qquad H^\epsilon_{iy}(s, w) = \begin{cases} \mu^\epsilon_s(y) - w_i(y) & \text{if } w_i(y) > w_j(y) \\ (\mu^\epsilon_s(y) - w_i(y))/2 & \text{if } w_i(y) = w_j(y) \\ 0 & \text{if } w_i(y) < w_j(y) \end{cases}$$

where the expected productivity at y is given by

$$(3.6) \qquad \mu^\epsilon_s(y) = \frac{\lambda s^\epsilon_1(y)}{\lambda s^\epsilon_1(y) + (1-\lambda)s^\epsilon_0(y)}$$

with $s^\epsilon_t(y)$ as in (3.4).

If firms do not make mistakes, the workers' payoffs are (up to a positive affine transformation, depending on the worker's own mistakes) exactly as in (3.1). Since affine transformations leave the solution invariant, we will consequently analyse the game in which the types of the worker have pure strategy set Y, the agents of the firms have strategy set W and in which the payoffs are as in (3.1), (3.5). This game will be denoted $\tilde{G}_\epsilon(Y, W, \lambda)$. Note that the perturbed game $\tilde{G}_\epsilon(Y, W, \lambda)$ is an ordinary agent normal form game of which we will analyse the Nash equilibria. There is no need to consider a more refined solution concept: All "trembles" have already been incorporated into the payoffs.

After having uniformly perturbed the game, the next step, in applying the HS theory consists in checking whether the game is decomposable, i.e. whether there exist (generalized) subgames, so called cells. (See the diagram on p. 127 of Harsanyi and Selten [1988]). It may be thought that, for each $y \in Y$, the agents $\{2y, 3y\}$ constitute a cell. Indeed these agents do not directly compete with an agent iy' with $y' \neq y$. However, they directly interact with both types of the worker and, hence, through the worker they indirectly compete also with agent iy'. To put it differently, the game does not contain any (proper) subcells, it is indecomposable. Therefore, we move to the next stage of the solution procedure, the elimination of inferior choices from the game.[4)]

4. Elimination of Inferior Strategies

A strategy is inferior if its stability set, i.e. the region of opponents' strategy combinations where this strategy is a best reply, is a strict subset of another strategy's stability set. It can be

[4] It will turn out that this step is actually redundant in the case at hand, i.e. the reader may move directly to Sect. 5 if he prefers. Since one of our aims is to illustrate the various aspects of the HS theory, we thought it best to include the discussion on inferior strategies.

verified that in our model, the inferior strategies are exactly those that are weakly dominated. Clearly, such strategies exist in our model: For any agent iy offering a wage greater than or equal 1 is weakly dominated. Furthermore, for the type t worker, choosing y with $y/(t+1) \geq maxW$ is dominated by choosing 0. HS require that one eliminates all such choices, and that one continues this process until there are no inferior choices left. (Along the way, one should also check whether the reduced game obtained contains cells, but this will never be the case in our model, see also Fn. 3). The following Proposition describes the irreducible game that results after all inferior strategies have been successively eliminated.

Proposition 4.1. *If ϵ is sufficiently small, then by iterative elimination of inferior strategies, the game $\tilde{G}_\epsilon(Y,W,\lambda)$ reduces to the game $\tilde{G}^r_\epsilon(Y,W,\lambda)$ in which the following strategies are left for the various agents:*

For type 0 : $Y_0^r = \{y \in Y;\ y < 1\}$
For type 1 : $Y_1^r = \{y \in Y;\ y < 3 - max(2\lambda, 1)\}$
For agent iy with $y < 1$: $W_{iy}^r = \{w \in W;\ w < 1\}$
For agent iy with $1 \leq y < 3 - max(2\lambda, 1)$: $W_{iy}^r = \{w \in W; w \in [\lambda^-, 1)\}$
For agent iy with $y \geq 3 - max(2\lambda, 1)$: $W_{iy}^r = \{\lambda^-\}$

Proof. Clearly $w \in W$ with $w \geq 1$ is inferior for any agent iy. Once these inferior strategies have been eliminated, choosing $y \geq 1$ becomes inferior for type 0 (it is dominated by choosing $y = 0$) and $y \geq 2$ becomes inferior for type 1. We claim that in the resulting reduced game an agent iy with $y < 1$ does no longer have any inferior actions. Namely, $w_i(y) = 0$ is the unique best response against $w_j(y) = 0$ if $s_0(y) = 1$ and $s_1(y) = 0$ (hence $\mu_s^\epsilon(y) < g$), so that offering zero is not inferior (Here we need that ϵ is small). Furthermore, if $w \leq 1^{--}$, then w is the unique best response of agent iy against $w_j(y) = w^-$ if $\mu_s^\epsilon(y) \approx 1$ (i.e. $s_0(y) = 0$ and $s_1(y) = 1$). Therefore, such wages also are not inferior. Finally, since both firms offering $w = 1^-$ at y is a strict equilibrium if $\mu_s^\epsilon(y) > 1^-$, also $w = 1^-$ is not inferior, which establishes the claim. Next, turn to an agent iy with $y \geq 1$. Since the type 0 worker cannot choose y voluntarily, we either have $s(y) = 0$, hence $\mu_s^\epsilon(y) = \lambda$, or the type 1 worker chooses y voluntarily, in which case $\mu_s^\epsilon(y) \approx 1$. Consequently, at y the firms play a Bertrand game for a surplus of λ or of 1. The standard Bertrand argument implies that, starting with $w = 0$, all wages with $w < \lambda$ will be iteratively eliminated. On the other hand, the same argument that was used for the case $y < 1$ establishes that wages strictly between λ^{--} and 1 cannot be eliminated. The reduction of the firms' strategy sets obtained in this way does not introduce new inferior strategies for the type 0 worker. Namely, if this worker expects that $y^* < 1$ will result in the wage 1^- while all other $y < 1$ yield wage 0, then the unique best response is to choose y^*, so that y^* cannot be inferior. If $\lambda > \frac{1}{2}$ the reduction leads to new inferior strategies for the type 1 worker, however. Namely, by choosing $y = 1$, he receives an income of at least $\lambda^- - \frac{1}{2}$ so that any education choice $y \geq 3 - 2\lambda$ becomes inferior. Finally, if y is inferior for both the type 0 and the type 1 worker, then firms at y play a Bertrand game for surplus λ and we already argued above that in this case only λ^- is not (iteratively) inferior. Hence, we have shown that $\tilde{G}_\epsilon(Y,W,\lambda)$ can be reduced to at least $\tilde{G}^r_\epsilon(Y,W,\lambda)$ as specified in the Lemma. Since the latter game is irreducible, the proof is complete.

□

The following Figure displays the result of Proposition 4.1. Along the horizontal y-axis, we indicate the noninferior actions of the worker. For each value of y, the shaded area displays the noninferior responses of the firms. The Figure is drawn for the case $\lambda > \frac{1}{2}$.

Figure 1. The reduced perturbed game $\tilde{G}^r_\epsilon(Y, W, \lambda)$.

By comparing Proposition 4.1 with Proposition 2 .2 it is seen that, at least if $\lambda > \frac{1}{2}$, some equilibria of the original game are eliminated by considering the reduced perturbed game. Specifically, separating equilibria in which type 1 invests very much in education as well as some equilibria involving randomization are no longer available in $\tilde{G}^r_\epsilon(Y, W, \lambda)$. All in all, however, this step of the HS solution procedure is not very successfull in cutting down on the number of equilibrium outcomes.

It is also instructive to compare Proposition 4.1 with the result that would have been obtained by iterative elimination of dominated strategies in the unperturbed game $G(Y, W, \lambda)$. In the latter, after wages $w \geq 1$ have been eliminated, choosing $y \geq 1$ becomes dominated for type 0, but not for type 1, hence, whenever such y is chosen, firms play a Bertrand game with surplus 1 so that all wage offers except 1^- are dominated. Hence, by choosing $y = 1$, the type 1 worker guarantees a utility of (about) $\frac{1}{2}$. In particular, choosing $y > 1$ is now dominated for this type. We see that the reduced game associated with the unperturbed game is strictly smaller than $\tilde{G}^r_\epsilon(Y, W, \lambda)$. The difference is especially dramatic if $\lambda < \frac{1}{2}$. Proposition 2.2 shows that in this case there is (essentially) only one equilibrium of the unperturbed game that remains in the reduced unperturbed game, viz. the separating equilibrium with $y_1 = 1$. Hence, if $\lambda < \frac{1}{2}$, the unperturbed game is dominance solvable [5)] , and indeed, the conventional analysis generates the efficient separating equilibrium ($y_0 = 0, y_1 = 1$) as the outcome when $\lambda < \frac{1}{2}$. The reduced perturbed game, however, does not force agents iy with $y \in (1, 2)$ to offer the wage of 1 since they may still think that they are reached by mistake. As a consequence, $\tilde{G}^r_\epsilon(Y, W, \lambda)$ still admits many equilibria even if $\lambda < \frac{1}{2}$, and the HS theory does not immediately yield efficient sepapration as the solution. Still we will show that, the risk dominance criterion of HS forces efficient separation as the solution in this case.

5. Uniformly Perfect Equilibria

As argued before, the multiplicity of equilibria in our model is caused solely by the imperfectness problem, i.e. by the fact that there necessarily exist unreached information sets at which the firms' beliefs are undetermined. At such education choices, firms may threaten to offer a wage of zéro which in turn implies that the worker will indeed not take such a choice. The final consequence is that there exist equilibria with unattractive payoffs for the worker.

The HS theory solves the game via its uniformly perturbed game to avoid the imperfectness problem. The natural question to ask is how successfull this step of the solution procedure is

---

[5] One could also perform this elimination in the extensive form. Since $y \geq 1$ is dominated for type 0, one eliminates the branch in which type 0 chooses such y. The resulting extensive game then has a subgame at each $y \geq 1$ and subgame perfectness forces the firms to offer the wage 1^-. This argument might suggest that in the perturbed game, for each $y \geq 1$ there is a cell consisting of the agents $1, 2y$ and $3y$. If this would hold, it would follow immediately that efficient separation is the Harsanyi/Selten solution if $\lambda < \frac{1}{2}$. However, it is not the case that the agents $1, 2y$ and $3y$ form a cell. Even though the cell condition (see Harsanyi and Selten [1988, p.95]) is satisfied for the agents $2y$ and $3y$, it is violated for the type 1 worker: His payoff depends in an essential way on how firms react at other values of y. The reduced perturbed game does not contain any cells, it is itself indecomposable.

to reduce the number of equilibria. At first glance it appears as if this step is very successfull, it seems that the uniformly perturbed game admits just a single equilibrium. Namely, assume that the equilibrium payoff of the type 0 worker would be strictly less than λ^-. Motivated by Proposition 2.3, it seems natural to assume that there exists y sufficiently close to zero that is not intentionally chosen by either type of worker. According to (3.5) and (3.6) firms will offer the wage λ^- at such y, but then the type 0 worker profits by deviating to y. The apparent contradiction shows that the type 0 worker should have an equilibrium payoff of at least λ^- and inspection of Proposition 2.3 shows that the unperturbed game has only one equilibrium that satisfies this condition, viz. the pooling equilibrium in which both types of workers choose $\overline{y} = 0$. It seems that only the Pareto best pooling equilibrium can be an equilibrium of the uniformly perturbed game, and that using uniform perturbations completely solves the selection problem.

The fallacy in the above argument is that there may not exist y close to zero with $s(y) = 0$. Even though it is true that in the unperturbed game at most 3 education levels can occur with positive probability in an equilibrium, this structural property no longer holds in the perturbed game. In most equilibria of the latter game, the type 0 worker is forced to randomize intentionally over many education levels including levels close to zero. Once, the type 0 worker chooses y intentionally, firms will not have unbiased beliefs at y and they may offer wages strictly below λ. In particular, the wage may be so low that the type 0 worker becomes indifferent between choosing y and taking any equilibrium education level, and in this case there is no reason why he should not intentionally choose y. To put it differently, the Pareto best pooling equilibrium of $G(Y, W, \lambda)$ is the only equilibrium that can be approximated by pure equilibria of $\tilde{G}_\epsilon(Y, W, \lambda)$. For later reference we list this result as Proposition 5.1.

Proposition 5.1. *Only the pooling equilibrium outcome in which both types of the worker do not invest in education can be approximated by pure equilibria of uniformly perturbed games.*

An equilibrium outcome of the unperturbed game $G(Y, W, \lambda)$ is said to be uniformly perfect if, for $\epsilon > 0$, there exists an equilibrium (s^ϵ, w^ϵ) of $\tilde{G}^r_\epsilon(Y, W, \lambda)$ that produces this outcome in the limit as ϵ tends to zero. Hence, Proposition 5.1 may be paraphrased as "the pooling equilibrium outcome at $y = 0$ is uniformly perfect". In the remainder of this section we first derive a condition that payoffs associated with uniformly perfect equilibria necessarily have to satisfy (Corollary 5.3), thereafter we show (Proposition 5.6) that this condition is sufficient for the outcome to be uniformly perfect as well. The overall conclusion (Proposition 5.7) will be that relatively many equilibrium outcomes of the original game are uniformly perfect.

The next Lemma states a lower bound on the wage that firms offer in an equilibrium of the uniformly perturbed game, as well as a derived lower bound on the utility of the type 1 worker.

Lemma 5.2. *If (s, w) is an equilibium of $\tilde{G}^r_\epsilon(Y, W, \lambda)$ and if u_t is the equilibrium payoff of the type t worker, then*

$$\text{(5.1)} \qquad \text{if} \quad y \geq \lambda^- - u_0\,, \quad \text{then} \quad w(y) \geq \lambda^-\,,$$

$$\text{(5.2)} \qquad u_1 \geq (\lambda^- + u_0)/2\,.$$

Proof. Assume $y \geq \lambda^- - u_0$ but $w(y) < \lambda^-$. Then $w(y) - y < u_0$, hence, the type 0 worker cannot choose y voluntarily. Therefore, $s_0(y) = 0$ and $\mu^\epsilon_s(y) \geq \lambda$. The Bertrand competition (Lemma 2.1), however, then forces firms to offer a wage of at least λ^-, hence $w(y) \geq \lambda^-$. The contradiction proves (5.1). To prove (5.2), assume (without loss of generality) that $\lambda^- - u_0 \in Y$.

By choosing, $y = \lambda^- - u_0$, the type 1 worker guarantees a wage of λ^-, hence, he can guarantee a payoff $\lambda^- - (\lambda^- - u_0)/2$. This establishes (5.2).

□

As a direct consequence of Lemma 5.2, we have

Corollary 5.3. *If (s, w) is a uniformly perfect equilibrium of $G(Y, W, \lambda)$ with u_t being the equilibrium payoff of the type t worker, then*

$$u_1 \geq (\lambda^- + u_0)/2 \ . \tag{5.3}$$

If (s, w) is a separating equilibrium with $s_1(y_1) = 1$, then $u_0 = 0$ and $u_1 = 1^- - y_1/2$, hence the above Corollary implies that $y_1 \leq 2^- - \lambda$. Consequently we have

Corollary 5.4. *If (s, w) is a uniformly perfect separating equilibrium and $s_1(y_1) = 1$, then $y_1 < 2 - \lambda$.*

It is easily checked that Corollary 5.3 does not allow us to eliminate any pooling equilibrium (in this case, (5.3) is always satisfied with equality), hence, let us turn to equilibria in which the worker randomizes. First, consider the case where only the type 1 worker randomizes, say between $\bar{y}$ and y_1 with $\bar{y} < y_1$. Then type 0 chooses $\bar{y}$ for sure, and $u_0 = w(\bar{y}) - \bar{y}$, $u_1 = w(\bar{y}) - \bar{y}/2$. Furthermore, $w(\bar{y}) < \lambda^-$ since $\mu_s^\epsilon(\bar{y}) < \lambda$ so that (5.3) is violated. Next, assume that type 0 randomizes, say between y_0 and $\bar{y}$ with $y_o < \bar{y}$. Then $y_0 = 0$ and type 1 also chooses $\bar{y}$ with positive probability. Furthermore, $u_0 = 0$ and $u_1 = \bar{y}/2$ since $w(\bar{y}) = \bar{y}$. Hence, (5.3) implies $\bar{y} \geq \lambda^-$. Therefore, $\bar{y} \geq \lambda$ in view of (2.8). We have shown:

Corollary 5.5. *Equilibria in which only the type 1 worker randomizes are not uniformly perfect. Equilibria in which the type 0 worker randomizes between $y_1 = 0$ and $\bar{y} > 0$ are uniformly perfect only if $\bar{y} \geq \lambda$, hence $u_1 \geq \lambda/2$.*

Figure 2. Uniformly Perfect Equilibrium Outcomes.

We now come to the main result of this section, which states that condition (5.3) is also sufficient for an equilibrium outcome to be uniformly perfect. We will give the formal proof only for separating equilibria. The reader may easily adjust the proof to cover the other classes of equilibria not excluded by Corollary 5.3 .

Proposition 5.6. *A separating equilibrium outcome (y_0, y_1) with $y_0 = 0$ and $y_1 < 2 - \lambda$ is uniformly perfect.*

Proof. The proof is by construction. Take ϵ small and define the strategies (s, w) by means of

$$s_0(y) = \begin{cases} \frac{\epsilon(\lambda - y - g)}{(1-\lambda)(1-|Y|\epsilon)(y+g)} & \text{if } \ 0 < y < \lambda \\ 0 & \text{if } \ y \geq \lambda \\ 1 - \sum\limits_{y \neq 0} s_0(y) & \text{if } \ y = 0 \end{cases} \tag{5.4}$$

$$s_1(y) = \begin{cases} 0 & \text{if } \ y \neq y_1 \\ 1 & \text{if } \ y = y_1 \end{cases} \tag{5.5}$$

$$(5.6)\qquad w_i(y) = \begin{cases} y & \text{if } 0 \leq y < \lambda \\ \lambda^- & \text{if } y \geq \lambda, y \neq y_1 \\ 1^- & \text{if } y = y_1 \end{cases}$$

Note that $s_0(y) \to 0$ for all $y \neq 0$ as $\epsilon \to 0$, so that s_0 is a well-defined strategy if ϵ is small. This strategy has been chosen so that

$$(5.7)\qquad \mu_s^\epsilon(y) = y + g \qquad \text{for } 0 < y < \lambda$$

(Direct verification using (3.4) and (3.6) is easy). Furthermore, if ϵ is small, then $\mu_s^\epsilon(0) < g$, so that due to (Lemma 2.1), firms' agents indeed bid equilibrium wages for $y < \lambda$. If $y \geq \lambda$ and $y \neq y_1$, then $\mu_s^\epsilon(y) = \lambda$, while $\mu_s^\epsilon(y_1) > 1^-$ if ϵ is small so that firms bid optimally also for these education choices. If wages are as in (5.6), the type 0 worker has the interval $[0, \lambda)$ as optimal education choices, while the optimal choice of the type 1 worker is y_1 since $y_1 < 2 - \lambda$. Hence, for ϵ small, (s, w) is an equilibrium of $\tilde{G}_\epsilon^r(Y, W, \lambda)$. Since (s, w) in the limit produces the separating outcome (y_0, y_1), this outcome is uniformly perfect.

□

The proof that other outcomes that are not eliminated by Corollary 5.3 are uniformly perfect proceeds similarly. The basic insight is that, if $y < \lambda^- - u_0$, then $y = w(y) - u_0$ for some $w(y)$ and the type 0 worker may choose y voluntarily in order to justify the wage offer $w(y)$ of firms at y. Furthermore, as long as the type 0 worker is indifferent, the type 1 worker will not have y as an optimal choice. Hence, condition (5.3) is sufficient for the equilibriumoutcome to be uniformly perfect, and we may state

Proposition 5.7. *The uniformly perfect equilibrium outcomes of $G(Y, W, \lambda)$ are*

(i) *all pooling equilibrium outcomes*

(ii) *the separating outcomes with $y_1 < 2 - \lambda$, hence, $u_1 \geq \lambda/2$*

(iii) *the equilibrium outcomes in which type 0 randomizes between $y_0 = 0$ and $\bar{y}$ with $\bar{y} \geq \lambda$, hence $u_1 \geq \lambda/2$.*

Note that all three classes of equilibria described in the above Proposition satisfy $u_1 \geq \lambda/2$. This condition is necessary for uniform perfectness (cf. Corollary (5.3)), but it is not sufficient: There exist equilibria in which only the type 1 worker randomizes that satisfy this condition and these are not uniformly perfect (Corollary 5.5).

6. Formations and Primitive Equilibrium Outcomes

Given that strict equilibria (i.e. pure strategy equilibria in which each player looses by deviating) are (at least at the intuitive level) more stable than non-strict ones, it frequently is more natural to select a strict equilibrium. (Harsanyi and Selten [1988, Sect.5.2]). Of course, strict equilibria do not always exist so that HS were led to search for a principle that generalizes (weakens) the idea of strictness as a selection criterion and that still helps to avoid those equilibria that are especially unstable. HS have come up with the concept of primitive formations. A formation of a game specifies for each agent a subset of his strategy set such that any best reply (in the original game) against any correlated strategy combination with support contained in the restricted game is again in the restricted strategy set. Primitive formations are sets that are minimal with respect to this property. HS (Lemma 5.2.1) have shown that primitive formations exist, and it is also true that any formation contains an equilibrium of the original game. If (s, w) is an equilibrium of $\tilde{G}^r_\epsilon(Y, W, \lambda)$, then we will write $F^\epsilon(s, w)$ for the primitive formation in $\tilde{G}^r_\epsilon(Y, W, \lambda)$ that contains (s, w). We will say that (s, w) spans $F^\epsilon(s, w)$. Note that $F^\epsilon(s, w) = \{(s, w)\}$ whenever (s, w) is a strict equilibrium. Hence, primitive formations are the smallest substructures with similar properties as strict equilibria. The HS theory favors the selection of equilibria which span primitive formations to retain as much as possible of the stability properties of strict equilibria. Specifically, HS consider as the natural solution candidates (the first candidate set) the set of all solutions to primitive formations. (See the flowchart on p.222 of Harsanyi and Selten [1988]). In this section, we determine the minimal formations of $\tilde{G}^r_\epsilon(Y, W, \lambda)$ and their solutions. We will proceed by constructing for each equilibrium (s, w) the minimal formation spanned by it, and then check whether there exists an alternative equilibrium (s', w') with

$$F^\epsilon(s', w') \underset{\neq}{\subset} F^\epsilon(s, w) \tag{6.1}$$

An equilibrium (s, w) belongs to a primitive formation if and only if no (s', w') satisfying (6.1) can be found. Such an equilibrium (s, w) will be called a primitive equilibrium. (This definition differs slightly from the HS definition, the results in this section however show that the equilibria that are primitive according to our definition are exactly the initial candidates in the HS sense.). Finally, an equilibrium outcome of the unperturbed game will be called primitive if it can be obtained as a limit of primitive equilibrium outcomes of perturbed games as the perturbations vanish.

We have already seen in Sect.4 that the efficient pooling equilibrium outcome of the unperturbed game can be approximated by pure equilibrium outcomes of the ϵ-uniformly perturbed game. Namely, if ϵ is small, the strategy combination

$$s_0(0) = s_1(0) = 1 \ , \ w_i(y) = \lambda^- \quad \text{for all} \quad y \in Y \ , \tag{6.2}$$

is an equilibrium of $\tilde{G}^r_\epsilon(Y, W, \lambda)$ which produces this outcome in the limit.[6] Note that the equilibrium (6.2) is strict, hence, primitive. Consequently, the pooling equilibrium outcome at $\bar{y} = 0$ is primitive as well.

Proposition 6.1. *The equilibrium outcome in which the workers are pooled at $\bar{y} = 0$ is primitive.*

To verify which other outcomes are primitive, the following Lemma is helpful.

Lemma 6.2. *Let (s, w) be an equilibrium of $\tilde{G}^r_\epsilon(Y, W, \lambda)$ and write $F^\epsilon_{iy}(s, w)$ for the strategy space of agent iy in the formation $F^\epsilon(s, w)$. For ϵ small, we have*

$$\text{if} \ \ y = 0 \ \text{or} \ \ s(y) = 0, \ \text{then} \ \ \lambda^- \in F_{iy}(s, w) \ . \tag{6.3}$$

Proof. If $s(y) = 0$ and workers play according to s, then firms play a Bertrand game for surplus λ at y. Remark 2.2 and the fact that each formation contains an equilibrium implies

[6] The reader may recall from Lemma 2.1 that the Bertrand game for a surplus of λ, has equilibria that differ from (λ^-, λ^-). However, these are not primitive, hence, at last we can justify Remark 2.2.

that $\lambda^- \in F^{\epsilon}_{iy}(s,w)$. Next, consider $y = 0$. The statement of (6.3) is clearly fulfilled for the equilibrium from (6.2). If (s,w) is an alternative equilibrium, then the argument from the proof of Proposition 5.6 implies that there exists an alternative best reply s' of the worker with $s'(0) = 0$. Since s' is a best reply $s' \in F^{\epsilon}(s,w)$ and $F^{\epsilon}(s,w) \subseteq F^{\epsilon}(s',w)$. The conclusion now follows from the first part of the proof.

□

The Lemma immediately implies that equilibrium outcomes of the unperturbed game with $u_1 < \lambda^-$ cannot be primitive. Namely, if (s,w) is an equilibrium of $\tilde{G}^r_{\epsilon}(Y,W,\lambda)$ with $u_1 < \lambda^-$, then both types of the worker have the education choice 0 as the best response whenever firms offer $w(y) \leq \lambda^- + y/2$ for $y \neq 0$ and offer λ^- for $y = 0$.(The Lemma implies that such a strategy of the firms belongs to $F^{\epsilon}(s,w)$). But if workers play this strategy, firms should offer $w = \lambda^-$ for all y and $F^{\epsilon}(s,w)$ contains the equilibrium (6.2), hence (s,w) is not primitive.

Corollary 6.3. *In any primitive equilibrium outcome, the payoff to the type 1 worker is at least* λ^-.

The above Corollary in turn implies

Corollary 6.4. *Equilibrium outcomes in which the workers are pooled at* $\bar{y} > 0$ *are not primitive.*

Next we show that the mixed equilibrium outcomes that were not yet eliminated in Sect.5 are not primitive.

Proposition 6.5. *Equilibrium outcomes in which the type 0 worker randomizes between 0 and* $\bar{y}$ *are not primitive.*

Proof. Let (s,w) be an equilibrium of $\tilde{G}^r_{\epsilon}(Y,W,\lambda)$ that produces in the limit an outcome as described in the Proposition. Then $s_0(\bar{y}) > 0$ and $s_1(\bar{y}) > 0$. Hence, the formation $F^{\epsilon}(s,w)$ also contains the strategy in which both workers choose $\bar{y}$ for sure. The firms' best response against this strategy is to offer λ^- for each education choice, and, if firms behave in this way, the workers should choose $y = 0$. Hence, $F^{\epsilon}(s,w)$ contains the equilibrium from (6.2) so that (s,w) is not primitive.

□

Figure 3 graphically illustrates the results obtained thus far.

Figure 3. Primitive Equilibrium Outcomes ($\lambda < \frac{1}{2}$).

Finally, let us turn to separating equilibrium outcomes. For the outcome to be primitive it is necessary that choosing $y = 0$ is not an alternative best response for the type 1 worker whenever firms offer the wage λ^- at $y = 0$, hence, $y_1 < 2(1-\lambda)$. (cf. Corollary 6.3). In the next proposition we show that this condition is not only necessary but that it is also sufficient for a separating outcome to be primitive.

Proposition 6.6. *A separating equilibrium outcome is primitive if and only if* $y_1 < 2(1-\lambda)$.

Proof. It suffices to show that the condition is sufficient. The proof is constructive. Let (y_0, y_1) be a separating equilibrium outcome with $y_0 = 0$ and $y_1 < 2(1-\lambda)$ and let the equilibrium

(s, w) of $\tilde{G}^{r}_{\epsilon}(Y, W, \lambda)$ that produces this outcome in the limit be constructed as in the proof of Proposition 5.6. It is easily seen that , for ϵ small, $F^{\epsilon}(s, w)$ contains the following restricted strategy sets

(6.4) $$Y_0 = \{y \in Y;\ y < \lambda\} \qquad Y_1 = \{y_1\}$$

(6.5) $$W_{iy} = \begin{cases} \{w \in W; w \leq \lambda^{-}\} & \text{if } y < \lambda, \\ \{1^{-}\} & \text{if } y = y_1, \\ \{\lambda^{-}\} & \text{otherwise.} \end{cases}$$

Furthermore, one may easily verify that the collection of strategies defined by (6.4) (6.5) is closed under taking best replies, hence, it is a formation and, therefore, exactly equal to $F^{\epsilon}(s, w)$. Since (s, w) is the unique equilibrium that is contained in this formation, (s, w) is primitive.

□

Since $y_1 \geq 1$ in any separating equilibrium outcome, we see from Proposition 6.6 that no such outcome is primitive if $\lambda \geq \frac{1}{2}$. By combining this observation with Proposition 5.6 and the previous results from this section, we, therefore obtain

Corollary 6.7. *(i) If $\lambda \geq \frac{1}{2}$, only the outcome in which the workers are pooled at $\bar{y} = 0$ is primitive.*
(ii) If $\lambda < \frac{1}{2}$, in addition to the efficient pooling outcome, also the separating equilibrium outcomes with $y_1 < 2(1 - \lambda)$ (hence $u_1 \geq \lambda$) are primitive.

7. Risk Dominance

In the previous section, we determined all primitive equilibria of $\tilde{G}^{r}_{\epsilon}(Y, W, \lambda)$. The set of all these equilibria is what HS call the first candidate set. HS propose to refine this set by a process of elimination and substitution until finally only one candidate, the solution, is left. Loosely speaking, this process consists in eliminating all candidates that are 'dominated' by other candidates and by replacing candidates that are equally strong by a substitute equilibrium. In our application, we will not need the substitution procedure. We will show that there exists exactly one equilibrium in the first candidate set that dominates all other equilibria in this set.

Attention will be confined to the case where $\lambda < 1/2$. If $\lambda \geq 1/2$, and ϵ is small, then the perturbed game $\tilde{G}^{r}_{\epsilon}(Y, W, \lambda)$ has just a single primitive equilibrium, and this induces the efficient pooling outcome, so that we have .

Proposition 7.1. *If $\lambda \geq 1/2$, then the HS solution of the Spence signaling game $\Gamma(Y, W, \lambda)$ is the pooling equilibrium in which both workers do not invest in education.*

Let e, e' be two different solution candidates of the perturbed game $\tilde{G}^{r}_{\epsilon}(Y, W, \lambda)$, write A for the set of agents participating in this game (with generic element a) and let $u_a(e)$ (resp. $u_a(e')$) be the payoff to agent a when e (resp. e') is played. Finally, denote by $A(e, e')$, the set of all agents for which $e_a \neq e'_a$, that is, who play differently in e and in e'. The strategy vector e payoff dominates e' if

(7.1) $$u_a(e) > u_a(e') \qquad \text{for all} \quad a \in A(e, e'),$$

The HS theory requires that one first eliminates all payoff dominated equilibria from the initial candidate set. In our case, the only initial candidates that could possibly be payoff dominated are those that approximate an equilibrium in which the type 1 worker is separated at an education level y' strictly above 1. Indeed, the type 1 worker prefers to be separated at a lower level y. However, condition (7.1) requires that one also considers the payoffs of the agents of the firms at y and y', and it cannot be the case that all these agents unanimously strictly prefer the type 1 worker to choose y: If agent iy strictly prefers the worker to choose y rather than y' , then agent iy' strictly prefers this worker to choose y'. Consequently, no initial candidate is payoff dominated, and we have

Proposition 7.2. *The criterion of payoff dominance does not reduce the initial candidate set.*

In case of multiple payoff undominated candidates the HS theory requires to compare equilibria using the risk dominance criterion. The notion of risk dominance is central to the HS theory. It tries to capture the idea that, in a situation where the players are uncertain about which of two equilibria should be played, the players enter a process of expectation formation that may finally lead to the conclusion that one of these equilibria is less risky than the other one, and that, therefore, they should play this less risky equilibrium. In the remainder of this section we will show that the best separating equilibrium, (i.e. type 1 chooses $y_1 = 1$), risk dominates all other solution candidates. Hence, this equilibrium is the HS solution if $\lambda < 1/2$. Before proving this main result, we introduce some notation and formally define the concept of risk dominance.

Assume it is common knowledge that the solution of the game will be either e or e' and let $A(e,e')$ be the set of those agents whose strategy in e differs from that in e'. The restricted game generated by e and e' is the game in which the player set is $A(e,e')$ and in which the set of strategy combinations is the smallest formation generated by $\{e,e'\}$ with all agents $a \notin A(e,e')$ being bound to use their $e_a = e'_a$-strategy. Let $a \in A(e,e')$ and assume this agent believes that his opponents will play e with probability z and e' with probability $1-z$. Then a will play his best response $b_a(z,e,e')$ against the (correlated) strategy combination $ze_{-a} + (1-z)e'_{-a}$. Harsanyi and Selten define the *bicentric prior* of agent a as $p_a = b_a(e,e') = \int_0^1 b_a(z,e,e')\,dz$. This bicentric prior may be interpreted as the mixed strategy which an outside observer (or an opponent of a) expects a to use. (Adopting the principle of insufficient reason, the outsider considers the beliefs of a to be uniformly distributed on [0,1]). Denote by p the mixed strategy combination $p = (p_a)_{a\in A(e,e')}$. This p represents the initial expectations of the players in this situation of common uncertainty. The tracing procedure transforms these preliminary expectations into final expectations. Formally, the *tracing procedure* is a map T that converts each mixed strategy combination p into an equilibrium $T(p)$. Risk dominance is defined by means of the tracing procedure. The equilibrium e is said to *risk dominate* e' if for the bicentric prior $p = b(e,e')$ generated by e and e' we have $T(p) = e$. We conclude this overview of definitions by briefly describing the operator T. In our case it turns out that the linear tracing procedure is well-behaved, so we only specify this one. Let G be a game with payoff function H, and for $t \in [0,1]$, let G_p^t be a game with the same strategy sets, but in which the payoffs are

$$H_a^t(\sigma) = tH_a(\sigma) + (1-t)H_a(\sigma_a, p_{-a}) \tag{7.2}$$

hence, for $t = 0$, one plays against the bicentric prior, for $t = 1$, one plays the original game. Let E_p^t be the set of equilibria of G_p^t. (In our case) it can be shown that $E_p = \{E_p^t; t \in [0,1]\}$

contains exactly one continous path connecting the unique equilibrium of G_p^0 with an equilibrium of $G_p^1 = G$. The tracing result $T(p)$ of p is the $t = 1$-endpoint of this continous path.

After this review of definitions, we turn to the results. We will first show that, for ϵ small, the efficient separating equilibrium risk dominates any other separating equilibrium. The intuition for this result is simple. Let e be the separating equilibrium in which type 1 chooses $y = 1$ and let e' be a separating equilibrium in which this type chooses $y' > 1$. The initial beliefs of the firms' agents at y and y' will be that the type 1 worker chooses both y and y' with a probability that is bounded away from zero. Since the type 0 worker only chooses y and y' by mistake, and mistakes are rare, firms will be willing to offer a wage $w = 1^-$ at y and at y'. Since y and y' garner the same wages, the type 1 worker strictly prefers to choose y as the cost incured there are lower. This reinforces firms at $y = 1$ to offer $w = 1^-$. On the other hand firms' agents at y' will gradually update their beliefs, they become more pessimistic and finally they will conclude that y' can only occur by mistake. Hence, ultimately they will offer $w = \lambda^-$: We end up at the separating equilibrium at $y = 1$.

The following Proposition makes this argument precise.

Proposition 7.3. *Let e' be an equilibrium of $\tilde{G}_\epsilon^r(Y, W, \lambda)$ in which the type 1 worker is separated at $y' > 1$. Then, if ϵ is small, there exists an equilibrium e in which the type 1 worker is separated at $y = 1$ that risk dominates e'. Hence, efficient separating equilibria risk dominate all other separating equilibria.*

Proof. Let e' be given. Modify e' such that the type 1 worker chooses $y = 1$ rather than $y' > 1$, and such that firms offer $w = 1^-$ at y rather than at y' (where they now offer λ^-). The resulting strategy combination e is an equilibrium. We will show that e risk dominates e'. Note that $A(e, e')$ consists of 5 agents, viz. the type 1 worker and the firms' agents at y and y'. The game relevant for the risk dominance comparison has strategy space $\{y, y'\}$ for the type 1 worker, whereas the firms may choose from $\{w \in W;\ \lambda^- \leq w \leq 1^-\}$ at y and y'. (Note that education levels $\hat{y}$ with $y < \hat{y} < y'$ are never optimal for the worker since $w^{-1}(\bar{y}) = \lambda^-$.) The payoff to the worker is as in (3.1), if agent iy overbids the opponent with wage w, then its payoff is

$$s^\epsilon(y)(\mu_s^\epsilon(y) - w) \tag{7.3}$$

where $\mu_s^\epsilon(y)$ is as in (3.6) (with $s_0(y) = 0$) and $s^\epsilon(y)$ is the probability that y is chosen by the worker

$$s^\epsilon(y) = (1 - \lambda)\epsilon + \lambda s_1^\epsilon(y)\ . \tag{7.4}$$

If agent iy bids lower than the opponent, its payoff is zero, if both agents at y bid the same wage w, they share the quantity from (7.3). Payoffs at y' are defined similarly.

We now compute the bicentric prior combination associated with e, e'. If the type 1 worker expects his opponents to play according to $ze + (1 - z)e'$, then his payoff if he chooses y is equal to

$$z1^- + (1 - z)\lambda^- - 1/2,$$

while if he chooses y' his payoff is

$$z\lambda^- + (1 - z)1^- - y'/2.$$

His best response is to choose y whenever

$$z > 1/2 - (y' - 1)/4(1 - \lambda).$$

Accordingly, his bicentric prior assigns the probability

$$p_1(y) = 1/2 + (y' - 1)/4(1 - \lambda). \tag{7.5}$$

to choosing y and the complementary probability to choosing y'.
Next, turn to agents of the firms. Let agent iy have beliefs $ze + (1 - z)e'$. If this agent chooses $w \notin \{\lambda^-, 1^-\}$, then he will only get a worker if e' is played and in this case his expected payoff is negative. Hence, such a wage cannot be optimal since by offering λ^- the agent guarantees a nonnegative payoff. In fact, λ^- yields an expected payoff of

$$(1 - z)\, \epsilon\, g/2\,. \tag{7.6}$$

On the other hand, if the agent chooses 1^- his expected payoff is

$$z\lambda^\epsilon\, g^\epsilon/2 + (1 - z)\, \epsilon\, (\lambda - 1^-)\,, \tag{7.7}$$

where λ^ϵ and g^ϵ are defined by

$$\left.\begin{aligned} \lambda^\epsilon &= s^\epsilon(y) \\ g^\epsilon &= \mu_s^\epsilon(y) - 1^- \end{aligned}\right\} \text{ with } s_1(y) = 1\,.$$

(Note that $(\lambda^\epsilon, g^\epsilon) \to (\lambda, g)$ as $\epsilon \to 0$). Comparing (7.6) and (7.7) we see that agent iy should choose λ^- whenever

$$z\lambda^\epsilon\, g^\epsilon/2 \;<\; (1 - z)\, \epsilon\, (1 - \lambda - g/2)\,. \tag{7.8}$$

Write δ^ϵ for the probability that (7.8) holds when z is uniformly distributed on $[0,1]$. Note that $\delta^\epsilon \to 0$ as $\epsilon \to 0$, hence, the prior strategy of agent iy chooses 1^- with probability close to 1 if ϵ is small. The computations at y' are identical to those at y, hence, we find for the prior strategies of the firms

$$w_{iy} = w_{iy'} = \begin{cases} 1^- \text{ with probability } 1 - \delta^\epsilon \\ \lambda^- \text{ with probability } \delta^\epsilon \end{cases}. \tag{7.9}$$

Given the prior strategy combination p as in (7.5), (7.9) it is easy to compute the equilibrium of G_p^0, that is, the starting point of the tracing path. Since the expected wage at y is the same as that at y', the worker chooses y since costs are lower there. Denote by $\mu^\epsilon(y)$ the expected prior productivity at y(i.e. $\mu^\epsilon(y)$ is determined by (7.4), (7.5) and (3.6)). Then $\mu^\epsilon(y) \to 1$ as $\epsilon \to 0$. The expected payoff from offering 1^- is at least equal to

$$(1 - \delta^\epsilon)(\mu^\epsilon(y) - 1^-)$$

whereas the expected payoff associated to any other wage is bounded above by

$$\delta^\epsilon\ \mu^\epsilon(y) - \lambda^-$$

Consequently, only 1^- is a best reply against the prior if ϵ is small. Hence, in any equilibrium of G_p^0, the firms offer $w_{iy} = 1^-$. The same argument also generates this conclusion at y'. Hence, G_p^0 has a unique equilibrium, and this is given by

$$s_1(y) = 1 \ , \quad w_{iy} = w_{iy'} = 1^- \tag{7.10}$$

Now consider $t > 0$. If firms do not change their wage offers, there is no reason for the worker to deviate from y, and if this worker stays at y, the agents $2y$ and $3y$ should not change their wage offers either. What about the firms' agents at y'? For t small, their (subjective) payoffs from (7.2) are still largely determined by their priors and they will find it optimal to offer $w = 1^-$. Consequently, for t small, the tracing path continues with the equilibrium from (7.10). However, if the worker remains at $y = 1$, then the expected productivity at y' decreases with increasing t and the payoffs to the firms' agents at y' would become negative for t close to 1 if these agents would remain at their wage offer of 1^-. Consequently, these agents will be the first to switch, and they will switch to lower wages. This, of course , reinforces the decisions of the worker and consequently of the firms' agents at $y = 1$: Along the tracing path, these agents will never switch, and the agents at y' have to adjust until finally an equilibrium of G is reached. In the end, these agents will, therefore, offer the wage λ^- and we see that the tracing path must end up in the equilibrium e. Hence, e risk dominates e'.

□

It is instructive to study in somewhat greater detail the adjustment process (i.e. the tracing path) that brings players from the prior p to the equilibrium e. As already remarked above only the agents at y' change behavior during the process. Take ϵ fixed and write $\tilde{G}_p^t$ for the game as in (7.2) played by these agents given that the type 1 worker chooses $y = 1$ for all t. $\tilde{G}_p^t$ resembles a standard Bertrand game for a surplus $\mu(t)$ where $\mu(t)$ is decreasing in t with $1^- < \mu(0) < 1$ and $\mu(1) = \lambda$. The only difference with an ordinary Bertrand game is that in $\tilde{G}_p^t$ each agent is committed to choose 1^- with probability $1 - t$. Nevertheless, the equilibria of $\tilde{G}_p^t$ can be read off from Lemma 2.1. It is easily seen that there exists $t_1 > 0$ such that only $(1^-, 1^-)$ is an equilibrium of $\tilde{G}_p^t$ for $t < t_1$. Furthermore, there exists $t_2 > t_1$ such that for $t \in (t_1, t_2)$ the game $\tilde{G}_p^t$ has three equilibria, viz. $(1^-, 1^-), (1^{--}, 1^{--})$ and a mixed one. If $t > t_2$, $(1^{--}, 1^{--})$ is still an equilibrium, but the other two are not, and they are replaced by two different equilibria, viz. (1-3g,1-3g) and a mixture of 1-2g and 1-3g. Hence, a switch of behavior has to occur at or before t_2. Since the tracing path must be continuous, it cannot jump from $(1^-, 1^-)$ to $(1^{--}, 1^{--})$ at t_1 or t_2, consequently it must bend backwards at t_2. Therefore, the initial segment of the tracing path looks as follows. From 0 to t_2 it consists of the equilibrium $(1^-, 1^-)$. At t_2 it bends backwards (i.e. t is decreasing), continuing with the equilibrium in which firms randomize between $1 - g$ and $1 - 2g$ (gradually increasing the probability of $1 - 2g$ from 0 to 1), at t_1 it bends forward again and continues with the equilibrium $(1 - 2g, 1 - 2g)$. This alternation between forward and backward moving segments continues while simultaneously lowering the wages until finally the path becomes stationary at (λ^-, λ^-).

We finally come to the risk dominance comparison of the pooling equilibrium with the best separating equilibrium. The main result of this section states that the separating equilibrium dominates the pooling one if $\lambda < \frac{1}{2}$ and ϵ is small. Again the intuition is simple. Consider a situation of mutual uncertainty concerning whether, in the unperturbed game, the separating equilibrium e or the pooling equilibrium e' should be played. The type 0 worker chooses $y = 0$ in both, hence, for him the situation is unproblematic. In fact, since the wage at $y = 1$ is at most 1^-, this type will always strictly prefer to choose $y = 0$. The prior uncertainty will however lead the type 1 worker to choose both $y = 0$ and $y = 1$ with a probability that is bounded away from zero. (Below we show that the probability is approximately 1/2). Hence, firms at $y = 1$ will infer that they face the type 1 worker and they will offer $w = 1^-$. On the other hand, firms at $y = 0$ infer that the expected productivity is below (and bounded) away from λ, hence, their wage offers will be below λ as well. Given that $0 < w(y) < \lambda$ for $y = 0$ and that $w(y) = 1$ for $y = 1$, the type t worker prefers to choose $y = t$. These choices reinforce firms to offer $w = 1^-$ at $y = 1$. At $y = 0$, however, firms gradually become convinced that they face the type 0 worker and this leads them to gradually lower their wage offers, until they finally offer 0. Hence, we end up at the separating equilibrium, the separating equilibrium risk dominates the pooling one.

Formalizing the above argument is, unfortunately, rather cumbersome since the HS theory requires working with the perturbed game. The difficulty is caused by the fact that the separating equilibrium can only be approximated by mixed equilibria of the perturbed game and the latter are rather complicated (cf. Proposition 5.6). Hence, there is an enormous number of switches along the tracing path. The difficulties are not of a conceptual nature, however, formally one just has to go through a large number of steps similar to the ones described in detail in the proof of Proposition 7.3. Since already in that case the notation became cumbersome and the formal steps were not particularly illuminating, we prefer to stick to the main ideas and make the risk dominance comparison in the unperturbed game. It can be shown that this shortcut does not bias the results. (The detailed argument for the perturbed game is available from the authors upon request).

Proposition 7.4. *If $\lambda < 1/2$, then in the unperturbed game, the best separating equilibrium risk dominates the best pooling equilibrium. The same dominance relationship exists in the perturbed game $\tilde{G}^r_\epsilon(Y, W, \lambda)$ provided that ϵ is small.*

Proof. The main advantage in working with the unperturbed game lies in the reduction in the number of agents involved in the risk dominance comparison: We do not have to consider the type 0 worker (he chooses 0 in both equilibria), nor the firms' agents at $y \neq 0, 1$. Let e denote the separating equilibrium in which the type 1 worker chooses 1 and let e' be the pooling equilibrium in which both workers choose 0. We first compute the bicentric prior. Let players have beliefs $ze + (1 - z)e'$. If the type 1 worker chooses $y = 0$ his expected payoff is $(1 - z)\lambda^-$, if he chooses $y = 1$, the expected payoff is [7] $z1^- + (1 - z)\lambda^- - 1/2$. Consequently, this worker will choose both $y = 0$ and $y = 1$ with a probability of approximately 1/2. Next, consider the agent of a firm at $y = 1$. This agent has to move only if the separating equilibrium is played, and in this case the agent of the competing firm offers 1^-. Hence, the only way in which this agent can make a profit is by also offering 1^-. Therefore, the bicentric prior is to offer 1^- for sure. Finally, consider a firm's agent at $y = 0$. It is easy to see that any wage $w \notin \{0, \lambda^-\}$ yields negative expected profits. If the agent offers the wage 0 the expected profit is 0, while the profit resulting from λ^- is equal to

$$-z(1 - \lambda)(\lambda - g) + (1 - z)g/2.$$

Consequently, the agent at 0 should offer the wage 0 if

$$z((1 - \lambda)(\lambda - g) + g/2) > g/2.$$

[7] The pooling equlibrium of the unperturbed game does not specify a unique wage at $y = 1$, we fix this wage at λ^-, that is, at the limit of the wages of approximating equilibria of the perturbed games.

Let δ be defined by

$$\delta = \frac{g/2}{(1-\lambda)(\lambda - g) + g/2} \quad , \tag{7.11}$$

then the bicentric prior of an agent at 0 chooses λ^- with probability δ and 0 with probability $1-\delta$. Since g can be chosen arbitrarily small, the initial expectation of the worker is, therefore, that the expected wage at 0 is close to 0. Consequently, for g small, the best response of the type 1 worker is to choose $y = 1$ for sure and this reinforces the firms to offer $w = 1^-$ at $y = 1$. We claim that in any equilibrium along the path followed by the tracing procedure, the type 1 worker chooses $y = 1$. Namely, this property can only fail to hold if the firms at $y = 0$ offer a sufficiently high wage (at least equal to 1/2) and this will never be the case since the expected productivity at 0 will always be below λ and $\lambda < \frac{1}{2}$. In fact along the tracing path the wage offer at $y = 0$ will never be above $\lambda/2$ since this is the highest expected productivity at $y = 0$ along the path. The exact tracing path can be determined by using an argument as the one that follows the proof of Proposition 7.3: For tracing parameter t, the agents at $y = 0$ play a Bertrand game for a surplus of (approximately) $(1-t)\lambda/2$ modified to the extent that each agent is committed to play his prior as in (7.11) with probability $1-t$. The unique equilibrium of this game for $t = 0$ is (y, y). As t increases, firms initially switch to higher wages (since for t small the expected surplus is large), however, for larger t, wages fall again since the surplus decreases. (Again the tracing path contains many backwards running segments). As t tends to 1 the surplus, hence, the wages tend to zero; along the way the wages never exceed the maximal surplus of $\lambda/2$. Consequently, the agents at 0 finally switch to the wage corresponding to the separating equilibrium. Since the other agents are already at this equilibrium from the beginning, the tracing path leads to the separating equilibrium. Hence, the separating equilibrium risk dominates the pooling one.

□

8. Conclusion

By combining the Propositions 7.1, 7.3 and 7.4, we obtain the main result of this paper.

Corollary 8.1. *The Harsanyi/Selten solution of the Spence signaling game $\Gamma(Y, W\lambda)$ is* [8)]

(i) the pooling equilibrium in which both workers choose $y = 0$ if $\lambda \leq \frac{1}{2}$,

(ii) the separating equilibrium in which the type t worker chooses $y = t$ if $\lambda > \frac{1}{2}$.

Even though we used the assumptions (2.6) - (2.10) to derive this result, it can be checked that at least for $\lambda \neq 1/2$, the statement of Corollary 8.1 remains correct if these assumptions are not satisfied. Hence, the result is independent of the discretization chosen. It is, therefore, justified to make a limiting argument and to talk about the HS solution of the continuum game. Hence, $g = 0$ and Y and W are continua, as in the usual specification of the Spence model found in the literature. Denote this game by $\Gamma(\lambda)$. The (limit as $g \to 0$ of the) solution found in Corollary 8.1 is known in the literature as the Wilson E_2-equilibrium (Wilson [1977]). It is that

[8] Actually, this Corollary only describes the outcome associated with the HS solution, it doesn't describe the wages firms intend to offer at education levels that are not chosen. From the previous section we know that $w_i(y) = \lambda^-$ for all y if $\lambda \leq 1/2$ and $w_i(y) = \min\{y, \lambda^-\}$ if $\lambda > 1/2$ and $y \neq 1$.

sequential equilibrium of $\Gamma(\lambda)$ that is best from the viewpoint of the type 1 worker. Hence, we have

Corollary 8.2. *For $\lambda \neq \frac{1}{2}$, the HS solution of the Spence signaling game $\Gamma(\lambda)$ is the Wilson E_2-equilibrium of $\Gamma(\lambda)$, i.e. it is the sequential equilibrium that is best for the type 1 worker.*

Given the ad hoc nature of Wilson's solution concept, it is to be expected that a coincidence as in Corollary 8.2 will not hold in general. However, the authors conjecture that for a broad class of signaling games that have similar structural properties as the game studied in this paper, the HS solution coincides with the solution proposed in Miyazaki [1977]. (The important properties are monotonicity and the single crossing condition, see Cho and Sobel [1977], it is conjectured that the number of types does not play a role). A detailed investigation into this issue will be carried out in a future paper.

The simple structure of our game enables a sensitivity analysis with respect to several assumptions made by Harsanyi and Selten that is difficult to carry out in general. We will not go into detail, but restrict ourselves to one issue. Some people have argued that the uniformity assumption made in the construction of the prior to start the tracing procedure is ad hoc. Consequently, one may ask how robust the results from Sect. 7 are with respect to this prior. The reader can easily convince himself that the outcome is very robust. Robustness especially holds for Proposition 7.4 : The separating outcome will result for $\lambda < \frac{1}{2}$ as long as the prior expectations of the players assign positive probability to this outcome (i.e. as long as the density of z is strictly positive on $[0, 1]$).

The theory of Harsanyi and Selten has both evolutionary and eductive aspects. (See Binmore [1987] for a general discussion of these notions). The preference for primitive equilibria is most easily justified by taking an evolutionary perspective, the tracing procedure most certainly is eductive in nature. It is interesting to note that the HS theory ranks evolutionary considerations prior to eductive ones. It seems that the ordering of steps has consequences for the final outcome since, at least in the unperturbed game, the separating equilibrium outcome with $y_1 = 1$ risk dominates any other equilibrium outcome for all values of λ. (cf the proofs of the Propositions 7.3 , 7.4). If the solution would be based on risk dominance considerations alone, the solution might always involve separation. Consequently, the solution that was initially proposed by Spence and that is defended and used in most of the subsequent literature may also be justified on the basis of risk dominance. Again we intend to investigate this issue more thoroughly in future work.

Finally, let us mention that recently related work has been done by Michael Mitzkewitz (Mitzkewitz [1990]). He computes the HS solution for signaling games in the following class: There are two players, player 1 has 2 possible types, and he can send 2 possible messages, to which player 2 can react in two different ways. Mitzkewitz does not assume the single crossing property, hence, he is forced to use an approach that differs from ours.

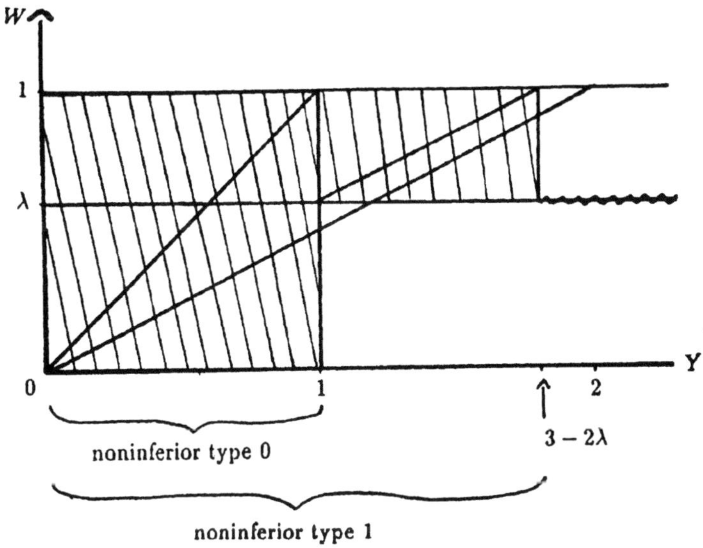

Figure 1. The reduced perturbed game $\tilde{G}^r_\epsilon(Y, W, \lambda)$.

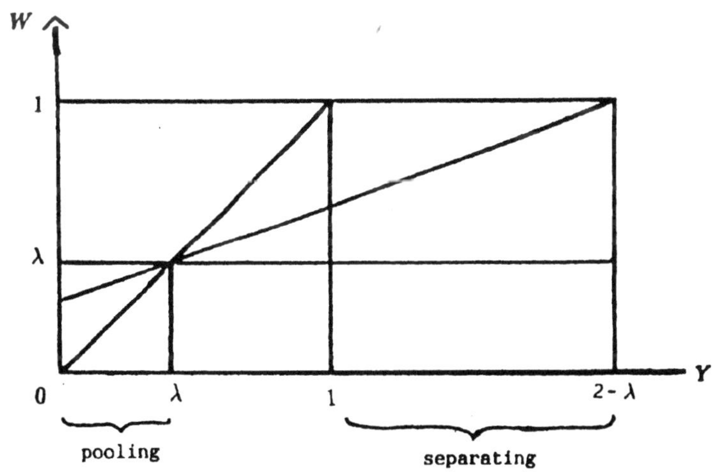

Figure 2. Uniformly Perfect Equilibrium Outcomes.

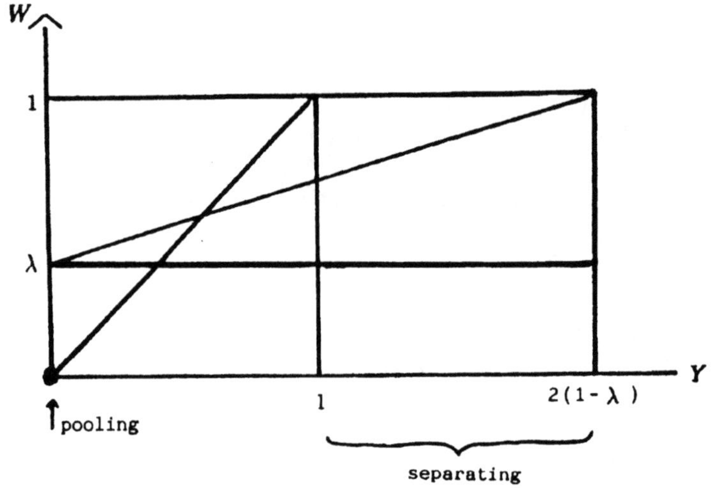

Figure 3. Primitive Equilibrium Outcomes ($\lambda < \frac{1}{2}$).

Appendix A: Dominance Solvability and the HS Solution

Consider the 2-person normal form game

	A_2	B_2	C_2
A_1	10, 10	0, 10	0, 0
B_1	10, 0	6, 6	0, 6
C_1	0, 0	6, 0	1, 1

Table 1: G

which admits three Nash equilibria, viz. $A = (A_1, A_2)$, $B = (B_1, B_2)$ and $C = (C_1, C_2)$.

We have that A_1 (resp. A_2) is dominated by B_1 (resp. B_2), while in the reduced game that results after A_1, A_2 have been eliminated, B_1 (resp. B_2) is dominated by C_1 (resp. C_2). Hence, the (unperturbed) game is dominance solvable, with solution C. (This is the unique stable equilibrium of the game). HS, however, do not analyse the unperturbed game, but rather a sequence of uniformly perturbed games. In the ϵ-uniformly perturbed game $G(\epsilon)$ of G each player, when he intends to choose the pure strategy X_i, will actually choose the completely mixed strategy $(1-2\epsilon)X_i + \epsilon Y_i + \epsilon Z_i$ with X_i, Y_i, and Y_i being pairwise different. Neglecting terms of order ϵ^2, the payoff matrix of $G(\epsilon)$ is given by

$10-30\epsilon$, $10-30\epsilon$	22ϵ, $10-24\epsilon$	11ϵ, 27ϵ
$10-24\epsilon$, 22ϵ	$6-8\epsilon$, $6-8\epsilon$	17ϵ, $6-17\epsilon$
27ϵ, 11ϵ	$6-17\epsilon$, 17ϵ	$1+2\epsilon$, $1+2\epsilon$

Table 2: $G(\epsilon)$

In $G(\epsilon)$, we have that A_1 (resp. A_2) is dominated by B_1 (resp. B_2). The reduced game in which these inferior strategies have been eliminated has two strict equilibria, viz. B and C. (There is a third equilibrium in mixed strategies but this is not primitive). Hence, the initial candidate set is $\{B, C\}$. Since B payoff dominates C, the HS solution of $G(\epsilon)$ is B. Consequently, B is the HS solution of G. Hence, perturbing first may make a difference since it can transfer non-strict equilibria into strict ones.

This example may lead the reader to think that the discrepancy between dominance solvability and the HS theory is caused by the fact that the latter makes use of Pareto comparisons and that a solution that is based only on risk dominance comparisons would always yield the stable outcome for dominance solvable games. Indeed in $G(\epsilon)$ the equilibrium (C, C) risk dominates (B, B) if ϵ is small. This issue is investigated further in Van Damme (1990).

References

Binmore, Ken (1987,1988) : Modeling rational players; *Journal of Economics and Philosophy. Part I,3 179-241, Part II,4 9-55.*

Cho, In-Koo and Kreps, David (1987) : Signalling Games and Stable Equilibria; *Quarterly Journal of Economics 102, 179-221.*

Cho, In-Koo and Sobel, Joel (1987) : Strategic stability and uniqueness in signaling games; *Mimeo, U.C. San Diego.*

Harsanyi, John and Reinhard Selten (1988) : A general theory of equilibrium selection in games; *MIT Press, Cambridge, MA.*

Mertens, Jean-François (1988) : Stable Equilibria - A reformulation; *Core DP 8838.*

Selten, Reinhard (1975) : Reexamination of the perfectness concept for equilibrium points in extensive games; *International Journal of Game Theory.*

Kohlberg, Elon and Jean-Francois Mertens (1986) : On the Strategic Stability of Equilibria; *Econometrica 54, 1003-1039.*

Mitzkewitz, Michael (1990) : Equilibrium Selection in signaling games; *Mimeo, Bonn.*

Miyazaki, Majime (1977) : The rat race and internal labor markets; *Bell Journal of Economics 8, 394-419.*

Spence, Michael (1974) : *Market Signalling: Informational Transfer in Hiring and Related Processes*; Harvard Univ. Press, Cambridge, USA.

Spence, Michael (1973) : Job Market Signalling; *Quarterly Journal of Economics 87, 355-374.*

Van Damme, Eric (1987) : *Stability and Perfection of Nash Equilibria*; Springer Verlag, Berlin.

Van Damme, Eric (1989) : *Stable Equilibria and Forward Induction*; Journal of Economic Theory 48, 476-496.

Van Damme, Eric (1990) : A note on risk dominance and dominance solvability; *Mimeo, Tilburg.*

Wilson, Charles (1977) : A Model of Insurance Markets with Incomplete Information; *Journal of Economic Theory, 167-207.*

INTERACTION BETWEEN RESOURCE EXTRACTION AND FUTURES MARKETS: A GAME-THEORETIC ANALYSIS[1]

by

Louis Phlips and Ronald M. Harstad

Abstract: This paper is an attempt to model three salient features of futures markets for exhaustible resources: oligopolistic control of the market for the basic commodity, the purely speculative nature of most of futures trading and apparent availability of unlimited funds to the traders.

Interaction between the extraction policy of duopolists and their futures trading is modelled with the help of a two-stage noncooperative game with incomplete information. The second stage is a two-period extraction game, which shows how futures positions taken by the producers affect their planned extraction rates and the resulting spot price of the resource. The first stage is a noncooperative futures game in which equilibrium net positions of and subgame perfect contracts between two producers and a representative speculator are determined.

The purely speculative part of futures positions depends mainly on the (known) structure of inconsistent prior beliefs about the level of uncertain market demand for the basic commodity. The hedging part is related to (known) levels of stocks of the resource available to the individual producers, and to their degree of risk aversion. The conditions under which each player goes net short or net long are worked out in detail.

0. Introduction

This paper is an attempt to model three features which, according to casual observation, characterize futures markets for exhaustible resources such as petroleum, tin or copper, and perhaps a number of other futures markets as well.

On the production side, the markets for the basic commodity are controlled by a few dominant producers: there is oligopolistic competition rather than perfect competition. These oligopolists determine the time path of their extraction rates. Simultaneously they can and do take positions on the futures markets for the commodity they extract. The profitability of these positions depends upon the difference between current futures price and the spot price that will materialize at maturity. This spot price, however, is to some extent under their control, since it depends at least in part upon their extraction policy. As a consequence, extraction

[1]This paper was written while the authors were visiting the Center for Interdisciplinary Research, Bielefeld, W. Germany. We are grateful for the Center's support, and for helpful comments from T. Brianza, E. van Damme, J.-F. Richard, Y. Richelle and R. Selten.

policies of the oligopolists are likely to influence profitability of their futures positions. Conversely, futures positions taken are likely to influence extraction policies and the time path of the spot price.

Secondly, the spectacular development of futures markets suggests that whoever wants to be active on these markets can find necessary funds without difficulty. The organizational set-up is such that only a small fraction of the value of futures transactions has to be available in cash. In addition, futures positions are closed out day by day, including the maturity date, so that actual payments are only a small fraction of the values involved. On top of this, banks apparently readily provide credit to finance these fractions -- until there is a crash, as exemplified by the 1985 tin crisis (see House of Commons (1986) and Anderson and Gilbert (1988)) or, more generally, by the October 1987 stock exchange crisis. This suggests that, for practical purposes, agents active on futures markets (be they producers, traders or speculators) can act (and can be modelled) as if they had no budget constraint.

A third feature which strikes the eye is the purely speculative nature of the majority of transactions on futures markets. Whether this is a recent phenomenon or not is immaterial. The point is that the traditional literature on futures is cast in terms of producers (competitive farmers are favorite examples) who use futures as a hedge against a possibly unfavorable spot price (at harvest time). It seems appropriate to focus analysis instead on the purely speculative part of futures positions, which are not related to producers' initial stocks of the natural resource.

We analyze the situation as a two-stage noncooperative game with inconsistent incomplete information. The production stage is modelled, in the pages that follow, as a Cournot duopoly game. The strategies are extraction rates of a homogeneous exhaustible resource (for which there can be only one equilibrium spot price in each time period). The game is played over the shortest possible number of periods, namely two. The second period is the maturity date of futures contracts made in the first period.

The equilibrium futures prices and the equilibrium futures positions are determined in a game played simultaneously in the first period. This is a three-person game between the two (risk-neutral or risk-averse) producers (call them A and B) and one representative (risk-neutral) speculator (S). All players know the rules of the extraction game played by the two producers. There is uncertainty for all about the level of aggregate demand for the exhaustible resource. The futures game is positive-sum (in terms of expected profit) whenever positions taken are sufficiently small relative to differences in prior beliefs about the level of aggregate demand. These beliefs are common knowledge.

The "futures game" thus leads at each of its "terminations" (sets of agreed-upon contracts) into an extraction game which is a proper subgame. Equilibrium payoffs in the extraction game vary with the futures positions taken, but all these subgames share a common structure, and a unique equilibrium extraction path is determined for each. This path is compared with a monopoly extractor's behavior when he speculates in the futures market (drawn from Brianza, Phlips and Richard (1987)).

We proceed in the standard way, backward, by considering the extraction game first, and then exploring in Section II the futures game leading to it.

I. The Extraction Game

Two producers, A and B, of an exhaustible resource have known initial stocks of s_{a0} and s_{b0}. They each face the intertemporal constraint that the initial stock has to be depleted in period 2, or

$$s_{a0} = q_{a1} + q_{a2} \tag{1.a}$$

$$s_{b0} = q_{b1} + q_{b2} , \tag{1.b}$$

$q_{it}(i = a,b;t = 1,2)$ being the extraction rate of producer i in period t. Each period will exhibit the same instantaneous inverse market demand curve

$$p_t(q_{at} + q_{bt}) = \frac{\alpha}{\beta} - \frac{q_{at} + q_{bt}}{\beta} \qquad t = 1,2 \tag{2}$$

(with $\alpha > 0$ and $\beta > 0$). In other words, any quantity extracted is immediately sold and consumed: there is no demand for speculative stocks. This admittedly strong assumption is made to simplify the analysis. (In real life, speculators of course not only buy or sell futures, but also buy the commodity in period 1 and carry it over to period 2 when they expect the spot price p_t to rise. Introducing this possibility would complicate the analysis unnecessarily.)

On the futures market, producers A and B have sold futures contracts to the speculator in (possibly negative) amounts f_{as} and f_{bs} at prices p_{as} and p_{bs}. In addition, at price p_{ab}, A has sold to B a quantity $f_{ab} = -f_{ba}$ on the futures market. We maintain the convention that $f_{ij} > 0$ if i has sold to j with $f_{ij} < 0$ if j has sold futures to i (i,j = A,B,S). Thus, $f_{ij} > 0$ indicates a short position for i and a long position for j, at least on this contract. (Also, note the convention that a doubly subscripted price is a futures price.) Details of the determination of these positions and prices are left to Section II. These positions relate to the producers' expectations α_a and α_b about the level of aggregate demand, α, which is the only unknown parameter in the extraction game. Planned extraction paths are determined at the beginning of period 1, before α is revealed.

To focus on futures market impact on extraction paths, we simplify by setting the interest rate to zero. Equilibrium extraction paths can then be readily compared to the constant extraction paths $\frac{s_{i0}}{2}$ obtained in the absence of a futures market. Producer A's intertemporal profit function is

$$\pi_a = p_2(q_{a2} + q_{b2})(q_{a2}-f_{ab}-f_{as}) + p_1(q_{a1} + q_{b1})q_{a1} + p_{ab}f_{ab} + p_{as}f_{as}, \quad (3.a)$$

which sums extraction sales and net profits of futures contracts. Similarly, producer B has the intertemporal profit function

$$\pi_b = p_2(q_{a2} + q_{b2})(q_{b2} + f_{ab} - f_{bs}) + p_1(q_{a1} + q_{b1})q_{b1} - p_{ab}f_{ab} + p_{bs}f_{bs}, \quad (3.b)$$

where the signs on f_{ab} terms are reversed because $f_{ab} > 0$ represents a short position for A, but a long position for B. Extraction costs are zero or constant and equal between the duopolists.

Payoffs in the extraction game are the certainty equivalent levels of profit:

$$W_a = E_a[\pi_a] - \frac{K_a}{2\beta^2} \operatorname{var}(\alpha_a)(s_{a0} - f_{ab} - f_{as})^2 \quad (4.a)$$

$$W_b = E_b[\pi_b] - \frac{K_b}{2\beta^2} \operatorname{var}(\alpha_b)(s_{b0} + f_{ab} - f_{bs})^2, \quad (4.b)$$

where $K_i \geq 0$ is the degree of constant absolute risk aversion. Equilibrium results from simultaneous maximization of (4.a) and (4.b) subject to (1.a), (1.b).

Notice the implications of depletion constraints (1.a) and (1.b) in this respect. Since there are only two periods, it is always optimal to extract what is left under the ground in period 2, whatever actual extraction rates were in period 1. There is thus no problem of time consistency. Second, no strategic commitments can be made for period 2 and so-called "path strategies" extending over the two-period horizon – in the terminology of Reinganum and Stokey (1985) – are ruled out. Third, backward induction is not necessary to find the subgame perfect equilibrium decision rules: it suffices to solve the (four equation) system of first-order conditions using $q_{a2} = s_{a0}-q_{a1}$ and $q_{b2} = s_{b0}-q_{b1}$. Equilibrium exists since each producer's marginal revenue declines as the extraction of the other producer increases (Novshek (1985)).

Maximization of Lagrangians corresponding to (4.a) and (4.b) with $p_t(q_{at}+q_{bt})$ specified as in Equation (2) gives

$$\frac{\alpha-2q_{a1}-q_{b1}}{\beta} = \frac{\alpha-2q_{a2}+(f_{ab}+f_{as})-q_{b2}}{\beta} = \eta_a \tag{5.a}$$

and

$$\frac{\alpha-2q_{b1}-q_{a1}}{\beta} = \frac{\alpha-2q_{b2}+(f_{bs}-f_{ab})-q_{a2}}{\beta} = \eta_b \tag{5.b}$$

where η_a and η_b are multipliers associated with (1.a) and (1.b). The unique equilibrium strategies are[2]

$$q_{a1} = \frac{s_{a0}}{2} + \frac{f_{bs}-f_{ab}}{6} - \frac{f_{ab}+f_{as}}{3} \tag{6.a}$$

$$q_{b1} = \frac{s_{b0}}{2} + \frac{f_{ab}+f_{as}}{6} - \frac{f_{bs}-f_{ab}}{3} \tag{6.b}$$

$$q_{a2} = \frac{s_{a0}}{2} - \frac{f_{bs}-f_{ab}}{6} + \frac{f_{ab}+f_{as}}{3} \tag{6.c}$$

$$q_{b2} = \frac{s_{b0}}{2} - \frac{f_{ab}+f_{as}}{6} + \frac{f_{bs}-f_{ab}}{3} \tag{6.d}$$

which sum to total extraction rates

$$q_1 = \frac{s_{a0}+s_{b0}}{2} - \frac{f_{as}+f_{bs}}{6} \tag{7.a}$$

$$q_2 = \frac{s_{a0}+s_{b0}}{2} + \frac{f_{as}+f_{bs}}{6} \quad . \tag{7.b}$$

According to Eqs. (6.a)-(6.d), taking a net short position $(f_{ab}+f_{as} > 0)$ implies that producer A pumps less in period 1 and more in period 2 than in the absence of a futures market. A long position carries the reverse implication. If A's competitor also has a net short position $(f_{bs}-f_{ab} > 0)$, this counteracts the reduction in A's extraction rate in period 1 and the increase in period 2. As the counteraction is less pronounced, contracts with the speculator impact on the industry's extraction path. Extraction is delayed if the speculator takes a net long position, and vice versa.

The effect on equilibrium prices is clear from the following equations:

$$p_1 = \hat{p} + \theta(f_{as}+f_{bs}) \tag{8.a}$$

$$p_2 = \hat{p} - \theta(f_{as}+f_{bs}) \tag{8.b}$$

---

[2]Equations (6) assume an interior maximum.

where $\hat{p} = \alpha/\beta-(s_{a0}+s_{b0})/2\beta$ and $\theta = 1/6\beta > 0$. A combined net short position increases p_1 and reduces p_2. In a world of increasing spot prices -- with a positive real interest rate -- this means that the variability of the spot price is reduced (or reversed). A combined net long position has the opposite effect.

At this point, a comparison with the monopoly case is in order. A monopolist with initial stock $(s_{a0}+s_{b0})$ and futures position $(f_{as}+f_{bs})$ extracts (taken from Brianza, Phlips and Richard, 1987):

$$Q_1 = \frac{s_{a0}+s_{b0}}{2} - \frac{f_{as}+f_{bs}}{4}$$

$$Q_2 = \frac{s_{a0}+s_{b0}}{2} + \frac{f_{as}+f_{bs}}{4}$$

and sets prices

$$P_1 = \hat{p} + \frac{3}{2}\theta(f_{as}+f_{bs})$$

$$P_2 = \hat{p} - \frac{3}{2}\theta(f_{as}+f_{bs}).$$

Notice that in a zero-interest-rate world without a futures market, duopoly achieves the same extraction and price paths as monopoly.

More importantly, with a futures market the impact of producers' futures activity upon *both* total extraction rates and prices is more pronounced in monopoly than in duopoly. In a world of increasing spot prices, *the variability of prices is larger under duopoly than under monopoly when the duopolists take a combined net short position.*

Of course, the duopolists expect to profit from the futures contracts they sign. Some of this profit is passed on to demanders of the resource being extracted. That is, relative to the extraction and price paths without a futures market, Eqs. (7) and (8) show some extraction moved from a (relatively) higher-priced to a (relatively) lower-priced period.

2. The Futures Game

We consider a noncooperative game with 3 players, producers A and B and the speculator S. At the beginning of period 1, a futures market is opened and contracts signed in accordance with rules specified as follows:

Step 1: A "commitment order", which is a permutation of the order {A,B,S}, is chosen randomly, each permutation with equal probability. This commitment order is common knowledge.

Step 2: Simultaneously, each player is allowed to place on the market ("announce") one sale contract offer and/or one purchase contract offer. Each offer consists of any nonnegative but finite (price, quantity) pair. Once announced, the offers are irrevocable.

Step 3: The player designated first in the commitment order irrevocably accepts or rejects each contract offered by another player.

Step 4: The player designated second in the commitment order irrevocably accepts or rejects each contract offered by another player (and not yet accepted).

Step 5: The last player irrevocably accepts or rejects each contract offered by another player (and not yet accepted). The futures market closes.

Step 6: The producers then play the extraction subgame.

Step 7: The demand level α is revealed.

While alternative rule configurations could be analyzed, this structure clarifies interaction between futures activity and production while maintaining simplicity.[3]

2.1 The Payoffs

Payoffs for producers A and B are given by the certainty equivalent profit functions (4.a) and (4.b). The speculator's payoff is

$$\phi_s = (p_2-p_{as})f_{as} + (p_2-p_{bs})f_{bs} \tag{9}$$

(recall $f_{is} > 0$ is a futures purchase by S from producer i).

We saw above that expectations about α do not affect the extraction subgame, because of the combined assumptions that α is the same in both periods and that given stocks must be depleted across the two periods. In the futures game, expectations play a central role, which can be clarified by expressing beliefs as expectations about $\hat{p}$ rather than α. Let the mean expectations be $\hat{p}_a$ and $\hat{p}_b$ for the producers, $\hat{p}_s$ for the speculator. Since an outside observer cannot separate risk aversion (K_i) from confidence of beliefs $(\mathrm{var}\ \hat{p}_i)$ in producer behavior, we simplify by setting $\mathrm{var}(\hat{p}_i) = \mathrm{var}(\hat{p}) = 1$.

Since the extraction game to follow is common knowledge, its unique subgame-perfect equilibrium strategies (6.a)-(6.d) and (8.a)-(8.b) can be incorporated in the

[3]Notice that intermediate steps after step 2 and before step 3 could be added, during which contracts offered could be irrevocably accepted or tentatively rejected. In the equilibria we consider, there would not be any acceptances "before the deadline" (steps 3-5). The possibility of improving one's offer (selling at a lower price, or buying at a higher price) could also be introduced between steps 2 and 3, and would not have any impact on subgame-perfect equilibria.

payoff functions:

$$W_a = s_{a0}\,\hat{p}_a + V_a \tag{10.a}$$

$$W_b = s_{b0}\,\hat{p}_b + V_b \tag{10.b}$$

$$\phi_s = (\hat{p}_s - p_{as})f_{as} + (\hat{p}_s - p_{bs})f_{bs} - \theta(f_{as} + f_{bs})^2 \tag{10.c}$$

where

$$V_a = (p_{ab} - \hat{p}_a)f_{ab} + (p_{as} - \hat{p}_a)f_{as} + \frac{\theta}{3}(f_{as} + f_{bs})^2 - \frac{K_a}{2}(s_{a0} - f_{ab} - f_{as})^2 \tag{11.a}$$

$$V_b = (\hat{p}_b - p_{ab})f_{ab} + (p_{bs} - \hat{p}_b)f_{bs} + \frac{\theta}{3}(f_{as} + f_{bs})^2 - \frac{K_b}{2}(s_{b0} + f_{ab} - f_{bs})^2 \tag{11.b}$$

(recall that $f_{ij} > 0$ implies i sells futures to j).[4] Because $s_{i0}\hat{p}_i$ is the expected value of the stock in the absence of a futures market, and is unaffected by futures trading, it is useful to focus on V_a, V_b and ϕ_s.

One striking way in which the futures and extraction markets interact shows up in Eqs. (11). Even if producer A chooses to be inactive in the futures market, its existence adds to his profit, as $V_a = (\theta/3)(f_{bs})^2$ when $f_{ab} = f_{as} = 0$. This is because a producer with a zero net short position rationally responds to his competitor's net position by shifting extraction to the period with a higher (discounted) price.

2.2. Beliefs and Attitudes Towards Risk

The payoffs are common knowledge; therefore players' beliefs about $\hat{p}$ are common knowledge. Under rational expectations, this common knowledge would make speculative futures trading impossible -- see the Groucho Marx theorem by Milgrom and Stokey (1982). We therefore abandon rational expectations and assume that the players have different prior beliefs, based on pure differences in *opinion*, in the terminology of Varian (1987). (Note that the differences in beliefs are not due to private information.) No player is inclined to alter his beliefs based upon the knowledge that rivals have different beliefs; each believes he can "outperform the market" because he is smarter than the others. The model thus forms a game of inconsistent incomplete information, as formalized in Selten (1982), Sections B.9-B.13.

[4]Equations (10) and (11) assume the extraction subgame has an interior maximum: Eqs. (6) are all positive. Each term involving θ becomes more complex otherwise.

Since beliefs do not affect equilibrium extraction rates, other than through futures positions, these differences in opinion are compatible with subgame-perfect equilibrium.

We assume throughout that the producers are risk-averse. Why this is necessary can be seen by examining the sums of the expected profits of two traders. The sums $V_i + \phi_s$, $i = a,b$, are positive (if the net positions are small enough relative to differences in beliefs) and quadratic in f_{is}, so that trading of finite amounts is possible between a producer and the speculator. This is true even under risk neutrality $(K_a = K_b = 0)$. However, the sum of the payoffs of the two producers is

$$V_a + V_b = (\hat{p}_b - \hat{p}_a)f_{ab} + (p_{as} - \hat{p}_a)f_{as} + (p_{bs} - \hat{p}_b)f_{bs} + \frac{2\theta}{3}(f_{as} + f_{bs})^2 \quad (12)$$

when $K_a = K_b = 0$. This sum is linear in f_{ab}, so that risk-neutral producers would trade unlimited amounts among themselves.

The implication of risk aversion is that the futures positions in our model will turn out to be related to the available stocks of the natural resource. Consequently, our futures market is not purely speculative in the sense of Kreps (1977) and Tirole (1982): the equilibrium futures positions will contain a speculative *and* a hedging component. This result contrasts with the finding of Brianza, Phlips and Richard (1987) that purely speculative trading is possible between a risk-neutral speculator and a risk-neutral monopolist with inconsistent prior beliefs.[5]

Purely speculative futures activity by producers does occur in our model, in the sense that trivializing the extraction subgame via zero resource stocks would not eliminate futures trading by A and B.[6] Because the differences in beliefs represent differences in opinion, the speculative components of futures contracts form a positive sum component of aggregate payoffs (for sufficiently small futures positions).

2.3 Net Futures Positions

Details of subgame-perfect equilibria will be provided in Section 2.4 below. Here, we focus on the net futures positions that will result from equilibrium contracts.

---

[5] Purely speculative trading would be possible with risk-neutral duopolists, as resource stocks would not appear in Eqs. (11). As mentioned, however, equilibrium with finite positions would require an artificial constraint on the size of contracts between the producers.
Under rational expectations, purely speculative futures trading with a monopolist is not possible in equilibrium -- see Anderson and Sundaresan (1984). In other words, with consistent prior beliefs the monopolist must view the futures market as a hedging instrument.

[6] We have made the speculator risk-neutral for simplicity. Note, though, that introducing risk aversion for S would reduce the size of hedging components relative to producers' speculative components in net positions.

Suppose equilibrium behavior will reach step 5 of the game (at the end of which the futures market closes), and for concreteness, S will be deciding whether or not to accept a futures contract offered by A, A having rejected any contracts offered by S. There will be some minimum profit level that S requires to be just indifferent between accepting or rejecting A's contract; this level may depend on the (p_{bs}, f_{bs}) contract.

Because A and S agree that their different beliefs and A's hedging options yield potential gains to trade, it will pay A to have announced a contract yielding S enough profit to assure acceptance. We avoid problems that sets of preferred outcomes are open by presuming that a player who otherwise could not trade accepts a contract when he is indifferent.

Thus, the equilibrium contract A should have announced is the (p_{as}, f_{as}) pair which maximizes his profit subject to S receiving his minimum acceptable profit, taking appropriate account of the f_{ab} and f_{bs} contracts. Indeed, once it is clear which contracts will be rejected, the appropriate f_{ab}, f_{as}, f_{bs} levels to announce result from simultaneous solution of the maximization of (V_a+V_b), (ϕ_s+V_a), and (ϕ_s+V_b) with respect to these f_{ij}'s. This system of equations is:

$$(K_a+K_b)f_{ab} + K_a f_{as} - K_b f_{bs} = \hat{p}_b - \hat{p}_a + K_a s_{a0} - K_b s_{b0} \tag{13.a}$$

$$-K_a f_{ab} - (K_a + \frac{4\theta}{3})f_{as} - \frac{4\theta}{3} f_{bs} = \hat{p}_a - \hat{p}_s - K_a s_{a0} \tag{13.b}$$

$$K_b f_{ab} - \frac{4\theta}{3} f_{as} - (K_b + \frac{4\theta}{3})f_{bs} = \hat{p}_b - \hat{p}_s - K_b s_{b0}\ . \tag{13.c}$$

Equation (13.c) is the sum of Eqs. (13.a) and (13.b): there is an indeterminacy. However, take two equations, for example (13.a) and (13.c), and rewrite these as

$$f_{ab} + f_{as} = \frac{1}{K_a}(\hat{p}_b-\hat{p}_a) + s_{a0} - \frac{K_b}{K_a} s_{b0} + \frac{K_b}{K_a}(f_{bs}-f_{ab}) \tag{14.a}$$

$$f_{as} + f_{bs} = \frac{3}{4\theta}(\hat{p}_s-\hat{p}_b) + \frac{3K_b}{4\theta} s_{b0} - \frac{3K_b}{4\theta}(f_{bs}-f_{ab}). \tag{14.b}$$

This two-equation system is in terms of the *net* futures positions of the players. Indeed, $f_{ab} + f_{as}$ is A's net position, $f_{bs} - f_{ab}$ is B's net position and $f_{as} + f_{bs}$ is the speculator's net position. Subtraction of (14.a) from (14.b) gives $f_{bs} - f_{ab}$ which can be substituted back, so that the net short positions are uniquely determined as follows:

for A:

$$f_{ab} + f_{as} = \sigma^{-1}[3K_b(\hat{p}_s-\hat{p}_a) + 4\theta(\hat{p}_b-\hat{p}_a) + (\sigma-\sigma_b)s_{a0} - \sigma_b s_{b0}] \quad (15.a)$$

for B:

$$f_{bs} - f_{ab} = \sigma^{-1}[3K_a(\hat{p}_s-\hat{p}_b) - 4\theta(\hat{p}_b-\hat{p}_a) + (\sigma-\sigma_a)s_{b0} - \sigma_a s_{a0}] \quad (15.b)$$

for S:

$$-f_{as} - f_{bs} = \sigma^{-1}[-3K_a(\hat{p}_s-\hat{p}_b) - 3K_b(\hat{p}_s-\hat{p}_a) - (\sigma-\sigma_a-\sigma_b)(s_{a0}+s_{b0})] \quad (15.c)$$

where $\sigma_i = 4\theta K_i$, $\sigma = \sigma_a + \sigma_b + 3K_aK_b$ (a negative amount indicates a net long position). Each net position is seen to be composed of a speculative part and a hedging part. The speculative component consists of the first two terms inside the brackets and is based on differences in beliefs, independent of stocks. The hedging component comprises the remaining terms which relate the futures positions to the available stocks of resources, and are independent of differences in beliefs.[7]

Pricing of speculative ventures is shown to be purely redistributive in a particularly striking way: Eqs. (15) do not contain the futures prices. In other words, contract curves in futures are vertical.

Note that a producer's net short position increases with his own stock and decreases with his competitor's stock of resources.[8] The speculator accommodates the hedging via a less short net position[9] when producers' stocks are increased. This hedging uses the futures market as an insurance or risk-sharing mechanism, and necessarily involves some futures sales to S.

Define the equilibrium net speculative positions to be the first two terms in (15.a), (15.b), (15.c). The signs and magnitudes of net speculative positions directly reflect differences in beliefs. The most optimistic player takes a net speculative long position, and vice versa: the player with the lowest $\hat{p}_i$ will be net speculative short. The player with median beliefs takes the smallest net speculative position. (Here, and throughout, we use "small" as an abbreviation for "small in absolute value", with "large" correspondingly used.) *Increasingly divergent beliefs lead to larger net speculative positions.*

---

[7]Equations (13)-(15) have assumed an interior maximum in the extraction subgame. The more complex equations that result when a producer finds nonnegative extraction a binding constraint do have his stock appearing in the speculative components.

[8]This suggests that the rival i to a dominant firm j does little hedging. Note, though, that if firm j has a larger stock but is also less risk-averse, this mitigates the effect. (In a many-period model with one depletion constraint, a dominant firm may not be constrained to a large depletion in any short time period, and may have a longer time horizon which could look like a lesser degree of risk aversion.)

[9]If the net position were long, "less short" would mean larger in absolute value.

Should a producer be more risk averse, in equilibrium he takes a smaller net speculative position (i.e., he moves closer to exact hedging of his stock), with both other players accommodating. (In particular, suppose $\hat{p}_a < \hat{p}_b < \hat{p}_s$ and B is net speculative short. Then increasing K_b leads A to a shorter and S to a less short net speculative position.) To complete comparative statics, a steeper demand curve (a higher θ) leads both producers to take less short net positions.[10]

Seemingly small changes in parameters can have notable effects upon net positions, extraction paths, and deviations from constant prices. We illustrate with some examples in the columns of Table 1. Each lists the parameters, followed by net

Table 1

$\hat{p}_a$		14.	14.	14.	14.	14.	14.	14.	14.
$\hat{p}_b$		15.	16.	15.	15.	15.	20.	20.	20.
$\hat{p}_s$		20.	20.	20.	20.	20.	15.	15.	15.
K_a		0.05	0.05	0.05	0.05	0.05	0.05	0.05	0.05
K_b		0.05	0.05	0.05	0.05	0.05	0.05	0.25	0.12
θ		0.15	0.15	0.75	0.15	0.15	0.75	0.15	0.15
s_{a0}		75.	75.	75.	40.	110.	75.	75.	75.
s_{b0}		75.	75.	75.	110.	40.	75.	75.	75.
net	A	30.56	39.44	14.51	-4.44	65.56	60.85	-8.45	18.
short	B	10.56	-0.56	-5.49	45.56	-24.44	-59.15	34.31	1.25
position	S	-41.11	-38.89	-9.02	-41.11	-41.11	-1.71	-25.86	-19.25
specul.	A	22.22	31.11	12.68	22.22	22.22	59.02	20.	33.
net sht.	B	2.22	-8.89	-7.32	2.22	2.22	-60.98	-20.	-36.25
position	S	-24.44	-22.22	-5.37	-24.44	-24.44	1.95	0.	3.25
q_{a1}		29.07	24.26	31.75	29.07	29.07	7.36	46.03	31.71
q_{b1}		39.07	44.26	41.75	39.07	39.07	67.36	24.66	40.08
$p_1 - \hat{p}$		6.17	5.83	6.77	6.17	6.17	1.28	3.88	2.88

[10] Increasing speculator risk aversion (as mentioned in footnote 6) has the same impact as increasing θ.

short positions (net long if negative) of the three players. Below that are the net speculative short positions, followed by each producer's equilibrium extraction rate in period 1 (which contrasts with $s_{i0}/2$ in the absence of a futures market). The last item is the price increase in period 1 relative to constant prices (p_2 is decreased by the same amount).

2.4 Subgame-Perfect Equilibria

The game described above has many equilibria. This section will provide the equations characterizing subgame-perfect equilibria, and then show how the characterizations arise in a detailed analysis of the commitment order.[11]

The subgame-perfect equilibria include equilibrium behavior in the extraction game, so Eqs. (6) are satisfied. For reasons mentioned in Section 2.3 and detailed below, the net positions satisfy Eqs. (15).

In equilibrium, a player offering a contract naturally prefers it to be accepted by another player to going unfulfilled. The player accepting the contract must prefer its terms to foregoing any trade with the contract offerer. The (weak) preferences for accepting contracts yield the following equations (where a boldface subscript indicates the player accepting the contract):

$$p_{a\mathbf{b}} \underset{<}{\geq} \hat{p}_a - K_a\left(s_{a0} - f_{as} - \frac{f_{ab}}{2}\right) \qquad \text{as } f_{ab} \gtrless 0 \quad (16.a)$$

$$p_{\mathbf{a}b} \underset{>}{\leq} \hat{p}_b - K_b\left(s_{b0} - f_{bs} + \frac{f_{ab}}{2}\right) \qquad \text{as } f_{ab} \gtrless 0 \quad (16.b)$$

$$p_{a\mathbf{s}} \underset{<}{\geq} \hat{p}_a - K_a\left(s_{a0} - f_{ab} - \frac{f_{as}}{2}\right) - \frac{\theta}{3}\left(f_{as} + 2\, f_{bs}\right) \qquad \text{as } f_{as} \gtrless 0 \quad (16.c)$$

$$p_{\mathbf{a}s} \underset{>}{\leq} \hat{p}_s - \theta\left(f_{as} + 2\, f_{bs}\right) \qquad \text{as } f_{as} \gtrless 0 \quad (16.d)$$

$$p_{b\mathbf{s}} \underset{<}{\geq} \hat{p}_b - K_b\left(s_{b0} + f_{ab} - \frac{f_{bs}}{2}\right) - \frac{\theta}{3}\left(f_{bs} + 2\, f_{as}\right) \qquad \text{as } f_{bs} \gtrless 0 \quad (16.e)$$

$$p_{\mathbf{b}s} \underset{>}{\leq} \hat{p}_s - \theta\left(f_{bs} + 2\, f_{as}\right) \qquad \text{as } f_{bs} \gtrless 0. \quad (16.f)$$

---

[11]An alternative game-theoretic approach to modelling a futures market would be cooperative, i.e., view the futures game as a bargaining problem. A variant of the bargaining set (Aumann and Maschler (1964)) could serve the role of resolving the indeterminacy in (13). The cooperative concept needed would be different from the bargaining set or other formalizations we have seen in several ways: it would have to separate coalition formation and contractual agreement, to allow for variable gains accruing to a producer who is not a member of a coalition as a result of that coalition's bargaining, and to obtain predictions about agreements in the face of inconsistent incomplete information. It would be preferable if the bargaining model produced prices or shadow prices without a presumption of a uniform price. (P.t.o.)

The direction of inequalities is simply that a seller prefers a higher, and a buyer a lower price. Equation (16.a), for example, is derived by solving

$$V_a - V_a\Big|_{f_{ab}=0} \geq 0$$

and (16.d) similarly from ϕ_s.

Equations (16) have to do with accepting each contract separately. A player also has the option of foregoing futures activity entirely, by refusing to offer or to sign any contracts. Thus, equilibrium must satisfy

$$V_a \geq \frac{\theta}{3}(f_{bs})^2 \tag{17.a}$$

$$V_b \geq \frac{\theta}{3}(f_{as})^2 \tag{17.b}$$

$$\phi_s \geq 0 \tag{17.c}$$

where the first two terms recognize that a contract between S and one producer benefits the other producer even if he is inactive in futures.

We discard two types of equilibrium as implausible because they are not subgame perfect. First, there are equilibria where (6.a)-(6.d) are not satisfied for all (f_{ab}, f_{as}, f_{bs}). In essence, these equilibria involve threats by one or both producers to punish deviations from the equilibrium path via altering their extraction policy to harm both themselves and the player being threatened. For example, there may be an equilibrium where S takes a net long position of 30 units. If (6.a)-(6.d) were satisfied everywhere, S would prefer a net long position of 45 units. However, A has threatened to extract 0 in period 1 and s_{a0} in period 2 (rather than profit maximizing) if S takes any net long position of more than 30 units. Notice that A does not have to carry out this threat in equilibrium, as S takes the 30-unit position.

Such threats are not credible and such equilibria are not sensible. The extraction game occurs after the futures market is closed, and now A and B derive higher certainty equivalent payoffs from satisfying (6.a)-(6.d) than from any unilateral deviation. They have no way to have committed themselves to any alternative extraction plan while the futures market was still open.

Development of such a cooperative solution concept is both beyond the scope and somewhat counter to the purposes of this paper. Bargaining seems necessarily to lead to outcomes which are not at all reflective of the structure of futures markets operations, and might naturally lead to joint agreements about contracts and extraction paths. Such joint agreements would yield little insight about the interaction between futures and cash markets.

Moreover, there are equilibria which fail to be subgame-perfect within the futures game itself. Equilibrium in itself does not take the commitment order into account: whenever another player ends up rejecting your contract offers, your best response is to accept his terms. But if you are deciding whether to accept or reject his terms during step 3, subgame perfectness expects you to consider what his rational behavior will be in the event that you reject his terms. In this event, the postulated equilibrium behavior that he rejects your terms during step 4 or 5 is not credible. We discard the equilibrium where you accept his terms in step 3, in favor of the subgame-perfect equilibrium where he accepts your terms after you have irrevocably rejected his.

We are left with infinitely many subgame-perfect equilibria, characterized as follows. Each satisfies (6.a) - (6.d), and, of course, (17.a) - (17.c). A player only accepts a contract if no alternative means of securing any gains from trade with the offerer remains. This property can be discerned by any contract offerer, who rationally announces contract terms which maximize his payoff subject to the contract being acceptable, i.e., subject to one of Eqs. (16). As discussed, such payoff-maximal contracts satisfy Eqs. (15), so *all subgame-perfect equilibria yield the same net positions*. Equilibrium contracts also yield the accepting player no increase in payoff, so the relevant equation in (16) is satisfied with equality.[12]

To describe the sequence of subgame-perfect actions which yield (15), (16) and (17), let i designate the player active during step 3, j the player active during step 4, and k the player active during step 5, and work backwards. Since at the end of step 5, the futures market closes, player k can do no better than to accept the contract offered by any player with whom k has not traded.

During step 4, player j rejects any contract offered by k, as j prefers the terms of his own contract to k's terms. In the event that step 3 saw i reject j's contract offer, j can do no better than to accept the contract (clearly intended for him) offered by i.

During step 3, i rejects a contract offered by j or k if i prefers that his contract terms be accepted. If i has a short position on the contract with j and a long position with k, or vice versa, then i rejects all contracts. Suppose, however, that i has a short position on both contracts, or long on both. Then he only had the opportunity to announce his preferred terms for one of these contracts during step 2. (Recall only a player taking a short position on one contract and a

[12] Notice that, although contracts are offered to the market and not identified as to the identity of the acceptor, in equilibrium the acceptor is known at announcement.

The equality in Eq. (16) can be thought of as the limit of a sequence of contract terms which are increasingly profitable to the offerer and all of which the player strictly prefers to accept. An alternative approach would be to introduce a smallest money unit, and require all prices to be integer multiples of this unit. No qualitative changes would result, save that acceptance need never be the result of indifference. Notation and calculation would be complicated considerably.

long position on his other can announce two contracts which could possibly be accepted.) He rationally would have chosen to announce his major contract (ij if $|(p_{i\mathbf{j}}-p_{\mathbf{i}j})f_{ij}| > |(p_{i\mathbf{k}}-p_{\mathbf{i}k})f_{ik}|$). Now, during step 3, he can do no better than to accept the terms dictated him on his minor contract.

During step 2, contract offers that will later be accepted are announced, in accordance with (15) and (16) as described. Formally, it does not matter whether contracts rejected in equilibrium satisfy (15) (though they must be acceptable to the offerer), or whether these contracts are even announced.

Averaging across commitment orders, a player who takes a short position in one contract and a long position in his other, in all commitment orders, obtains the gains from exchange in each of the two contracts 5 times out of 6 (if B is this player, for example, only in order ASB does S dictate terms to B, and only in order SAB does A dictate to B). The gains from exchange between a player who only takes short positions and one who only takes long positions, across commitment orders, accrue to each 3 times out of 6.

3. Obtaining a Unique Set of Contracts

Differences in beliefs about the uncertain level of resource demand, producers' stocks of resources and levels of risk aversion, and the slope of the resource demand curve suffice to determine uniquely the traders' net positions, the extraction rates, and the divergence from constant prices. Yet these parameters leave open an infinity of equilibrium options for the prices and quantities written on futures contracts, and for the distribution of payoff levels across players.

None of the subgame-perfect equilibria strike us as particularly implausible, so reducing to a unique prediction is not an equilibrium refinement task in the usual sense.[13] A unique prediction can be attained by a revision of the rules described in Section 2, without altering any of the analysis of Section 3.

Specifically, we remove the word "simultaneously" from the description of contract offer announcement in step 2, and substitute a rule that the player chosen (at random) to be first in the commitment order also is allowed to be first in announcing any contract offers.[14]

This rule alteration yields a unique subgame-perfect equilibrium. Being first to commit, player i is also first to announce. Let Y_i be the function obtained when

---

[13]Without extensive exploration, it appears to us that most refinements discussed in the literature (see van Damme (1987) for a clear exposition) leave us with an infinite set. One approach could be to follow the equilibrium selection methods of Harsanyi and Selten (1988); to do so would require alteration of the game to finite strategy sets. Our approach here is simpler and more transparent, but may not be robust to changes in the structure of the game.

[14]Other methods of breaking announcement simultaneity may also yield a unique prediction in a less straightforward manner.

the relevant equations in (15) and (16) are substituted for prices and all but one futures position in i's payoff function (V_a, V_b or ϕ_s). Y_i is a quadratic function, mapping any one variable, f_{im}, into the payoff i obtains if he announces the contract offer pair $(f_{im}, p_{i\blacksquare})$ derived from an equality in (16), and if all succeeding announcements are in accordance with (15) and (16), and contract acceptance behavior is as described in Section 2.4. If Eqs. (17) are satisfied, the path in the subgame following announcement $(f_{im}, p_{i\blacksquare})$ which leads to payoff $Y_i(f_{im})$ for i is a subgame-perfect path. (If i takes short positions (or long positions) in both contracts, set f_{im} to be his major contract.)

All players can discern that i will reject any contract offers by m in step 3 forcing m to accept $(f_{im}, p_{i\blacksquare})$, that is, to accept i's terms. Accordingly, unilateral deviations from the prices and quantities derived from (15) and (16) cannot increase payoffs.

Thus, the $f_{i\blacksquare}$ which maximizes Y_i subject to (17) is i's best choice for the first announcement.[15] With divergent beliefs, $0 < \theta < \infty$ and at least one risk-averse player, contracts with both other players represent potential gains to trade for player i, so he cannot prefer an $f_{i\blacksquare}$ failing to satisfy (17). In this sense, his announcement choice is unconstrained.

4. A Concluding Perspective

In the preceding computations, the true level of aggregate demand and the correct forecast of the spot market price do not appear. Each contract has its own equilibrium futures price. Consequently, there is no sense in which "the" futures price could be a predictor, unbiased or biased, of the spot price to materialize at maturity. With inconsistent beliefs, rather than private information, each player has his own immutable opinion about how the spot price will behave in period 2; he does not look to the futures price to provide any information to update this opinion. As such, "the market" does not "aggregate" information in the sense these words are usually applied in futures market models. Whether a strength or a weakness, this is a difference of our approach.

Some differences of a game-theoretic approach to futures market analysis are clearly strengths from an industrial organization point of view: game theory allows us (a) to take account of the strategic interdependence characteristic of oligopoly in the interaction between cash and futures markets, and (b) to introduce explicitly institutional arrangements that characterize operation of futures markets and trace out their impact on market outcomes. While some of our simplifications were more to

[15] It is straightforward, though tedious, to show that equations (17) bound $f_{i\blacksquare}$. Only if Y_i happens to achieve a global minimum exactly at the midpoint of upper and lower bounds will the best $f_{i\blacksquare}$ fail to be unique.

clarify than to describe, the empirical relevance of basic noncooperative elements of the structure should be clear.

Indeed, a formalization of the problem as a two-stage game and resulting characterization of subgame-perfect equilibrium outcomes appears a natural way to model the interaction between oligopolistic control of spot markets for natural resources and manipulation of corresponding futures markets. We find that the existence of futures markets can be profit-enhancing to oligopolists in general, and that profitable speculation (as well as hedging) transforms the temporal evolution of extraction rates and the spot price. In particular, the model shows how a futures position taken by a competitor can be a source of profit for an individual producer, whether he is active in the futures market or not. Equilibrium outcomes include the striking and perhaps descriptive possibilities that a risk-averse producer can take a net long futures position (his net speculative long position being larger than his hedging position) and even a net long hedging position (when a competitor has a larger stock to deplete).

The futures market is modelled as an oral double auction ("open cry") in rough description of operating rules of such markets as the London Metal Exchange. While the particular commitment order rule we use is more transparent than some complex but realistic alternatives, it does capture the serious phenomenon that minor details of timing can have major impacts on profit distributions. A reassuring feature of equilibrium outcomes in this model is that only profit distributions depend upon commitment order; many equilibria exist, but equilibrium net futures positions and extraction paths are uniquely determined by stocks of resources, risk aversion, demand responsiveness and differences in beliefs.

The oral double auction avoids the untenable assumption of a single uniform price in a futures market. Thus, our simple story reflects more naturally the multiplicity of prices (often among the same traders) during one trading day on, say, the London Metal Exchange. This is one of a number of realistic divergences from the standard capital asset pricing model in vogue in the futures markets literature. The futures market in subgame-perfect equilibrium is efficient, in the sense that net positions taken maximize the sum of payoffs, but no efficient markets assumption is employed. Finally, the usual rational expectations assumption (which precludes examining interactions between producers' net speculative positions and interdependent extraction policies) is abandoned in favor of inconsistent prior beliefs. We are pleased to discover that the differences in beliefs play an essential role in determining the sign and size of net speculative positions.

The presumption that the underlying commodity is a depletable resource played a role for which other technological relations or constraints could be substituted. In many markets trading in futures or other financial instruments, we see underlying oligopolistic industrial structures, a high volume of essentially speculative trading, and participants with apparent access to unlimited credit. A game-theoretic approach to analyzing these markets should be fruitful as well.

References

Anderson, R.W., ed. (1984). The Industrial Organization of Futures Markets. Lexington, Mass: D.C. Health and Company.

Anderson, R.W. and C.L. Gilbert (1988). Commodity Agreements and Commodity Markets - Lessons from Tin. The Economic Journal **98**: 1-15.

Anderson, R.W. and M. Sundaresan (1984). Futures Markets and Monopoly, In: R.W. Anderson (ed.): The Industrial Organization of Futures Markets. Lexington, Mass.: D.C. Health and Company, pp: 75-106.

Aumann, R.J. and M. Maschler (1964). The Bargaining Set for Cooperative Games. In: M. Dresher, L.S. Shapley and A.W. Tucker (eds.): Advances in Game Theory. Annals of Mathem. Studies No 52. Princeton, N.J.: Princeton University Press, pp. 443-476.

Brianza, T., L. Phlips and J.-F. Richard (1987). Futures Markets, Speculation and Monopoly". Revised version of CORE Discussion Paper 8725, Louvain-la-Neuve.

Harsanyi, J.C. and R. Selten (1988). A General Theory of Equilibrium Selection in Games. Cambridge, Mass.: MIT Press.

House of Commons (1986). The Tin Crisis. Second Report from the Trade and Industry Committee, Session 1985-86. London: HMSO.

Kreps, D.M. (1977). A Note on 'Fulfilled Expectations' Equilibria. J. Econom. Theory **14**: 32-43.

Milgrom, P.R. and N.L. Stokey (1982). Information, Trade and Common Knowledge. J. Econom. Theory **26**: 17-27.

Novshek, W. (1985). On the Existence of Cournot Equilibrium. Rev. Econom. Studies **52**: 85-98.

Reinganum, J.F. and N.L. Stokey (1985). Oligopoly Extraction of a Common Property Natural Resource: the Importance of the Period of Commitment in Dynamic Games. Intern. Econom. Rev. **26**: 161-173.

Selten, R. (1982). Einführung in die Theorie der Spiele mit unvollständiger Information. Information in der Wirtschaft, Schriften des Vereins für Socialpolitik, Gesellschaft für Wirtschafts- und Sozialwissenschaften, N.F. Bd. **126**: 81-147, Berlin: Duncker & Humblot.

Tirole, J. (1982). On the Possibility of Speculation Under Rational Expectations. Econometrica **24**: 74-81.

van Damme, E.E.C. (1987). Stability and Perfection of Nash Equilibria. Berlin: Springer-Verlag.

Varian, H.R. (1989). Differences in Opinion in Financial Markets. In: C. Stone (ed.): Financial Risk: Theory, Evidence and Implications. Amsterdam: Kluwer Academic Publishers.

A FRAMING EFFECT OBSERVED IN A MARKET GAME

Wulf Albers and Ronald M. Harstad

Introduction

A wide variety of markets transacting via auctions share this characteristic: when a bidder selects his strategy, the value to him of the auctioned asset is uncertain. This characteristic lends the common–value auction model (Wilson [1977], Milgrom and Weber [1982]) much of its richness. It also makes the bidder's strategic problem more complex: laboratory observations of common–value auctions have suggested that subjects exhibit substantially more difficulty approaching equilibrium profit levels, and indeed avoiding losses, in common–value auctions (Kagel and Levin [1986], Kagel, Levin and Harstad [1988]) than in the more artificial setting of private–values auctions (where each bidder knows for sure the asset's value to him, cf. Kagel, Harstad and Levin [1987], Cox, Roberson and Smith [1982]).

In a canonical common–value auction, asset value is a random variable v. Each bidder privately observes a signal informative about v, typically these signals are independently and identically distributed given v. Based on this information, bidders compete under known auction rules. A symmetric equilibrium analysis of the auction has each bidder rationally take into account when selecting bids that, if he wins, all rivals will have estimated the asset to be worthless; the symmetric assessments yield this outcome because each rival observed a lower signal than the winner. Field observations of offshore oil lease auctions led Capen, Clapp and Campbell [1971] to suggest instead that bidders systematically fail to take into account the likelihood that the winner is the bidder who most overestimated the asset's value. This "winner's curse" is alleged to affect bidding and profitability in publishing (Dessauer [1981]), professional sports (Cassing and Douglas [1980]) and corporate takeovers (Roll [1986]). The laboratory experiments cited have found indications of the winner's curse to be widespread and persistent: more experience in a particular auction setting seems to be required for subjects to cope with this adverse selection problem than to cope with nearly any other game–theoretic issue presented in controlled laboratory environments.

We report here on a series of oral or "English" auction experiments designed to allow observation of the key characteristic of standard common–value auctions in a somewhat more simplified setting. The simplification arises principally in making subjects' information about asset value unconditionally independent. We conduct several replications of the following auction. Each of five bidders privately observes his "draw", a random variable drawn from a uniform distribution on the set {0, 5, 10, ..., 195, 200}. With this information, the five bid for an asset worth precisely the mean of the five draws. The draws are not signals, in that the asset value v is determined (nonstochastically) from the aggregate information contained in the draws, not vice versa. Presumably, the underlying statistical properties of this auction are more readily comprehended than in the canonical common–value model; the notions of a "low draw" or a "high draw" can be given unconditional meanings, and straightforward numerical interpretations.

The idea behind this approach was to reduce the internal variance of the model: the value of the asset does not contain any additional error term, and the sum of the information of all subjects together gives the correct value of the asset. Moreover, in symmetric equilibrium, at the end of every auction the asset value is precisely known to the winner. It can be calculated from rivals' dropout prices and his own draw.

As described in Section 2, the auction was conducted by having an auctioneer call out prices in ascending order, with subjects announcing when they wished to drop out, i. e., to cease competing. The asset was awarded to the last remaining competitor at the price at which his last rival dropped out of the bidding. The symmetric equilibrium model of this auction does have the bidder with the highest draw winning, and all bidders taking proper Bayesian account of this feature throughout the bidding. Bidders may fail to incorporate such Bayesian calculations, so a winner's curse hypothesis offers an alternative prediction of behavior; given the dynamic nature of English auctions, two formulations of the winner's curse are specified in Section 3.

Additionally, a treatment is incorporated that allows observing a pure framing effect in the market. The term framing effect comes from psychological studies of individual decision making under uncertainty, which suggest that two logically equivalent problems can lead to systematically different evaluations and decisions when questions are framed in different ways (Tversky and Kahneman [1981]). Economists have questioned whether markets as institutions discipline agents so as to eliminate or seriously constrain such irrational behavior (see, e. g., Knez and Smith [1987]). These experiments provide another vantage point on this issue, by observing whether a framing effect matters to market outcomes and to individual behavior in markets, in a setting presenting market participants with a task which may be at or beyond their rational capabilities.

The framing effect is introduced as follows. Experiments 1–5 were conducted as described above, under what we label the mean condition. Experiments 6–9 were conducted under the sum condition: each of five bidders privately observes his draw from a uniform distribution on the set {0, 1, 2, ..., 39, 40}. With this information, the five bid for an asset worth precisely the sum of the five draws. Notice the transformation of one condition into the other is completely transparent (multiplication of draws by 5, and mean instead of sum), and ought not affect rational behavior. Section 4 describes our observations as to whether it did.

1 Theoretical Predictions

Except where we explicitly consider differences, all discussion of theoretical predictions and of observations will be couched in terms of the mean condition draws. Thus, a bid associated with a draw of 35 should take into account that 35 is a relatively low draw (out of 200, corresponding to a sum–condition draw of 7 out of 40).

1.1 Symmetric Equilibrium

The underlying symmetry of the auction leads a game–theoretic analysis to predict a symmetric equilibrium solution. Here symmetry means that any two bidders with the same draw, observing the same history of dropout prices so far, will plan to continue competing up to exactly the same price. Equilibrium means that, given the behavior of rival bidders, no bidder can gain by a unilateral change of strategy. Techniques in Levin and Harstad [1986] (including showing that symmetric equilibrium must prescribe monotonicity, i. e. higher bidding at higher draws) can readily be adapted to the current auctions to show that symmetric equilibrium is unique. [1])

Symmetry and monotonicity imply that a bidder will be outbid if any rival has a higher draw. If a bidder is outbid, he does not care how high he bids. So the game–theoretic model has each bidder assuming that no rival observed a higher draw, an assumption that will (in equilibrium) be harmless when incorrect. This assumption guides plans as to how

---

[1]) It is well known that there exist asymetric equlibria which are degenerate, in the sense that one bidder always wins. Bikhchandani and Riley [1989] show that the symetric equilibrium is the only nondegenerate equilibrium, under a rather opaque condition that is satisfied in this model. Neither our observations or those of Harstad, Kagel and Levin [1989] lend any support to the degenerate equilibria.

far to compete before dropping out, with the effect that each bidder incorporates from the beginning information that would be available to him if he were to be the winning bidder. If a rival has a lower draw than yours, he will drop out sooner than you plan to drop out. So in essence, each bidder in equilibrium assumes that any rival still competing observed the same draw the bidder himself observed; this assumption is updated whenever a rival drops out, revealing that the rival actually had a lower draw.

It clarifies presentation and discussion to distinguish two rank orders. Let p_1, p_2, p_3, p_4 denote the nondecreasing sequence of prices at which a 1st, 2nd, 3rd, and eventually 4th bidder ceases competing ("dropout prices"). Thus, the market price paid by the winner is p_4, and the winner's profit is $v-p_4$. Let x_1, x_2, x_3, x_4, x_5 denote the draws observed by the bidders when ordered to correspond to the order in which they cease competing. That is, x_1 is the draw seen by the first bidder to drop out. Under symmetric monotonic behavior, x_1 is also the lowest of the five draws; otherwise it need not be. Let $x(1)$, $x(2)$, $x(3)$, $x(4)$, $x(5)$ represent the draws ranked in ascending order. Thus, $x(5)$ is the highest draw, while x_5 is the draw observed by the winning bidder.

The basic incentive structure of English auction rules is price–taking: your decision regarding how long to continue competing will determine whether you win, but the price you pay if you win will be determined by your rivals' decisions about how long to compete. The assumptions mentioned lead the bidder with the lowest draw to assume each rival shares his draw, thus v equals his draw, so $p^*_1 = x_1$. (Throughout, an asterisk will designate an equilibrium prediction. Recall that this prediction translates into a sum condition prediction $p^*_1 = 5x_1$.)

Note the implication of symmetry: each bidder initially plans to continue competing until the price reaches his draw, so long as no other bidder has dropped out. When they learn of p_1, each of the 4 remaining bidders re–evaluates his plans, estimating v at $0.2p_1 + 0.8x$ for the x he observed. Each plans to continue in the bidding up to this price, so $p^*2 = 0.2p_1 + 0.8x_2$. Once again, p_2 reveals to each of 3 remaining bidders that they must re–estimate v. Table 1 provides formulas and illustrates this logic for the mean draws and sets of draws used in the questionnaire.

While v is viewed as uncertain by each individual bidder, in equilibrium, for the winner this uncertainty is resolved as soon as his last rival drops out. At this point, v is inferentially nonstochastic, equal to the average of p_4 and the price at which the winner was planning to drop out $(0.1p_1+p_2/6+p_3/3+0.4x_5)$. Equilibrium profit is thus $0.2(x_5-x_4)$, which also becomes nonstochastic when p_4 becomes known. Ex ante, for x drawn uniform on $\{0, 5, ..., 200\}$, average equilibrium profit is 20/3.

1.2 Winner's Curse Models

The winner's curse phenomenon has not been given as precise a definition as equilibrium has. The fundamental distinction is that bidders falling prey to the winner's curse select bids under a presumption that there is no particular rank–order relationship between their draw and rivals' draws. In the auctions observed here, a bidder who recognizes that his draw of 20, say, is unlikely to be a median draw may nonetheless bid under a dangerous assumption in line with the winner's curse. That is, he may rely upon the independence underlying his draw and rivals' draws, assuming that he can bid as if rivals' draws are unrelated to his. A specific formulation of this view is what we label

> the **unresponsive winner's curse model**: assuming that rivals' draws each average 100, are independent of a bidder's draw, a bidder with draw x does bid up to $0.2x + 0.8*100$.

If all bidders were to adopt this strategy, they would be behaving symmetrically, and the winner would be "cursed" to discover that his rivals' draws are related to his draw, which is an upper bound on their draws. (Even if four rivals' draws will average 100, the four lowest of five draws will average less.) The unresponsive winner's curse model has later dropout prices unrelated to earlier dropout prices; earlier dropout prices do contain the complete information about their draws (the bidder's draw is 5 times as far from 100 as his dropout price), but this information is hypothesized not to be used in this model. The unresponsive winner's curse model predicts an overall average profit of –20/3.

A bidder can begin the auction naively assuming there is no relationship between his draw and others, and still see advantages to behaving differently when rivals drop out early than when they compete longer. A version of a winner's curse effect remains if he assumes draws of rivals still competing are unrelated to his own draw. This consideration gives

> the **responsive winner's curse model**: (Counterfactually) rivals who have already dropped out are assumed to have dropped out at their draws (i. e., using $p(x) = x$), and rivals remaining are assumed to have draws averaging 100.

Table 1 provides the resulting predictions for mean draws and for sets of draws in the questionnaire. When a dropout price below 100 is observed, this model lessens the winner's curse by assuming an overestimate of the dropper's draw, but less of an overestimate than 100 would be. It still leaves bidders in serious jeopardy whenever $x<100$, as remaining

Table 1: Drop-out Prices in Equilibrium Bidding , and Responsive Winner's Curse Model for Three Examples of Auctions (general formulas at the bottom)

	Equilibrium Model				Responsive Winner's Curse Model			
	rank of dropout				rank of dropout			
	first	second	third	fourth	first	second	third	fourth
	dropout prices of bidders who dropped out before:							
	—	p1=33.33	p1=33.33	p1=33.33	—	p1=86.67	p1=86.67	p1=86.67
			p2=60.00	p2=60.00			p2=90.67	p2=90.67
				p3=80.00				p3=95.46
draw:	dropout price (if next to drop out):				dropout price (if next to drop out):			
x= 33.33	p1= 33.33*				p1= 86.67*			
x= 66.67	p1= 66.67	p2= 60.00*			p1= 93.33	p2= 90.67*		
x=100.00	p1=100.00	p2= 86.67	p3= 80.00*		p1=100.00	p2= 97.33	p3= 95.46*	
x=133.33	p1=133.33	p2=113.33	p3=100.00	p4= 93.33*	p1=106.67	p2=104.00	p3=102.13	p4=101.27*
x=166.67	p1=166.67	p2=140.00	p3=120.00	p4=106.67	p1=113.33	p2=110.67	p3=108.90	p4=107.89
	dropout prices of bidders who dropped out before:							
	—	p1=20	p1=20	p1=20	—	p1=84	p1=84.	p1=84
			p2=36	p2=36			p2=84.8	p2=84.8
				p3=48				p3=85.76
draw:	dropout price (if next to drop out):				dropout price (if next to dropout):			
x= 20	p1= 20*				p1= 84*			
x= 40	p1= 40	p2= 36*			p1= 88	p2= 84.8*		
x= 60	p1= 60	p2= 52	p3= 48*		p1= 92	p2= 88.8	p3= 85.76*	
x= 80	p1= 80	p2= 68	p3= 60	p4= 56*	p1= 96	p2= 92.8	p3= 89.76	p4= 86.91*
x=100	p1=100	p2= 84	p3= 72	p4= 64	p1=100	p2= 96.8	p3= 93.76	p4= 90.91
	dropout prices of bidders who dropped out before:							
	—	p1=120	p1=120	p1=120	—	p1=104	p1=104	p1=104
			p2=136	p2=136			p2=108.8	p2=108.8
				p3=148				p3=114.56
draw:	dropout price (if next to dropout):				dropout price (if next to drop out):			
x=120	p1=120*				p1=104*			
x=140	p1=140	p2=136*			p1=108	p2=108.8*		
x=160	p1=160	p2=152	p3=148*		p1=112	p2=112.8	p3=114.56*	
x=180	p1=180	p2=168	p3=160	p4=156*	p1=116	p2=116.8	p3=118.56	p4=121.47*
x=200	p1=200	p2=184	p3=172	p4=164	p1=120	p2=120.8	p3=122.56	p4=125.47
drop-out price (as function of own signal x, others' draws xi / others' dropout prices pi) by rank of dropout:								
first	p1= x				p1= .2x + .8			
second	p2=.2x1+.8x =.2p1+.8x				p2= .2p1 + .2x + .6			
third	p3=.2x1+.2x2+.6x =.15p1+.25p2+.6x				p3= .2p1 + .2p2 + .2x + .4			
fourth	p4=.2x1+.2x2+.3x3+.4x=p1/10+p2/6+p3/3+.4x				p4= .2p1 + .2p2 + .2p3 + .2x + .2			

* Dropout prices of bidders who actually drop out next (if the draws are as given in the left column) are marked by an asterix.

rivals' draws are overestimated when it matters (i. e., when the bidder wins). For x>100, however, once two or three rivals have dropped out, 100 will be on average an underestimate of rivals' draws, so this model can readily observe profits when v>100. Overall, it predicts an average profit of −1.23.

The two formulations used to yield predictions of a winner's curse story are unaffected by the framing effect of switching between mean and sum conditions. Psychologists' arguments for the relevance of framing effects are based upon a decision process where subjects adopt an anchor and then systematically underadjust their estimation away from that anchor. The way a question is framed may alter the anchor selected. In this context, it is not clear to us how differential predictions of the winner's curse may be built by hypothesizing that the mean and sum conditions lead to subjects adopting different anchors. We do not rule out the possibility; we simply are unsure how to model it.

2 The Experiment

The subjects were recruted from a microeconomics course at the University of Bielefeld. They were mainly business administration students. The experiment was run in the same room as the course, but the participation was voluntarily. 45 students participated, i. e. about one third of the students enrolled. The subjects were distributed to 9 different tables in a way that students who were acquainted with each other sat at different tables.

The instructions were supported by a one page leaflet containing all essential information. (The modifications for the sum condition are given in brackets.)

> **The Problem:** An asset will be auctioned among 5 people in an English auction. The **Value** of the **asset is the mean (sum) of 5 individual values** each of which is known to one of the 5 persons. – The 5 individual values are **independent draws** from the interval 0 to 200 (0 to 40). Every individual value can adopt the values 0,5,10,15,...,195,200 (0,1,2,3,.. .,39,40), and each of these values has the same probability. – Example: individual values 135,75,110,170,45 (17,25,22,34,9) give a mean (sum) of 107, which is the value of the asset.
>
> **The English auction:** An auctioneer counts aloud the numbers 0,1,2,3,... to 200. A bidder does not act as long he is willing to pay the announced number as the price of the asset. As soon as the auctioneer says a number which is higher than a bidder (for instance bidder A) is willing to pay, the bidder says "A off". This means that Bidder A is not bidding any more from then on. – The auction is done, as soon as

only one bidder is still bidding. He is the buyer. The number, at which the last of the other bidders said "X off", is the price of the asset. (profit = value – price)

Remarks to the rules of the auction: The decision "X off" is irrevocable. – If several bidders say "X off" at the same price, that one receives the asset who said "X off" last. (This decision is made by the auctioneer, if necessary, by flipping a coin.)

Remark concerning the execution: The individual values have been taken in advance as independent draws, and assigned to a separate list for every player. Do carefully pay attention, to pick up your individual values for the correct rounds. (It seems to be helpful, to erase each value after transferring it to the auction form.)

Aim (German: Zielsetzung): Experienced bidders can conclude from their individual number and the bidding behavior of the others on the mean (sum), and can behave accordingly.

These Instructions were read aloud to all subjects simultaneosly. – Before separating the groups and reading the instructions the experimentor shortly motivated the two different framing conditions by two different cover stories:

Mean: Five firms are interested in the rights to strike oil from a certain area. They have different estimates of the value of these rights, given by the five best experts in this field. Every firm does only know the estimate of one expert. The correct value is precisely the mean of the expert opinions.

Sum: Five firms are interested in getting the rights to strike oil from a certain area. The area can be subdivided into five subareas. In each of these subareas one of the firms drilled down a testwell, giving the precise value of this subarea. The result of this test is not known to the other firms. The value of the total area is precisely the sum of the values of the subareas. The area is auctioned in one piece.

The groups ran between 48 and 60 auctions in a time of around 4 hours including instructions. Five groups were run under the mean, four under the sum condition.

All subjects (and the supervisors = auctioneers) had identical forms in which they filled in the dropout prices of all players, the true value of the asset and the calculated profit for each single auction, so that it can be taken for sure that every player really perceived these data. The auctioneer had a reservation price which was 20 units below the value of the asset in the respective round. When the last dropout price did not meet the reservation price, the asset was not sold and the subjects were not informed about the exact value in this round. Otherwise the true value of the asset was given by the auctioneer after every auction. Information about the individual values was not given at any time.

After every sixth auction, before the value of the asset was reported, the subjects were asked to guess the value.

After every twelfth auction, a separate questionnaire was distributed asking for the dropout prices in different situations:

> Question 1: You are bidder 1. Your individual value is x_1 (=20 or 120). At which number p_1 do you say "X off", if no one else stops bidding before you?
>
> $x_1 = 20 \longrightarrow p_1 = \ldots$
> $x_1 = 120 \longrightarrow p_1 = \ldots$
>
> Question 2: You are bidder 2. Your individual value is x_2 (=40 or 140). At which number p_2 do you say "X off", if no one else stops bidding before you? (Fill in p_1 from above.)
>
> $p_1 = \ldots,\ x_2 = 40 \longrightarrow p_2 = \ldots$
> $p_1 = \ldots,\ x_2 = 140 \longrightarrow p_2 = \ldots$

Questions 3 and 4 gave the same type of question for the fourth and the third bidder who dropped out, where the individual values were (60 or 160) for the third and (80 or 180) for the fourth. Thereby two increasing sequences of individual values were asked, namely 20,40,60,80 and 120,140,160,180, with the second 100 points higher than the first. Each of the corresponding questions was based on "observed" dropout prices $p_1, p_2, ..$ of the dropout responds of the same subject in the preceding questions, and on the individual information of the respective question (x_1 or x_2 or..). In this framework the players had the opportunity to give a complete set of decisions in all positions. It might even be suggested that he was motivated to decode the preceding dropout signals, in order to reach a reasonable result. (In advance we feared that the decoding might be done just by looking at the individual values of the preceding questions. But the subjects did apparently not behave like that.)

The value of each point was 1 DM (over 1/2 Dollar). This meant that if they played the equilibrium dropout prices, they could win 300 DM per table. If they cooperated, and did all drop out at low prices, they could have even made a profit up to 100 DM per game. To prevent such cooperation the subjects were told that every asset had an announced reservation price, below which it would not be sold. (The reservation prices were 20 points below the values, but this was not known to the subjects.)

The subjects were told that they did have an initial endowment of 60 points, and that in similar experiments the most successful players received total profits of about 150 DM (about 75 Dollars) to 200 DM (about 100 Dollars). (Anyway the subjects did not suspect in advance that they might make losses in the auctions.) The players were instructed that no one would lose money in the experiment, but that they should contact the experimentor, if high losses in the preceding sessions caused, that they did not feel motivated any more. (One subject did so, and we raised his aggregated profit at that moment from –300 up to –50.) During the experiment, facing the high losses by the winners curse, the players were told that other players were also making losses, and that it might be that we would give some additional endowment to all of them. – The final payoffs of the experiment were based on an initial endowment of DM 60. The most successful players received DM 175, 121, 111, 105, and 84. The sum of all payoffs was 1215 DM (about 610 Dollars).

Table 2: Mean Nash–Equilibrium Profit, and Mean Observed Profit *)

Exp.	# aucts	mean NE profit all auctions	mean observed profit all auctions	mean observed profit auctions 1-48	mean observed profit auctions 37-48
1M	72	6.97 (0.79)	-3.74 (2.24)	-2.73 (2.71)	-9.92 (5.86)
2M	60	6.53 (0.72)	-2.46 (1.91)	-0.43 (2.04)	-2.25 (3.90)
3M	60	6.48 (0.63)	-8.20 (2.65)	-9.44 (2.74)	-13.33 (3.93)
4M	60	7.18 (0.73)	-6.28 (1.96)	-4.94 (2.28)	-4.00 (3.84)
5M	54	5.63 (0.75)	0.19 (2.79)	-0.13 (3.06)	6.25 (5.34)
1S	60	6.75 (0.85)	-1.98 (2.40)	-2.25 (2.49)	-1.17 (6.21)
2S	48	6.58 (0.75)	-3.50 (2.02)	-3.70 (2.03)	1.67 (3.11)
3S	56	6.85 (0.64)	-7.42 (2.58)	-7.10 (2.80)	-5.25 (4.89)
4S	48	7.40 (0.83)	-3.69 (2.39)	-3.69 (2.39)	-7.42 (6.23)
1M-4M	252	6.80 (0.36)	-5.13 (1.12)	-4.40 (1.25)	-7.38 (2.32)
1S-4S	212	6.85 (0.39)	-4.09 (1.20)	-4.19 (1.23)	-3.04 (2.68)

*) The terms in brackets give the respective standard deviations

3 Results

3.1 Observations of Market Prices

Aggregate outcomes are compiled in Table 2. Experiments are identified by a number and letter M for mean condition or S for sum condition. Experiments with the same number had identical realizations for all draws. Rows labelled 1M–4M and 1S–4S combine observations for the indicated experiments. The number of auctions completed in each experiment is shown in column 2. Equilibrium predicted mean profits are shown in column 3. Mean observed profits are reported in columns 4, 5 and 6, for all auctions, for auctions 1–48, and for auctions 37–48 (the latest dozen completed in all experiments). Each mean profit entry is followed by its standard error in parentheses. Table 3 reports two frequencies of interest: the frequency with which nonnegative profit levels were observed, in column 2, and the frequency with which all five bidders were still competing when the price exceeded the value of the asset, in column 3.

Figures 1–4 diagram observed and predicted profit for the sets of matching experiments. Time progresses along the horizontal axis. The profit that would have been observed in symmetric equilibrium in each auction is indicated as a dollar sign; the actual profit observed in experiment 1M is indicated as an open square, and that observed in experiment 1S as a plus sign.

The overall impression is clearly an impact of some version of the winner's curse which is widespread and persistent. Average profit over the last 12 auctions only reaches a nonnegative level in experiments 5M and 2S. Both levels and frequencies of losses are not significantly different between the mean condition and the sum condition (mean profit for experiments 1M–5M is –4.18). The following null hypotheses cannot be rejected by either parametric differences in means tests or two–sample nonparametric rank tests:

a. Levels of profits are drawn from the same distribution in mean and sum conditions.

b. Signs of profits are drawn from the same distribution in mean and sum conditions.

c. Levels of profits are drawn from the same distribution in auctions 4–27 as in either auctions 37–48 or the last dozen auctions.

Figures 5–9 display profit as a function of value (centered on its mean of 100), both as scatter plots and as a regression line. Figures for experiments not shown are similar, as the tendency to observe losses whenever value was below its mean level of 100 and typically profits when value was above its mean level is robust across experiments and experience. Market prices show a mean reversion tendency sufficiently strong to generate this profit pattern. To understand this tendency, it is necessary to examine observed strategies as indicated by dropout prices.

3.2 Observed Dropout Prices

As indicated above, the symmetric equilibrium predictions begin with the subject with the lowest draw x(1) dropping out when the price reaches his draw (or 5 times his draw in the sum condition). 44 of the 45 subjects continued to compete past this predicted level on at least 90% of the occasions when they held the lowest draw.

The excessively high levels of p_l, the first dropout price, are significantly different for the two treatment conditions. While a hypothesis of homogeneous behavior when dropping out first can be rejected in all experiments (via f–tests on fixed–effects regressions, $p < 0.1$), bidders in the sum condition exhibit significantly higher ($p < 0.001$) first dropout prices, both in nonparametric paired comparisons and in fixed–effects regressions of p_l on the draw of the bidder dropping out first. Combining within conditions, OLS regressions of p_l on the draw held yield $p_l = 56.9 + 0.33$ draw, for the mean condition, and $p_l = 73.4 + 0.25$ draw (that is, + 1.25 times an [unadjusted] draw ranging from 0–40), for the sum condition. Recall that the equilibrium model predicts a zero intercept and a slope of 1; either winner's curse model predicts a 0.2 slope and an intercept of 80.

Table 3: Auctions with Positive Profits, and Auctions where the First Dropout Price p_l, was Greater that the Value of the Asset (Proportions within the First 48 Auctions

Exp.	% of auctions with profit>0	% of auctions with pl>value	Exp.	% of auctions with profit>0	% of auctions with pl>value
1M	50.0	22.9	1S	43.8	29.2
2M	46.8	0.0	2S	43.8	29.2
3M	33.3	20.8	3S	40.9	29.5
4M	45.8	22.9	4S	45.8	31.2
5M	50.0	16.7			
1M-4M	43.9	16.8	1S-4S	42.7	29.2

This difference also shows up in columns 3 and 6 of Table 3, where the first drop out has already ensured a loss in 17% of the cases for mean condition experiments, but 29% of the cases for sum condition experiments. It is not easy to provide a thorough explanation for this difference; clearly subjects fall prey to the framing effect in some manner. Possibly the mean and sum conditions lead to anchoring and adjustment decisions with different anchors (as if first dropouts in the mean condition were calculating how far above their draw to drop out, while first dropouts in the sum condition were calculating how far below 100 to drop out).

Fixed–effects regressions of p_1 on the draw held by the first person to drop out, with or without time trends or lagged profit, explain between 50% and 60% of the variance in p_1 (except for 80% in 4M and only 37% in 3S). Between the small coefficients on the draw, and the remaining 40–50% unexplained variance, there is little opportunity to infer the draw from knowledge of p_1. This difficulty by itself must bear considerable responsibility for losses occurring in 57% of the auctions, and explain why mean condition auctions cannot take greater advantage of relatively less overbidding by the first bidder to drop out.

Behavior of the second, third and fourth bidders to drop out appears as if they are somehow disciplined to keep the impact of the framing effect upon p_1 from influencing p_4, the market price. Fixed–effects regressions of p_2 on x_2 are notably less different across the mean/sum treatment than the comparable regressions of p_1 on x_1: only two of the four paired comparisons of p_2 regressions exhibit average absolute values of subject dummy variable coefficients that are different across mean/sum conditions at 0.1 significance. Corresponding fixed–effects regressions of p_3 on x_3 and p_4 on x_4 or on x_5 (no 5th dropout price is ever revealed) show no differences at all across the framing effect.

When one or two bidders have dropped out, a suspicion that they have remained competing at prices above their draws should lead bidders still competing to drop out earlier than they otherwise would. However, this tendency should show up as an intercept adjustment in a linear bid function—it should not reduce the slope below 0.2 (the prediction of the equilibrium and both winner's curse models), as a higher draw still impacts positively on value. To examine this, for each experiment, p_3 (and separately p_4) was regressed on earlier dropout prices, a time trend, and x_3 (separately, x_4). Individual intercept coefficients for each subject were estimated, as usual for fixed effects estimates; individual slope coefficients were also estimated. Of the 45 slope coefficients on x_3, no estimate exceeded 0.15. Of the x_4 coefficient estimates, one was 0.08 and significantly positive ($p < 0.2$, 2–tailed), another was (counterintuitively) –0.10 and significantly negative; the remaining 43 slope estimates were below 0.12 and insignificant. The tendency of p_1 to be well in excess of x_1 clearly left little room for maneuvering to avoid losses when $v < 100$.

Some of the explanation for profits observed to increase with v for $v > 100$ relates to p_3 and p_4 behavior which provides virtually no information about x_3 and x_4.

Without depending upon the distributional assumptions underlying fixed–effects regressions, it is possible to see how well various rules of thumb organize the data. Across all sum condition auctions, the unresponsive winner's curse model (drop out at $.2x + .8*100$) has an average error of 0.16 in predicting p_2. The responsive winner's curse (drop out at $.2x + .2p_1 + .6*100$) does a much better job of organizing p_2 observations in mean condition auctions.

Notice that every subject regularly observes outcomes where $v < p_1$, outcomes where v is so slightly in excess of p_1 as to make the likelihood that $p_1 > x_1$ prohibitively high, and outcomes where he himself dropped out first at a price above his draw. So each should be aware of the need to consider p_1 as an upwardly biased estimator of x_1, and possibly an even more biased estimator of x(1). The evidence suggests they do not attempt this crucial task: several rules of thumb which would take into account $x_1 < p_1$ when determining p_2 do not organize the data as well as the two rules of thumb just mentioned. Yet the unresponsive winner's curse sets p_2 at the asset's expected value given x_2 and assuming the other four draws (including x_1) average 100, taking no account of any information inferable from p_1. The responsive winner's curse sets p_2 at the asset's expected value given x_2, assuming (counterfactually) $p_1 = x_1$ and assuming the other four draws average 100.

While the unresponsive winner's curse predicts p_3 on average 0.5 points higher than observed in sum condition auctions, and 2.3 points lower than observed in mean condition auctions, it is an extremely high variance predictor in either. Variance accounted for (via measuring root prediction squared error), among simple rules of thumb, the best we have found at predicting p_3, in either treatment, is that the bidder drops out as soon as legally permitted after the previous dropout. The same rule of thumb is the best simple predictor of p_4 observations in both treatments. (Both the unresponsive and the responsive winner's curse fare very badly as predictors of p_4 behavior.) In fact, respectively in mean and sum conditions, p_4 falls within a point of tying p_3 with 49.7% and 60.1% frequency; of these ties, 38.9% and 33.6% happen to be cases where $x_4 < x_3$, and 50.5% and 54.9% happen to yield losses. Allowing for overlap, 68.4% and 69% of ties end up being cases where $x_4 > x_3$ and a profit resulted. Notice that the most workable simple predictors of p_3 and p_4 take no account even of the information the bidder dropping out had: his own draw.

There are some indications that the ties observed may have related to rational attempts to cope with not entirely hopeful situations (cf. Table 4). Consider the second bidder to drop out: in mean condition experiments he ties the first dropout 17 % of the time. These ties were cases where $x_2-x_1=-7$ on average, while cases where $p_2>p_1$, x_2-x_1 averaged 21. In the sum condition, where p_1 reflected substantially more overbidding, p_2 tied p_1 with 33 % frequency, on average x_2-x_1 was 10 for ties, 35 otherwise. These distinctions for cases of ties are essentially duplicated for p_3. The last bidder to drop is substantially more likely to tie in the mean condition (41 %) and somewhat more likely in the sum condition (53 %). For ties, x_4-x_3 averaged 13 (mean), 29 (sum). When the fourth bidder competed longer, x_4-x_3 averaged 12 (mean), 53 (sum). Overall the ties were occurring when the tying bidders signal was low relative to the current price, which is consistent with essentially more frequent ties in the sum condition.

Table 4: Mean Draws, Mean Drop–out Prices, Mean Values for Different Drop–out Ranks (First 48 Auctions)

	value = mean of individual draws					value = sum of individual draws				
rank of dropout	% ties	mean draw *)	mean dropout price *)	mean value o.asset	value./. dropout price	% ties	mean draw *)	mean dropout price *)	mean value o.asset	value./. dropout price
first	—	51	74	98	24	—	45	85	99	14
second ties	17	61(-7)	92(—)	101	9	35	56(10)	94(—)	94	0
no ties		68(21)	89(19)	97	8		79(35)	95(15)	102	7
all		67(13)	89(15)	98	8		71(26)	94(10)	99	5
third ties	29	69(-10)	99(—)	93	-6	46	77(6)	96(—)	93	-3
no ties		99(38)	96(10)	100	4		105(32)	101(8)	104	3
all		91(24)	97(8)	98	1		92(20)	99(5)	99	0
fourth ties	41	103(13)	95(—)	90	-5	53	118(29)	99(—)	95	-4
no ties		143(12)	107(9)	103	-4		143(53)	108(8)	103	-5
all		127(12)	102(5)	98	-4		132(41)	103(4)	99	-4
winner		154(28)	—	98			153(21)	—	99	
		mean price payed:		102	-4		mean price payed:		103	-4

*) The terms in brackets give the mean additional draw (with respect to the draw of the player who dropped out before), and the mean additional dropout price (with respect to the same player).

Both regression analysis and comparative evaluation of rules of thumb, then, lead to the conclusion that the framing effect noticeably alters the behavior of the first bidder to drop out, but quicker following dropouts in sum conditions keep this effect from impacting on market prices, profits and p_3. There is little evidence that much valuable information can be gleaned from the prices at which the first couple bidders drop out, and much less evidence that bidders incorporate in their behavior in any systematic information from preceding dropout prices.

3.3 Information and Learning by Punishment

The idea that adequate bidding behavior is a matter of learning sounds reasonable, and can help to explain the observed results. Since the only player who can be punished for inadequate behavior is the winner of the asset, learning should – at least in the first phase of the learning process – mainly address the dropout price of the last player who drops out (p_4). If he drops out too early, the asset is left to a rival, at "too low" a price (that is, a price where the winning rival on average makes a nonnegligible profit, while of course the last bidder to drop out makes zero profit). – A basic adjustment of the intercept of the bidding function can then serve to keep overall mean losses reasonably small (nearly zero), but such simple adjustment would not take into account of information that may be available from the earlier dropout prices.

Of course, there are very simple rules, by which an overall profit of 0 can be reached. For instance if all bidders drop out at 100. But such a behavior would not be in equilibrium, since by dropping out later if own draws are above the mean and dropping out earlier at lower own draws gives additional profits. Moreover the dropout prices of the others can be used to get additional information about the asset. As we mentioned above, in equilibrium bidders do get the total information about the rivals' draws, but this was clearly not observed in the experimental behavior.

This raises the question, how much of the information about the individual draws was aggregated by the auction. It seems reasonable to take the dropout price p_4 (which is the price) as a conservative estimate of the aggregated information. This assumption can be motivated in two ways:

(1) Bidders see that they incur losses, when they drop out above asset value, and that they leave profits to others, if they drop out at prices below the value. This suggests that they – in the mean – should just drop out at the value.

(2) A more detailed analysis gives the result that this bidder (bidder 4), when deciding on his drop out price, has – explicitly or implicitly – (a) to make an estimate of rivals' draws, (b) to add his own draw, and (c) to estimate the last competing bidders' draw and add it. Although it is not unreasonable that such a behavioral idea guided the intentions of the players, the exact analysis gives that under normal conditions the bidder should learn to drop out earlier, namely at the price obtained by adding his own draw x_4 to (a) and (b). (Assume a monotonic

dropout behavior, i.e. that a bidder with a higher draw drops out at a higher price: he will be punished, if he drops out at a higher price, since if he then wins the asset, then his price is the higher one and the draw of the remaining bidder is overestimated. He also is punished for a lower dropout, since he afterwards can observe that the remaining bidder makes still good profits, a part of which he could have received had he won the auction.) – This means that, at the highest possible level of information, the dropout price of the price setting bidder can only reach $(x_1+x_2+x_3+x_4+x_4)/5$. I.e. p_4 is for (x_5-x_4) below the value $(x_1+x_2+x_3+x_4+x_5)/5$.)

Of course approach (2) is more correct, the remaining part of information, x_5-x_4, is not available to the subjects, and any estimate of this can theoretically only shift up the reaction curve by a constant. However, it cannot be excluded, that in a situation with personal contact part of this remaining information x_5-x_4 can even be transposed to the other (for instance by observing the trembling of fingers, etc.). This argument makes it reasonable to take p_4 as a slightly conservative indicator on how much of the information $x_1+x_1+x_3+x_4+x_5$ has been incorporated.

As can be inferred from figures 5–9 observed behavior was less informative about bidders' draws. Since profit is $(x_1+x_2+x_3+x_4+x_5)/5-p_4$, one minus the slope of the regression line in figures 5–9 is a conservative estimate of the amount of information about the draws which is aggregated in p_4. So the observed behavior on average only aggregated 50.79 % (in sum condition) of this information, leading bidders who were clearly seeking profit to the level shown in these figures. (The regression line in figures 5 and 6 intercept $v=100$ at about $p_4=104$.)

The intercept is a result of the overall learning behavior of the subjects over all games, avoiding expected losses, where the level of 104 shows that they did not yet learn adequately. This level has in the average been reached already after the first 12 rounds, and it seems that no more essential learning took place after that with respect to the mean level of losses.

Clearly, a bidder would also have reasons to deviate from his own most reasonable estimate into a more conservative dropout price. However this would – if done willingly – only effect the (slope for) high values, and not the unsuccessful behavior at low values v. The fact that the figure shows a linear shape of the in section suggests that a splitted behavior did not take place, and that the subjects everywhere used all the information they perceived. (Systematic underestimations or overestimations do only effect the intercept, not the slope.)

The data show that different groups performed essentially different in information aggregation. It seems surprising that the two groups on tables 2M and 2S (with identical draw data) performed best, and that the groups of the corresponding tables 3M and 3S (again with identical draw data) performed best. This slightly suggests that good information transferring can be motivated by the history in which signals are presented.

3.4 Bidders' Estimates of Value

Recall that every sixth auction involved an extra step: after the four dropout prices were known, the bidders were asked to estimate the value, before its announcement. Subjects appeared to take this request seriously. The dasta of table 5 show that the bidders, who dropped out first, second or third (on average) underestimated the value of the asset (after knowing all four dropout prices). Players who dropped out fourth, and "actively" set the price (not tying), on average underestimated the value by 3 points (mean), and 7 points (sum), and thereby implicitly estimated losses of 3 and 7 points. Winning bidders, however, overestimated the value by 8 points (mean), and 4 points (sum), and thereby implicitly estimated profits of the same amounts. The actual average losses were 5 points (mean), and 0 points (sum) in the respective rounds. Overall the data illustrate that on average the prrocedure selects just those players as winning, who do oversestimate the asset.

A question is, why the winning bidder was willing to bid higher than he thought the asset was worth? We initially suspected an answer comes from the frequently closely bunched occurrences of p_2, p_3 and p_4: in such cases, when (unknown to the bidders) the auctioneer was calling out a price equal to the asset's value, there may have been 3 or even 4 bidders still competing, all thinking at the moment that the value is higher. When a couple bidders drop out shortly thereafter, the price setter may have rationally viewed these bunched dropouts as suggesting a lower value than he thought before hearing such close dropouts. In fact the (implicitly) estimated losses of the 4th to dropout were on average for 7 points (mean), and 6 points (sum) higher in cases of ties than otherwise: Tying fourth dropouters are more cautios in evaluating the asset than those who do not tie. However, the winner on average overestimates the asset in both cases, in the sum condition even for the same amounts.

The price setter's estimates may become slightly more accurate in later auctions than earlier, but overall there is little sign of improvement in estimating accuracy as bidders gather experience. In view of the regression analyses which suggested that draws x_4 play virtually no role in determining p_4, it may be surprising that the winner regularly estimates the value about 10 points higher than the price setter, and this more or less matches the average of $x_5 - x_4$, the difference in their draws (9 points). This compares averages, however, of two sets of observations with extremely high standard errors. Of course, we have no corroborative evidence in the bidding: there is no way to know at how much higher a price the winner would have wanted to cease competing.

Table 5: Mean Draws, Mean Drop–out Prices, Mean Values, Mean Estimated Values for Different Drop–out Ranks (Every 6th of the First 48 Auctions)

	value = mean of individual draws					value = sum of individual draws				
rank of dropout	% ties	mean draw 1)	mean dropout price 1)	mean val.of asset2)	value./. dropout price 2)	% ties	mean draw 1)	mean dropout price 1)	mean val.of asset2)	value./. dropout price 2)
first	—	69	86	107(103)	21(17)	—	53	88	108(94)	20(6)
second ties	17	50(-57)	103(—)	114(101)	11(-2)	42	63(6)	98(—)	101(92)	3(-6)
no ties		79(18)	100(18)	106(98)	6(-2)		112(62)	102(21)	113(104)	11(2)
all		74(5)	100(15)	107(98)	7(-2)		91(38)	100(12)	108(99)	8(-1)
third ties	14	64(0)	98(—)	87(94)	-11(-4)	54	93(0)	99(—)	104(96)	5(-3)
no ties		107(31)	108(7)	111(106)	3(-2)		104(15)	111(9)	112(114)	1(3)
all		101(26)	106(6)	107(104)	1(-2)		98(7)	104(4)	108(104)	4(0)
fourth ties	38	77(13)	104(—)	97(106)	-7(+2)	73	119(25)	103(—)	104(100)	1(-3)
no ties		157(61)	116(9)	114(111)	-2(-5)		159(49)	119(11)	117(110)	-2(-9)
all		127(26)	112(5)	107(109)	-5(-3)		129(31)	108(3)	108(103)	0(-5)
winner		166(40)	—	107(115)	(3)		167(38)	—	108(112)	(4)
		mean price payed:		112	-5		mean price payed:		108	0

1) In data–columns 2 and 3 the terms in brackets give the mean additional draw (with respect to the draw of the player who dropped out before), and the mean additional dropout price (with respect to the draw of the player who dropped out before).
2) In data–columns 4 and 5 the terms in brackets give the mean estimated value (by questionnaire), and the differene of this estimated value to the dropout price (i.e. the "estimated profit").

3.5 Responses to Questionnaires

In addition to providing estimates every six auctions, recall that the subjects were asked to fill out a questionnaire every twelfth auction. (Details are in the last part of Section 2.) The average responses to the questionnaire are presented in Table 6. Averages from mean and sum conditions are combined, as they are virtually identical. Overall, there is no clear pattern showing improvement in bidding on questionnaires as subjects acquire more experience (experiment 2M clearly improves in $q_1(20)$, and 1S improves in $q_4(180)$, but these are balanced by disimprovements in other experiments). The spread $q_4(80) - q_1(20)$ is systematically narrower than the spread $q_4(180) - q_1(120)$ in individual responses.

The questionnaire provided a means of observing hypothetical choices of the same subject in all four dropout positions of a low–draws and then a high–draws auction, just shifted by a constant of 100. It could also have served to suggest behavioral patterns for bidding, by pointing to the notion of bidders dropping out in the order of their draws. There is no evidence it actually served this purpose. Apparently subjects responded as we had hoped, not with separate calculations, but more with the same behavioral patterns they were using for the bidding.

Table 6: Average Responses to Questionnaires in a Low Draws and a High Draws Auction (Equilibrium and Responsive Winner s Curse Predictions in Parentheses)

rank of dropout	first		second		third		fourth	
questionnaires, low draws auction:								
own draw	x1	20	x2	40	x3	60	x4	80
avg dropout price	q1	72	q2	79	q3	84	q4	88
(equilibrium)	(p*1)	(20)	(p*2)	(36)	(p*3)	(48)	(p*4)	(56)
(resp.winner's curse)	(p*1)	(84)	(p*2)	(85)	(p*3)	(86)	(p*4)	(87)
questionnaires, high draws auction:								
own draw	x1	120	x2	140	x3	160	x4	180
avg dropout price	q1	110	q2	119	q3	126	q4	132
(equilibrium)	(p*1)	(120)	(p*2)	(136)	(p*3)	(148)	(p*4)	(156)
(resp.winner's curse)	(p*1)	(104)	(p*2)	(109)	(p*3)	(115)	(p*4)	(121)
observed behavior:								
avg own draw	x1	48	x2	69	x3	91	x4	129
avg dropout price	p1	79	p2	92	p3	98	p4	102
(equilibrium)	(p*1)	(33)	(p*2)	(60)	(p*3)	(80)	(p*4)	(93)
(resp.winner's curse)	(p*1)	(87)	(p*2)	(91)	(p*3)	(95)	(p*4)	(101)

If the pattern of higher dropout prices relating to higher draws is presumed to persist, then the value could be expected to fall in the range [56,80] and average 68 for the low–draws auction, and in the range [156,160] and average 158 for the high–draws auction. This would mean that most subjects respond to the low–draws questions with behavior that has 3 subjects still competing when the price has gone beyond the highest possible value of 80. They respond to the high–draws questions with behavior that does not even attempt to compete for profits (with $x_2 = 140$, inferring $x_1 = 120$, and $x_3, x_4, x_5 > 140$, the 2nd, 3rd and 4th bidder all know $v > 136$, yet all three drop out before that). With the high draws exceeding the low draws by 100, a value 20 points higher is ensured, so responses should be 20 points higher only if there is no reason to assume rivals' draws are higher. There is reason to assume this, and responses are more than 20 points higher in 90% of the cases. However, if rivals still competing are assumed to have as high a draw as yours, and higher draws are inferred for rivals who dropped out at much higher prices, responses 100 points higher are justified. The observed tendency to compete only 40 points higher shows an assumption that rivals have higher draws, but little confidence in the inferring from higher dropout prices. (The observed increases from low–draws to high–draws questions are larger than predicted by either winner's curse model.)

Responses are not only less responsive to the 100 point difference than would seem appropriate, but the responsiveness to higher draws (20 above the previous draw) is insufficient. It is well above (except for $q_4(80) - q_3(60) = 4$) the 4 points justified if you assume no rival's draw is higher, but averages roughly half the responsiveness warranted if you assume all rivals still competing also have draws 20 points higher (which would imply $q_2 = q_1 + 16$, $q_3 = q_2 + 12$, $q_4 = q_3 + 8$). Nonetheless, this incorporates information about draws x_3 and x_4 to a substantially greater degree than observed in the actual dropout prices p_3 and p_4.

While average responses to the questionnaire are very close to identical between mean and sum conditions, a systematic difference in the distributions underlying these averages may illuminate behavior patterns. The whole distributions of sum condition responses to $q_1(20)$, $q_1(120)$ and $q_4(180)$ each form only a part of the corresponding distribution of mean condition responses. This is shown in Figures 10–12.

The remaining quarter of the mean condition responses form a behaviorally (and perhaps attitudinally) distinct mode. The distinct responses to $q_4(180)$ lie above 145, apparently "aggregating" the information of the draws 120, 140, 160, 180, as if these subjects have no fear to compete to high prices when warranted. The distinct responses to $q_1(20)$ lie below 45, perhaps "hesitating" appropriately when the only information they have is pessimistic. The distinct responses to $q_1(120)$ lie above 125, who "reveal" their supposition that other bidders still competing have higher draws and who are willing to bid high on this basis. Such revealers do not exist in the bidding data, where none of the 32 cases with $x_1 > 120$ have $p_1 > 120$. Notice that when x_1 is low, someone assuming rivals' draws are higher than x_1 cannot be distinguished from someone simply assuming that rivals' draws are typical (i. e., average 100). Similarly, hesitators cannot be separated in cases where $x(1)$ is high, and aggregators cannot be distinguished when x_4 is low.

Behavior which combines "hesitating" and "aggregating" appears significantly more rational than most subjects' behavior observed in the bidding. Responses to questionnaires showing both attitudes occur only for 3 subjects in experiment 2M. The other 6 aggregators are also in the distinct $q_1(120)$ mode, responding as if willing to compete to high prices whether justified or not.

4 Indications

Despite the apparent difficulty subjects faced in making sufficient adjustments for the adverse selection problem in these auctions, we have observed the framing effect of mean versus sum condition nullified in impact on market outcomes. The story across most subjects is one of the framing effect notably altering the first dropout price, but what could be loosely called market forces constraining later dropout prices so as to wipe out this initial impact.

The two formulations for the winner's curse model presented in Section 2 both come closer than the symmetric equilibrium model in predicting the observed average level of profit. The responsive winner's curse model also is consistent with the observed regularity of profits occurring when the asset is more valuable than average, and losses when it is less valuable than average. None of these three models is at all an adequate predictor of observed patterns of individual competition. In particular, all three predict continued impact of private information (the draw) upon dropout prices, when the data clearly reject this hypothesis for the 3rd and 4th bidders to drop out.

The simpler statistical properties of this independent information variant of the canonical common–value auction appear if anything to be less important to the rate of subject learning than the disconcerting role of 100 as a natural anchor. There are enough differences to prevent clear and direct statistical comparisons, but the signs are that subjects in the independent information variant may take even longer to learn to deal with the winner's curse than in the canonical variant (cf. Harstad, Kagel and Levin [1989]).

References

Bikhchandani, Sushil and John Riley, "On Unique Equilibria in Open Auctions," working paper, Department of Economics, University of California, Los Angeles, 1989.

Capen, E. C., R. V. Clapp and W. M. Campbell, "Competitive Bidding in High–Risk Situations," Journal of Petroleum Technology, 1971, v. 23, 641–653.

Cassing, James and Richard W. Douglas, "Implications of the Auction Mechanism in Baseball's Free Agent Draft," Southern Economic Journal, 1980, v. 47, 110–121.

Cox, James C., Bruce Roberson and Vernon L. Smith, "Theory and Behavior of Single–Object Auctions," in Vernon L. Smith (Ed.), Research in Laboratory Economics, v. 2, Greenwich: JAI Press, 1982.

Dessauer, John P., Book Publishing, New York: Bowker, 1981.

Harstad, Ronald M., John H. Kagel and Dan Levin, "Revenue Comparisons in Open– and Sealed–Bid Auctions, with Control for Extraneous Sources of Variation," working paper, Department of Economics, Virginia Commonwealth University, 1989.

Kagel, John H., Ronald M. Harstad and Dan Levin, "Information Impact and Allocation Rules in Auctions with Affiliated Private Values: A Laboratory Study," Econometrica, 1987, v. 55, 1275–1304.

Kagel, John H. and Dan Levin, "The Winner's Curse and Public Information in Common Value Auctions," American Economic Review, 1986, v. 76, 894–920.

Kagel, John H., Dan Levin and Ronald M. Harstad, "Judgment, Evaluation and Information Processing in Second–Price Common Value Auctions," working paper, Department of Economics, University of Pittsburgh, 1988.

Knez, Mark and Vernon L. Smith, "Hypothetical Valuations and Preference Reversals in the Context of Asset Trading," pp. 131–154 in Alvin E. Roth (Ed.), Laboratory Experimentation in Economics: Six Points of View, Cambridge: Cambridge University Press, 1987.

Levin, Dan and Ronald M. Harstad, "Symmetric Bidding in Second–Price, Common–Value Auctions," Economics Letters, 1986, v. 20, 315–319.

Milgrom, Paul R. and Robert J. Weber, "A Theory of Auctions and Competitive Bidding," Econometrica, 1982, v. 50, 1089–1122.

Roll, Richard, "The Hubris Hypothesis of Corporate Takeovers," Journal of Business, 1986, v. 59, 197–216.

Tversky, Amos and Daniel Kahneman, "The Framing of Decisions and the Psychology of Choice," Science, Jan. 1981, 453–458.

Wilson, Robert, "A Bidding Model of Perfect Competition," Review of Economic Studies, 1977, v. 44, 511–518.

Figure 1: Experiments 1M and 1S
Profit, Observed and Predicted

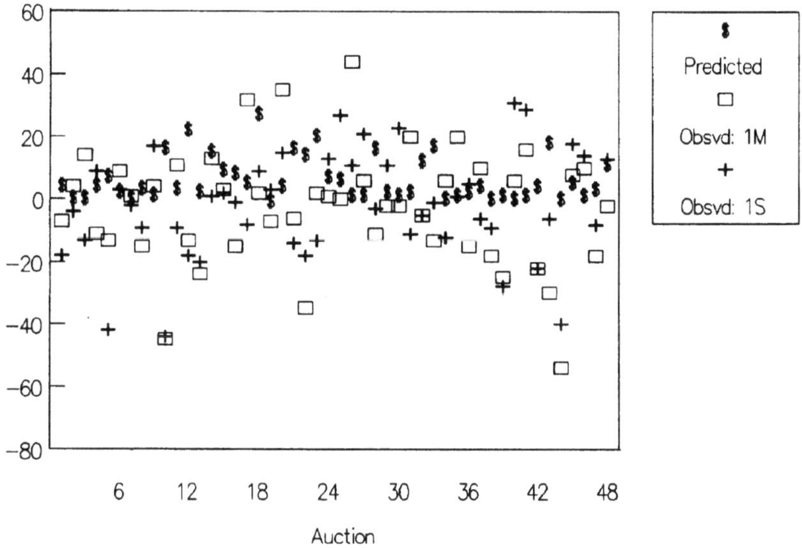

Figure 2: Experiments 2M and 2S
Profit, Observed and Predicted

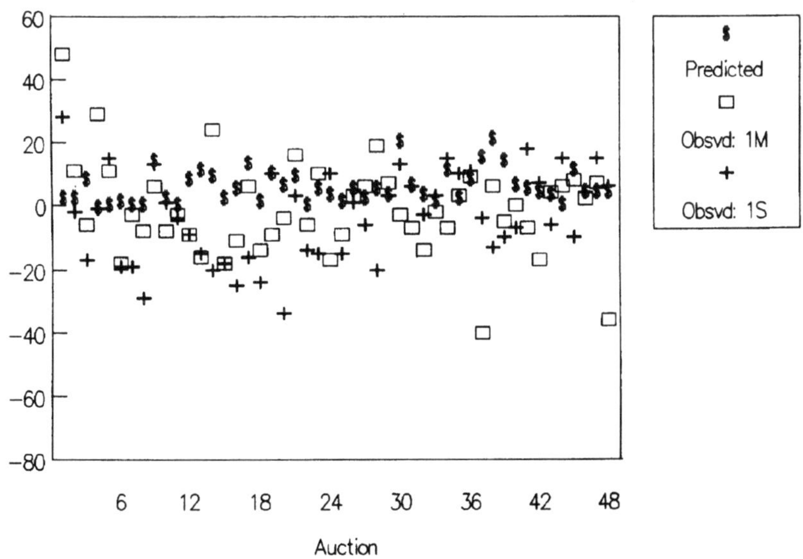

Figure 3: Experiments 3M and 3S
Profit, Observed and Predicted

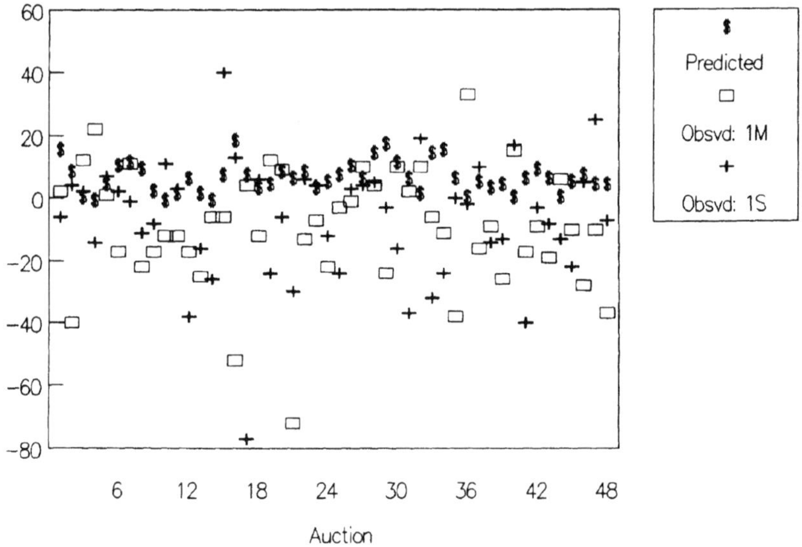

Figure 4: Experiments 4M and 4S
Profit, Observed and Predicted

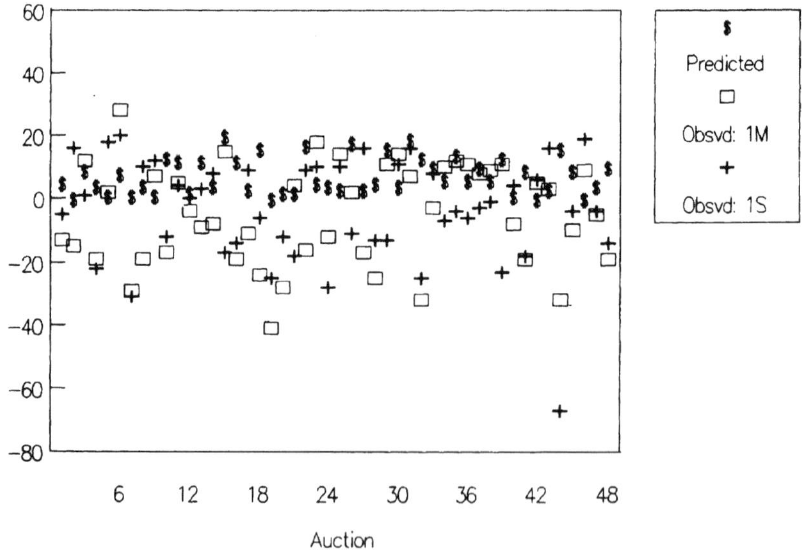

Figure 5: Mean Condition
Profit as a Function of Value

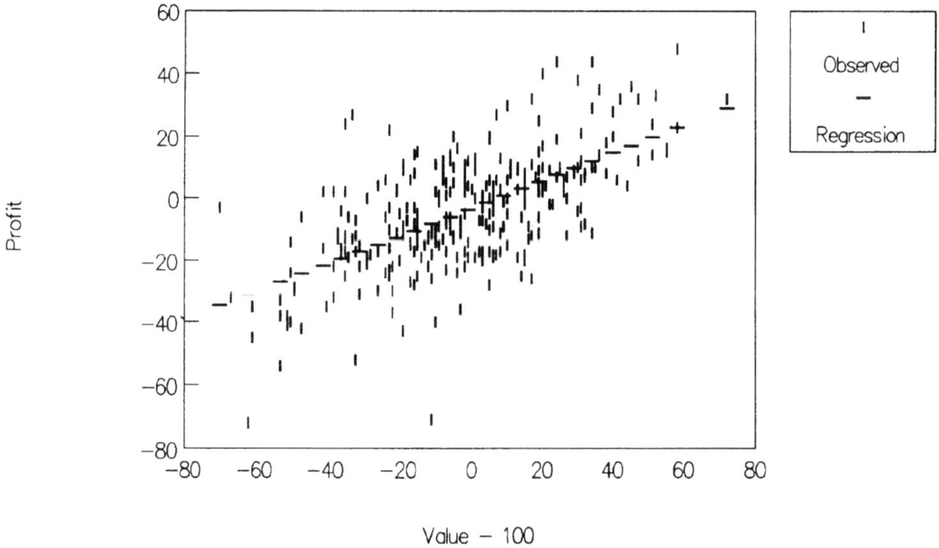

Figure 6: Sum Condition
Profit as a Function of Value

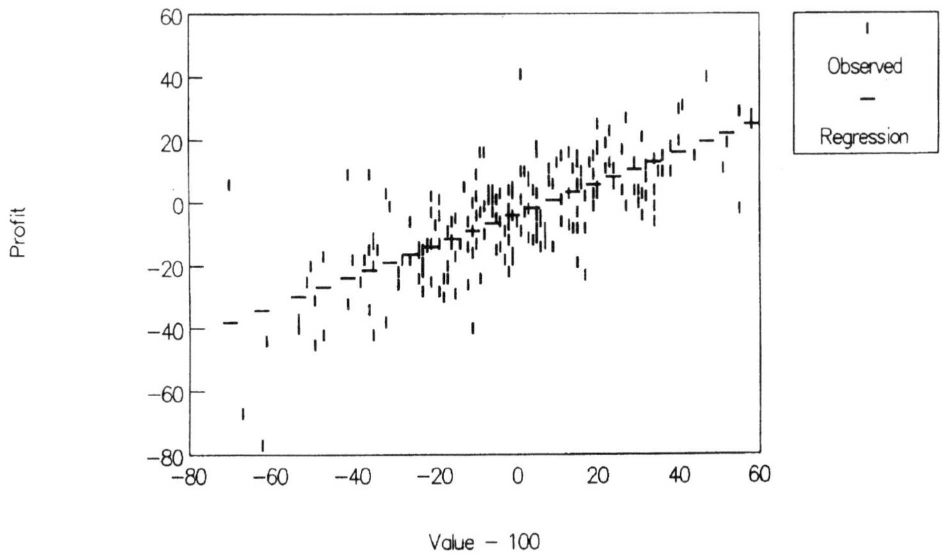

Figure 7: Experiment 2M
Profit as a Function of Value

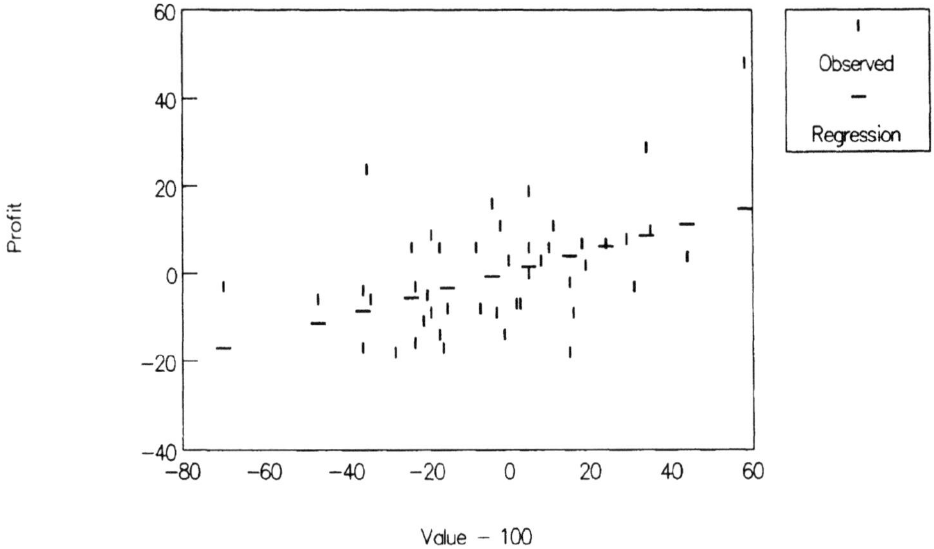

Figure 8: Experiment 3S
Profit as a Function of Value

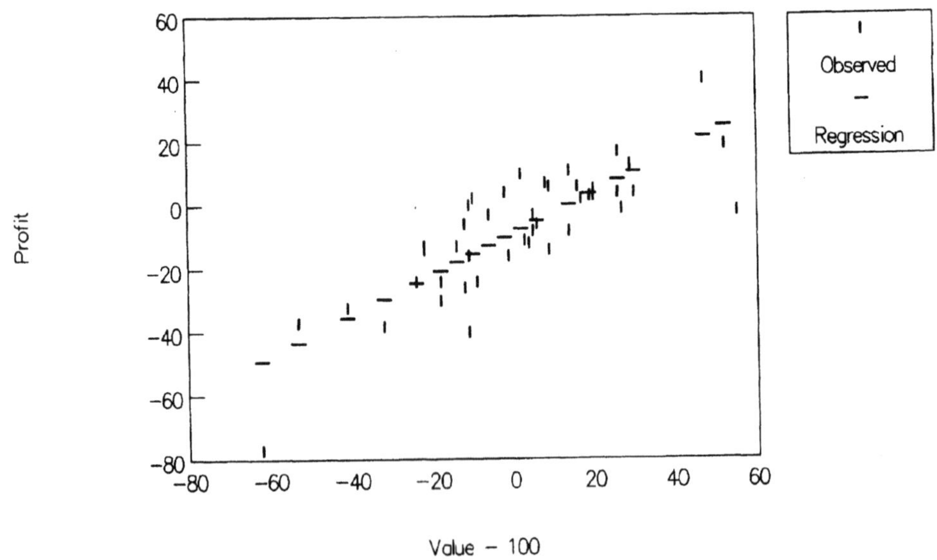

Figure 9: Experiment 4S
Profit as a Function of Value

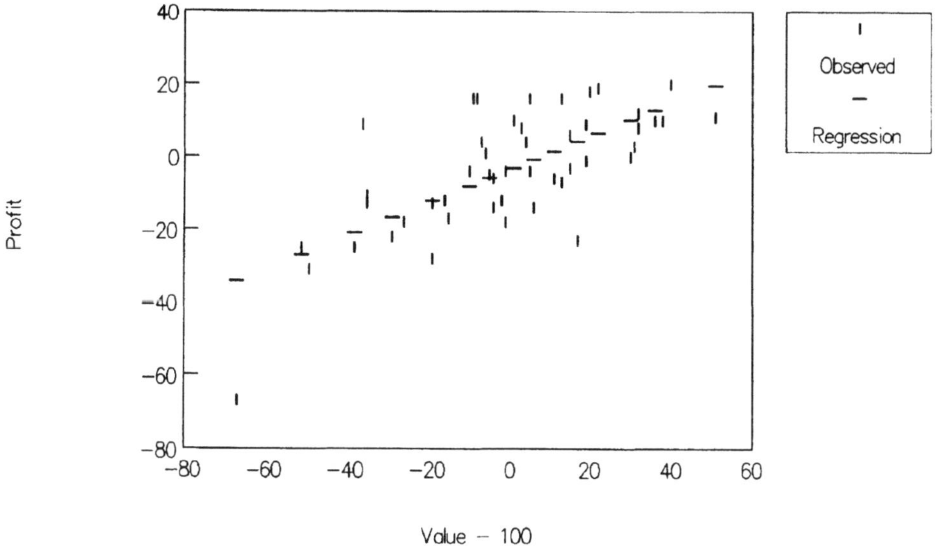

Figure 10: Response Frequencies
1st Drop–Out Price, Draw = 20

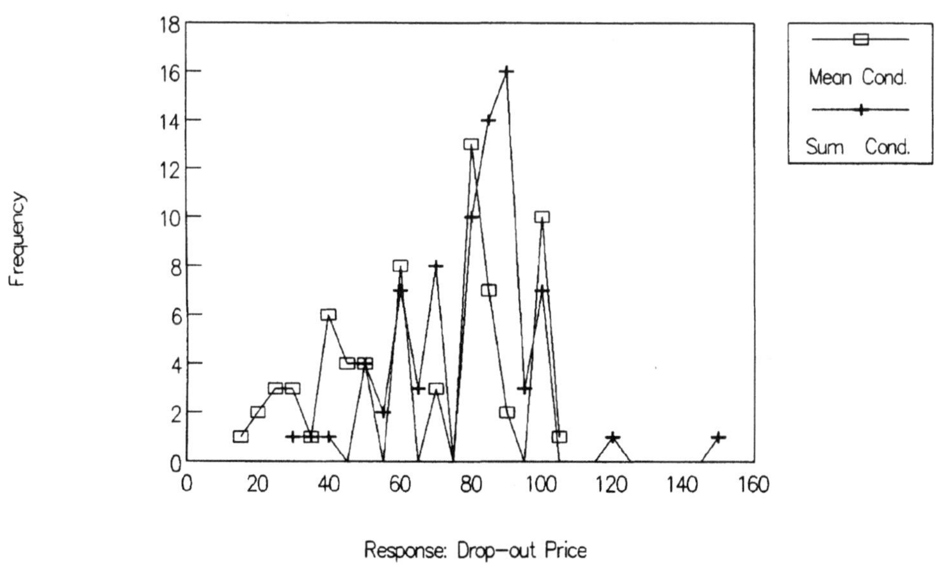

Figure 11: Response Frequencies 1st Drop-Out Price, Draw = 120

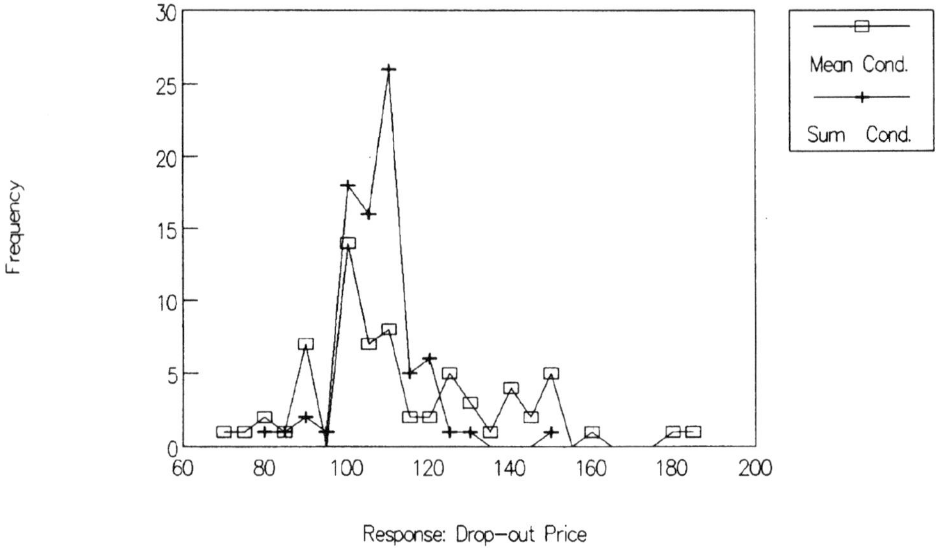

Figure 12: Response Frequencies 4th Drop-Out Price, Draw = 180

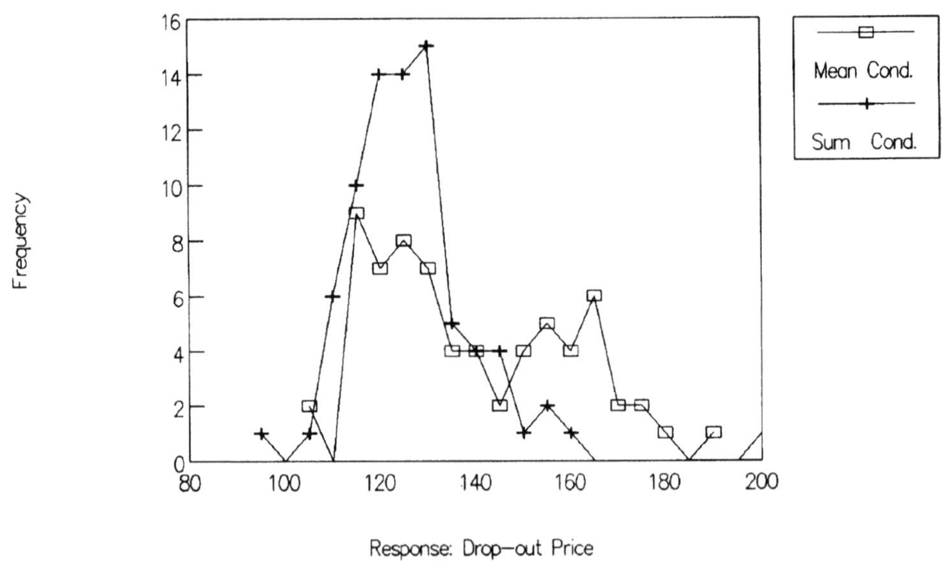

RENT DISSIPATION AND BALANCED DEVIATION DISEQUILIBRIUM IN COMMON POOL RESOURCES: EXPERIMENTAL EVIDENCE[1]

by

James M. Walker, Roy Gardner, and Elinor Ostrom

Abstract: This paper examines resource allocation in an open access common-pool resource. focus on the predictive strength of a Nash equilibrium model which predicts that users of open access common-pool resources will appropriate units from the resource at suboptimal rates near full rent depletion. We present evidence from laboratory experiments designed to investigate the accuracy of the rent dissipation prediction for such environments.

I. INTRODUCTION

There exists considerable theoretical and field work examining optimal resource allocation for resource environments known as open access common-pool resources (CPRs).[2] A generally accepted premise is that, void of imposing private property rights or a central planner, users of open access CPRs will overinvest in appropriation from the resource, leading to the complete dissipation of economic rents and possible resource destruction.[3] The reasoning leading to this result is not game theoretic. Indeed, for a small number of players, complete rent dissipation may contrast sharply with the outcomes predicted by any Nash equilibrium strategy, as we show below.

Using laboratory experiments, this paper investigates the choice behavior of agents in a decision environment designed to capture the properties assumed in standard

---

[1]A preliminary version of this research was presented at the Conference on Equilibrium Models, Bielefeld, Germany, June 1988. We wish to thank the participants of this conference for their helpful comments. Any errors remain the responsibility of the authors. Financial support from the National Science Foundation (Grant # SES-8619498) is gratefully acknowledged. All data are stored on permanent PLATO disk files and are available on request. Send inquiries to Professor James M. Walker, Department of Economics, Ballantine 901, Indiana University, Bloomington, Indiana 47405.

[2]Following Gardner, Ostrom, and Walker (1988) we define an open access CPR to be a natural or man-made resource system from which a flow of subtractable (rival in consumption) resources are available over time and: (1) are sufficiently larg that it is costly (but not necessarily impossible) to exclude potential beneficiaries from appropriating resource units, or (2) in which the property rights in use are such that potential appropriators cannot be legally excluded. See Bromley (1982) and Ciriacy-Wantrup and Bishop (1975) for a discussion of the distinction between open access CPRs and common property CPRs.

[3]The <u>necessity</u> of an external actor imposing a change in property rights on those using a common-pool resource is not empirically supported. Many users of common-pool resources have implemented their own property rights (see E. Ostrom, 1988, and many of the individual cases reported in McCay and Acheson, 1987, and National Research Council, 1986.) In related work, we are analyzing the conditions associated with changes in institutional arrangements from open-access to various other forms of property relationships.

models that focus on this dilemma.[4] Our investigation is designed to test the accuracy and robustness, at the aggregate and individual levels, of the standard prediction of complete rent dissipation versus the predictions from a Nash equilibrium model.

The formal modelling and description of rent dissipation in open access common-pool resources traces its roots to the seminal paper by Scott Gordon (1954). Prior to Gordon's analysis, discussions of appropriation from open access resources focused primarily on the biological issues related to the size of the resource stock. Gordon's work redirected the focus to the problem of optimal appropriation, with a goal of economic efficiency. Briefly, the behavioral hypothesis described by Gordon can be summarized as follows. Individuals appropriate resource units in a resource environment where marginal changes in individual appropriation levels have external effects on the production relationship for appropriation faced by all other users. Specifically, Gordon assumes that increases in levels of appropriation by individual users lower the marginal physical product to investment by all users. Given the external nature of this effect and the lack of well-defined property rights, individual users are assumed to ignore the effects and focus only on average returns from investment. Note that an immediate consequence of this assumption is that, for a finite number of players, the resulting behavior will not in general constitute a Nash equilibrium. If individuals follow a behavioral strategy that focuses only on average returns, investments in appropriation from the resource will continue to a level at which average revenue product of increased appropriation equals marginal opportunity costs. The resourc is "mined" at a level beyond the economically efficient point at which marginal revenue product equals marginal opportunity cost and all economics rents are dissipat d. It does not immediately follow that such behavior leads to the secondary issue of destruction of a renewable resource. This latter prediction requires further constraints on the specific relationship between the reproductive nature of the resource and the level of appropriation. This paper focuses only on the issue of overappropriation.

The theoretical framework laid out by Gordon has been extended by others such as Smith (1968) and Clark (1976). There have also been numerous "case studies" directed

---

[4]Following Blomquist and Ostrom (1985), we distinguish between the resource units that can be withdrawn from a resource and the actual resource system itself. Following Plott and Meyer (1975), we call the process of obtaining these resource units "appropriation" to distinguish this process from that of producing or distributing the resource units. As discussed in Gardner, Ostrom, and Walker (1988), the CPR dilemma may actually be composed of numerous "separable" allocation problems. The work here focuses on one subset of these choice problems, efficient "appropriation." While the resource system may be jointly used, the actual resource units are not subject to joint use. In appropriation problems, the allocation problem to be solved focuses on how to allocate the yield from the CPR in an economic and equitable fashion. For analytical purposes, appropriation questions can be separated from "provision" questions which r late to creating the resource, maintaining or improving the production capabilities of the resource, or avoiding the destruction of the resource.

toward estimating the degree to which open access resources are mined at a level predicted by the standard theor tical models (e.g., Alexander, 1982; Agnello and Donnelly, 1975; Bell, 1972; and Johnson and Libecap, 1982). However, field studies are limited by the extent to which the optimal level of appropriation and actual appropriation are calculable and observable. Further, such studies are limited by the extent to which the observer is able to infer the consequences of parametric changes in the behavioral environment. Prior field research has not always made a clear distinction among the diverse property rights and regimes that are used in relation to CPRs (see Schlager and E. Ostrom, 1988). The choice environments described in this experimental study are designed to test the behavioral accuracy of CPR theoretical models and to investigate the linkages between individual behavior and environmental (parametric) conditions.[5] By creating an environment where no institutional configuration exists to monitor or limit the amount of investment, the r sulting strategy space captures the essence of an open access property rights regime.

The paper proceeds as follows. Section II describes the actual choice environment faced by experimental subjects. Given this basic framework, the next three sections focus on predicted and actual behavior according to rent dissipation and Nash equilibrium theories for specific experimental parameterizations. In section III, we report results for a design for which the common-pool resource has multiple Nash equilibria, all close to each other in payoff space, and a positive opportunity cost. In section IV, the parameterization still leads to multiple Nash equilibria, but the opportunity cost is zero. One consequence of this change is to increase the differ nce in payoffs that result at the Nash equilibrium relative to those that occur at zero rents. In section V, the parameterization is altered to yield a unique Nash equilibrium. Our working hypothesis was that this change would reduce the behavioral complexity of reaching the Nash equilibrium.

Our essential findings in all three treatments are that: (1) Nash equilibrium predictions outperform the full rent dissipation hypothesis at the level of average individual behavior; but (2) individual play does not confirm Nash equilibrium play. The resulting state of affairs we term a balanced deviation disequilibrium. The conclusion offers possible explanations for the balanced deviation disequilibria we observe, as well as suggestions for further research.

---

[5]We are not aware of any experimental work that has focused specifically on the appropriation problem and rent dissipation as described in this paper. Th r has been previous experimental work, however, that focuses on behavioral choices in a shared resource environment. Jorgenson and Papciak (1980) and Messick and McClelland (1983) are illustrative of the focus of this prior research. See Samuelson and Messick (1988) for a survey. Also, see Millner and Pratt (1988) for an interesting experimental study of rent-seeking activities in two-person lottery games. If one considers profit dissipation in an oligopoly market as analogous to rent dissipation in a CPR, then experiments such as Alger (1986) and Morrison and Kamarei (1989) reveal broad tendencies consistent with ours.

II. EXPERIMENTAL ENVIRONMENT

A. Subjects and the Experimental Setting

The experiments reported in this paper were conducted using subjects drawn from the undergraduate population at Indiana University. Students were volunteers recruited primarily from principles of economics classes. Prior to recruiting, potential volunteers were given a brief explanation in which they were told only that they would be making decisions in a "economic choice" environment and that the money they earned would be dependent upon their own investment decisions and those of the others in their experimental group. All experiments were conducted on the PLATO computer system at Indiana University. The use of the computer facilitates the accounting procedures involved in the experiment, enhances across experiment control in experimental procedures, and allows for minimal experimenter interaction. In the experiments reported in this paper, all subjects are experienced in the decision environment. They had all previously participated in an experiment utilizing the same decision environment (not necessarily the same parameters). Subjects were randomly drawn from the pool of inexperienced subjects. No experimental session represents a group held intact from a previous session.

B. The Choice Environment

At the beginning of each experimental session, subjects reviewed the experimental instructions. They were reminded that they would be making a series of investment decisions, that all individual investment decisions were anonymous to the group, and that at the end of the experiment they would be paid privately (in cash) their individual earnings from the experiment. Subjects then proceeded to go through, at their own pace, the instructions that described the investment decisions.[6]

Subjects were instructed that each period they would be endowed with a given number of tokens (e_i) and that each period they were to invest their endowment between two markets. Market 1 was described as an investment opportunity in which each token yielded a fixed (constant) rate of output and that each unit of output yielded a fixed (constant) return. Market 2 (the CPR) was described as a market that yielded a rate of output per token dependent upon the total number of tokens invested by the entire group. The rate of output at each level of investment for the group was described in functional form as well as tabular form. Subjects were informed that they would receive a lev l of output from market 2 that was equivalent to the percentage of total group tokens they invested. Further, subjects knew that each unit of output from market 2 yielded a fixed (constant) rate of return. Figure 1 displays the actual information

---

[6]A complete copy of the instructions is available from the authors on request.

subjects saw as summary information in the experiment (illustrated for the payoff schedule used in our Design I experiments). The instructions were written to describe each level of information displayed in Figure 1. Subjects knew with certainty the total number of decision makers in the group, total group tokens, and that endowments were identical. They did not know the actual number of decision periods that would constitute the experiment. Subjects were separated by blinders and were not allowed to communicate. Subject's decisions were made simultaneously in any given decision period.

UNITS PRODUCED AND CASH RETURN FROM INVESTMENTS IN MARKET 2
commodity 2 value per unit = $ 0.01

Tokens Invested by Group	Units of Commodity 2 Produced	Total Group Return	Average Return per Token	Additional Return per Token
8	181	$ 1.81	$ 0.23	$ 0.23
16	323	$ 3.23	$ 0.20	$ 0.18
24	427	$ 4.27	$ 0.18	$ 0.13
32	493	$ 4.93	$ 0.15	$ 0.08
40	520	$ 5.20	$ 0.13	$ 0.03
48	509	$ 5.09	$ 0.11	$-0.01
56	459	$ 4.59	$ 0.08	$-0.06
64	371	$ 3.71	$ 0.06	$-0.11
72	245	$ 2.45	$ 0.03	$-0.16
80	80	$ 0.80	$ 0.01	$-0.21

The table shown above displays information on investments in Market 2 at various levels of group investment. Your return from Market 2 depends on what percentage of the total group investment is made by you.

Market 1 returns you one unit of commodity 1 for each token you invest in Market 1. Each unit of commodity 1 pays you $ 0.05.

Figure 1: Instructions: Summary Information

In designing any experiment the question of the most acceptable way to operationalize the theoretical environment must always be examined. Most theoretical models are incomplete with respect to the exact form and level of information that should be available to decision makers, as well as the specific institutional arrangements for translating subject's choices into outcomes. In fact, it is such variables that can be explored in examining the robustness of theoretical predictions. Our design can be viewed to some extent as a "boundary" experiment for investigating the notion of rent dissipation in CPR environments. That is, from a parametric point of view, groups are not extremely large (N = 8) and subjects are given explicit information on the marginal effects of investment in the CPR.

If the behavioral results we obtain are contrary to predictions of rent dissipation then we would conclude that the environment necessary for theoretical confirmation might be obtained by altering one of our key experimental controls (e.g., group size, openness of the resource, investment information, etc.). However, if our results support an hypothesis of significant rent dissipation, they are suggestive of a behavioral prediction that is quite robust. Further, one should note that our environment (parallel to Gordon's analysis) abstracts away from an environment in which investment decisions are time dependent. The complexity of this time dependence did not seem appropriate for this study. The complexity of intertemporal choice would hav been confounded with our goal to investigate the behavioral hypothesis that individuals will tend to ignore marginal production externalities imposed on the group. Finally, as discussed in Gardner, Ostrom, and Walker (1988), it is quite useful analytically to s parate CPR decision problems that are static in nature (appropriation) from thos that are dynamic in nature (provision).

III. DESIGN I EXPERIMENTS

The parametric conditions for our initial set of four "Design I" experiments are shown in Table 1. Parametric conditions were constant within a given experiment. All xperiments were conducted for at least 20 decision periods (no more than 25).

Experiment Type	Design I
Number of Subjects	8
Individual Token Endowment	10
Production Function: Mkt.2*	$25(\Sigma x_i)-.30(\Sigma x_i)^2$
Market 2 Return/unit of output	\$.01
Market 1 Return/unit of output	\$.05
Earnings/Subject at Group Max.	\$.92
Earnings/Subject at Nash Equil.	\$.67
Earnings/Subject at Zero Rent	\$.50

*Σx_i = the total number of tokens invested by the group in market 2. The production function shows the number of units of output produced in market 2 for each level of tokens invested in market 2.

Table 1: Experimental Design I - Parameters for a Given Decision Period

A. Theoretical Predictions

Given our experimental design, we can proceed to examine possible theoretical predictions for this choice environment. The predictions are classified into three principle areas: (1) group optimality (maximum rents), (2) Nash equilibria, and (3)

complete rent dissipation. Figure 2 illustrates group behavior that would be consistent with these alternative theoretical predictions. As noted in Figure 2, a group investment of 33 tokens yields a level of investment at which MRP = MC and thus maximum rents (denoted T1). Conversely, a group investment of 67 tokens yields a level of investment at which ARP = MC and thus zero rents from Market 2 (denoted T2).

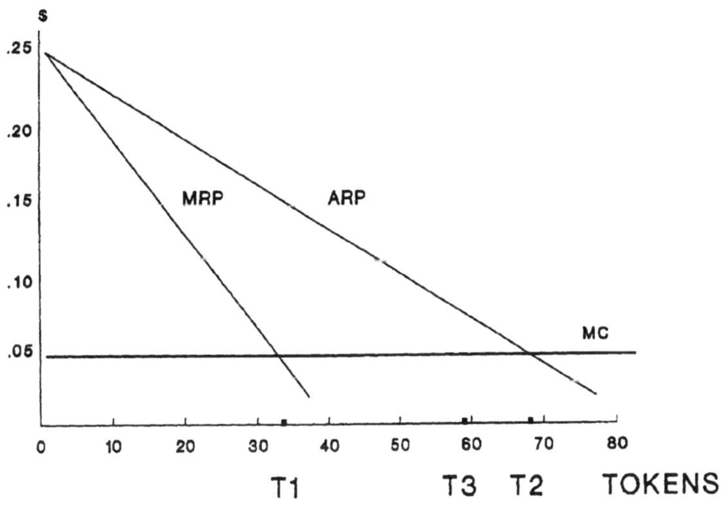

Figure 2: Theoretical Predictions: Design I

Given the nature of the production function for market 2 and the fact that subjects have multiple tokens to invest, there exist a set of Nash equilibria that differ from either rent maximization or minimization. The derivation of the Nash equilibria proceeds in the following manner. Let $f_i(x)$ be the payoff function for player i, when the vector of strategies is x. At a Nash equilibrium x^*, one has:

$$\partial f_i(x^*)/\partial x_i = 0 \tag{1}$$

for every i, for the case where tokens are continuously divisible. In our experiments with 8 players, the payoff function is of the form:

$$f_i(x) = a(10-x_i) + (x_i/\Sigma x_i)(b\Sigma x_i - c(\Sigma x_i)^2). \tag{2}$$

Since we know that $\Sigma x_i>0$ (if no tokens were invested in market 2 by other players, then player i would be best off investing all his/her tokens in market 2), we can write (2) as:

$$f_i(x) = a(10-x_i) + x_i(b-c\Sigma x_i). \tag{3}$$

Differentiating (3) and substituting into (1), we have

$$0 = -a + (b-c(x_i+\Sigma x_i)) \quad \text{for } i=1,2,3,\ldots,8. \tag{4}$$

The system of equations in (4) has a symmetric solution in which x_i^* is the same for all i. Denoting this common value by x_i^*, we have from (4) that $0 = -a + (b-9cx_i^*)$ which can be rewritten as:

$$x_i^* = (b-a)/9c. \tag{5}$$

Multiplying x_i^* by 8 yields the aggregate tokens at equilibrium, namely $(8/9)(b-a)/c$.

Now if tokens are not infinitely divisible, all equilibria are in the vicinity of this one. All of these equilibria converge to (5) as tokens become more divisible. Let k and k+1 be the two integers nearest the aggregate token equilibrium given by (5). One of these will correspond to a set of asymmetric equilibria. Take m an integer, and solve the equation:

$$mn + (m+1)(8-n) = k. \tag{6}$$

In (6), m is the number of tokens, and n is the number of players contributing m tokens. Now test whether a player contributing m tokens is maximizing profits given that k tokens are being invested. Test similarly for a player contributing m+1 tokens. If both tests pass, then k is an aggregate investment for any asymmetric equilibrium with m tokens contributed by n players, and m+1 contributed by 8-n players. If both tests don't pass, then try (6) with k+1 on the right hand side. This test will pass, yielding asymmetric equilibria on that side.

Now look at the system of equations in mixed strategies. Since this game is symmetric, it will have a symmetric equilibrium. Due to the law of large numbers, this symmetric equilibrium will have an aggregate contribution very close to (5), with probability mixture replacing infinite divisibility of tokens. Suppose m and m+1 are the token levels at the asymmetric equilibria just found. Then a mixed strategy equilibrium is approximately (to the nearest 10^{-1} for our designs) a symmetric Nash equilibria , i.e., invest m with probability p and m+1 with probability 1-p {satisfying $[pm + (1-p)(m+1)]$ = expected contribution = x_i^*}. The complete set of equilibria defined above constitute all equilibria for the indivisible token case.

Now consider the specific parameterization for Design I experiments. For this parameterization, the strategy set for each player is $x_i \in \{0,1,2,\ldots,10\}$, where x_i denotes the number of tokens in market 2. The payoff for player i $h_i(x)$, in cents, is:

$$h_i(x) = \begin{cases} 50 & \text{if } x_i = 0 \\ 5(10-x_i) + (x_i/\Sigma x_i)(25\,\Sigma x_i - .3(\Sigma x_i)^2) & \text{if } x_i > 0 \end{cases}$$

where $x = (x_1,\ldots,x_8)$ is the vector of strategies of all players. This symmetric game has 57 Nash equilibria. 56 of these are asymmetric and in pure strategies, with

$\Sigma x_i = 59$ (approximately 41% of rents possible from market 2).[7] Th se are generated by having 5 players play $x_i = 7$ and 3 players play $x_i = 8$. The game also has a symmetric Nash equilibrium in mixed strategies, with $E(\Sigma x_i) = 59$ and $Var(\Sigma x_i) = 1.7$ (denoted as T3 in Figure 2). This equilibrium is generated by each player playing $x_i = 7$ with probability .62 and $x_i = 8$ with probability .38. It is this symmetric mixed strat gy that equilibrium selection theory (Harsanyi and Selten, 1988) predicts will be played.

B. Experimental Results - Design I

We begin our interpretation of the experimental observations with a descriptive look at the data across each experiment. Our primary focus is: (a) the extent to which this environment leads to an inefficiency in resource allocations and (b) the nature of this allocation across individuals. In Figure 3, we present period by period observations on the ARP and MRP across experimental decision periods. Contrary to the prediction of zero rents, we do not find our experimental markets stabilizing at full rent dissipation. Instead, we observe a general pattern across experiments where rents decay toward zero then rebound as subjects reduce the level of investment in the common-pool resource. In Table 2, we present summary information with regard to the aggregate tendencies of these four experiments. On average (pooling across all experiments), we find average rents equal to only 29% of optimum. Thus, although rents do not stabilize at the zero rent prediction, they fall far short of maximum. Further, from Figur 3, we see that there is some tendency for rents to decrease with repetition of the d cision proc ss.

DESIGN I EXPERIMENTS

		Length
1xhp:	22.6, 15.5, (-14 to 58)	n = 25
2xhp:	24.1, 22.6, (-14 to 65)	n = 20
3xhp:	23.9, 24.0, (-49 to 61)	n = 20
4xhp:	40.5, 20.6, (-14 to 83)	n = 20
Pooled:	29.0, 19.6, (-49 to 83)	

*All decision periods are used in calculating the descriptive statistics.
xhp = experienced, high pay

Table 2: Descriptive Statistics: Percentage of Maximum Rents Accrued (mean - standard deviation - range)*

[7]Rents accrued as a percentage of maximum = (Return from market 2 minus the opportunity costs of tokens invested in market 2)/(Return from market 2 at MR=MC minus the opportunity costs of tokens invested in market 2). Opportunity costs equal the potential return that could have been earned by investing the tokens in market 1.

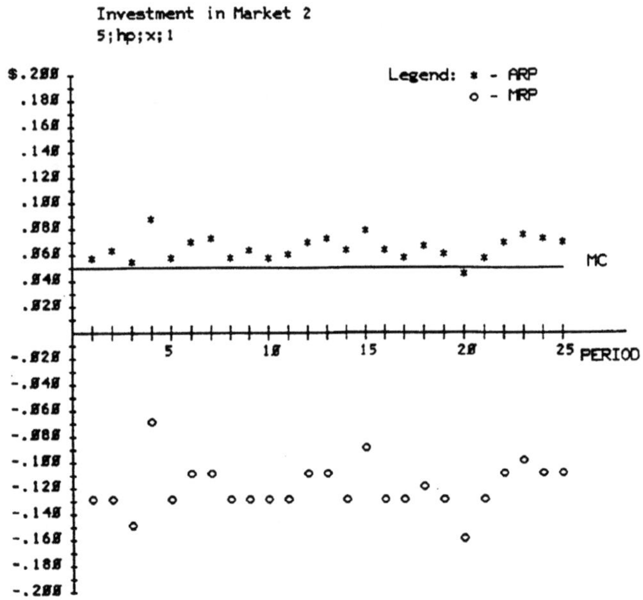

FIGURE 3a: DESIGN I - ARP AND MRP BY PERIOD - EXPERIMENT 1

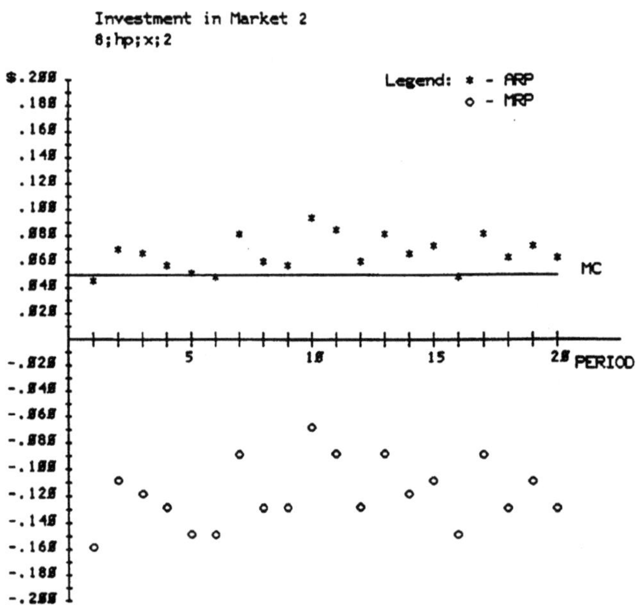

FIGURE 3b: DESIGN I - ARP AND MRP BY PERIOD - EXPERIMENT 2

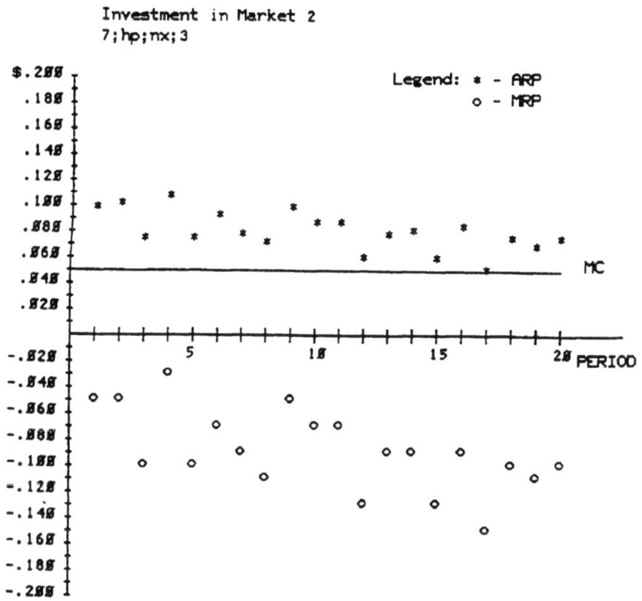

FIGURE 3c: DESIGN I - ARP AND MRP BY PERIOD - EXPERIMENT 3

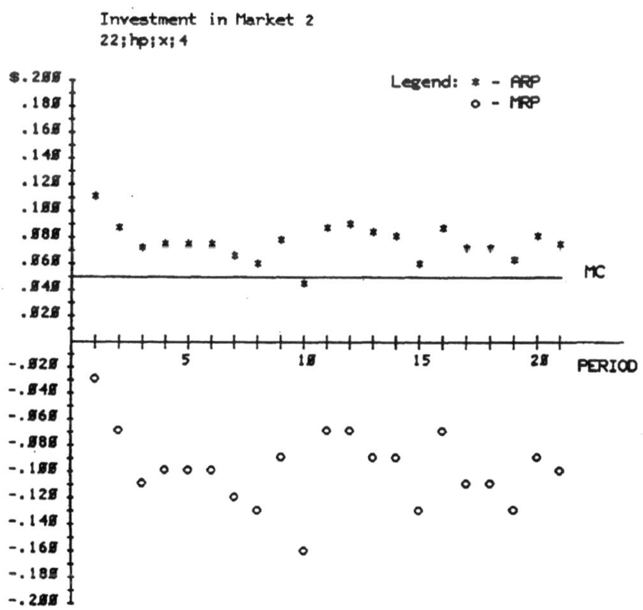

FIGURE 3d: DESIGN I - ARP AND MRP BY PERIOD - EXPERIMENT 4

Recall that the Nash equilibrium prediction of 59 tokens invested in market 2 implies a level of rents equal to 41% of optimum. At the aggregate 1 vel, the Nash prediction is in fact doing the best job of predicting subject behavior. The Nash prediction is, however, a prediction about <u>individual</u> behavior. How does our data conform to the individual predictions for the mixed strategy equilibrium for this environment? Recall that this symmetric game actually has 57 Nash equilibria. 56 of these are asymmetric and in pure strategies, with $\Sigma x_i = 59$. These are generated by having 5 players play $x_i = 7$ and 3 players play $x_i = 8$. The game also has a symmetric Nash equilibrium in mixed strategies, with $E(\Sigma x_i) = 59$. This equilibrium is generated by each player playing $x_i = 7$ with probability .62 and $x_i = 8$ with probability .38.

In Table 3, we present frequency counts across experiments that describe the extent to which individual decisions match those predicted by the possible Nash equilibria. Specifically, the data in Table 3 are classified with regard to two principal research questions.

Q1 - To what extent do period by period observations meet the criteria of 59 tokens allocated to market 2?

a) the number of periods in which 59 tokens were allocated in market 2.

b) if 59 tokens were contributed to market 2, the number of periods in which all subjects invested 7 or 8.

c) if 59 tokens were contributed to market 2, the number of periods in which there were 5 equal to 7 and 3 equal to 8.

Q2 - What is the frequency of periods in which investments of 7 or 8 were mad ?

a) periods in which all investments were 7 or 8.

b) periods in which all but 1 investment was 7 or 8.

c) periods in which all but 2 investments were 7 or 8.

d) periods in which all but 3 investments were 7 or 8.

e) periods in which all but 4 investments were 7 or 8.

f) periods in which all but 5 investments were 7 or 8.

g) periods in which all but 6 investments were 7 or 8.

h) periods in which all but 7 investments were 7 or 8.

i) periods in which no investments of 7 or 8 were made.

Experimental	Q1			Q2									Numb r of
Condition	a	b	c	a	b	c	d	e	f	g	h	i	Periods
HPX	6	0	0	0	0	3	7	27	24	21	5	0	87

<u>Table 3</u>: Periods in which Individual Investments were Consistent with Nash Design I Experiments

The results are quite interesting. Unlike the results reported on average rents, the data on individual investments lend little support to the Nash prediction. With regard to question 1 (Q1), we found only 6 of 87 periods in which actual market 2 investments equaled the Nash prediction of 59. Further, there were no periods in which 59 tokens were invested and all investments were 7 or 8. The frequency counts relat d to question 2 (Q2) illustrate the degree to which the individual decisions differ from the Nash prediction. One can see from Table 3 (columns Q2e through Q2i) that in 77 of 87 periods at least one half of the players (4) do not play the Nash strategy of investing 7 or 8 tokens in market 2. Although the results are not reported, we investigated questions Q1 and Q2 for each individual experiment. No individual experiment stands out from the others. That is, at the individual experiment level, no single experiment approaches the pattern of individual investment decisions predicted by the Nash equilibrium. We emphasize that if individual decisions do not match the Nash prediction, rational (income maximizing) behavior predicts that groups will not remain at a contribution decision that only matches the Nash prediction in the aggregate. Thus, it is not surprising that we find groups continuing to alter investment decisions in those periods that match Nash (approximately) in the aggr gate but not in an individual sense.

An alternative approach for investigating individual behavior is to focus on behavior over time for an individual. Did individuals consistently make the same investment choices? Did the pattern of choices vary across payoff conditions or experience? Was there evidence of learning (strategies in later periods that were consist ntly different than those of earlier periods)? In Table 4 we present fr quency counts that can be used to shed light on individual choice patterns. The data are separated into subjects whose investments of tokens in market 2 always equal: (1) 9 or 10; 8 or 7; 6 or 5; 4 or 3; 2 or less and (2) 10 to 6 or 5 to 0. The latter division was chosen to investigate the frequency with which subjects consistently played a "non-cooperative" strategy (10 to 6) or a "cooperative" strategy (5 to 0).

Panel A displays data using all periods of a given experiment, Panel B uses only periods 1-5 of a given experiment, and Panel C uses only periods 16-20 of a given experiment. Several observations are drawn from the data. First, most subjects did not adopt a particular (narrow) strategy that they used across all experiments. Only 2 of the 32 subjects in this series of experiments followed a strategy always within on of the token intervals arrayed in Table 4. Second, 15 of the 32 subjects adopted strategies that varied across decision periods between either a "non-cooperativ " (10-6) or a "cooperative" (5-0) strategy. The remaining 17 subjects (53%) consistently adopted the non-cooperative strategy of investing in the (10-6) range. Third, Pan ls B and C suggest that there is little "learning" of strategies for experienced subjects. The frequency data in periods 16-20 compare closely to that for periods 1-5.

PANEL A - ALL PERIODS

Experimental Condition	9,10	Number of Individuals Always Investing 8,7	6,5	4,3	2,1,0	10 - 6	5 - 0
HPX	2	0	0	0	0	17	0

PANEL B - PERIODS 1-5

Experimental Condition	9,10	Number of Individuals Always Investing 8,7	6,5	4,3	2,1,0	10 - 6	5 - 0
HPX	2	1	0	0	0	25	0

PANEL C - Periods 16-20

Experimental Condition	9,10	Number of Individuals Always Investing 8,7	6,5	4,3	2,1,0	10 - 6	5 - 0
HPX	2	2	0	0	0	24	1

Table 4: Individual Data: Consistency in Investment Patterns Design I Experiments

In summary, our initial design yields aggregate levels of rent that are consistent with the Nash equilibria predictions. At the individual decision level, however, we find very little support for the mixed strategy Nash equilibrium. Our next two designs employ parametric changes designed to investigate the robustness of these results.

IV. DESIGN II - ZERO MARGINAL COST EXPERIMENTS

The parametric conditions for our three "Design II" experiments are shown in Table 5. As with Design I, parametric conditions were constant within a given experiment. All experiments were conducted for at least 20 decision periods (no more than 25). One should note (relative to the Design I experiments) there are several key comparisons in the parameterizations. In Design II: (a) the payoff function for market 2 was identical to that used in Design I, (b) the payoff from market 1 was reduced to zero, (c) individual token endowments were increased to 15 tokens, and (d) subjects started the experiment with an initial capital endowment of \$5.00. The reasons for the increased token endowments and the use of an up front capital endowment will be clear after we investigate the theoretical properties of this design.

Experiment Type	Design II
Number of Subjects	8
Individual Token Endowment	15
Production Function: Mkt.2*	$25(\Sigma x_i)-.30(\Sigma x_i)^2$
Market 2 Return/unit of output	$.01
Market 1 Return/unit of output	$.00
Earnings/Subject at Group Max.	$.65
Earnings/Subject at Nash Equil.	$.26
Earnings/Subject at Zero Rent	$.00

*Σx_i = the total number of tokens invested by the group in market 2. Th production function shows the number of units of output produced in market 2 for each level of tokens invested in market 2.

Table 5: Experimental Design II - Parameters for a Given Decision Period

A. Theoretical Predictions

As with the Design I parameterizations, Design II allows for three non-cooperative theoretical predictions. Figure 4 illustrates group behavior that would be consistent with these alternative predictions. As noted in Figure 4, a group investment of 42 tokens yields a level of investment at which MRP = MC and thus maximum rents (denoted T1). Conversely, a group investment of 83 tokens yields a level of investment at which ARP = MC and thus zero rents from market 2 (denoted T2). As with Design I, ther exist a set of Nash equilibria that differ from either rent maximization or minimization. Consider the specific parameterizations for market 1 and market 2. The strategy s t for each player is $x_i \in \{0,1,2,\ldots,15\}$, where x_i denotes the number of tokens in market 2. The payoff for player i $h_i(x)$, in cents, is:

$$h_i(x) = \begin{cases} 0 & \text{if } x_i = 0 \\ 0(10-x_i) + (x_i/\Sigma x_i)(25\Sigma x_i - .30(\Sigma x_i)^2) & \text{if } x_i > 0. \end{cases}$$

where $x = (x_1,\ldots,x_8)$ is the vector of strategies of all players. This symmetric game has 57 Nash equilibria. 56 of these are asymmetric and in pure strategies, with Σx_i = 74 (approximately 40% of rents possible from market 2). These are generated by having 6 players play $x_i = 9$ and 2 players play $x_i = 10$. The game also has a symmetric Nash equilibrium in mixed strategies, with $E(\Sigma x_i) = 74$ (denoted as T3 in Figure 4). This equilibrium is generated by each player playing $x_i = 9$ with probability .74 and $x_i = 10$ with probability .26.

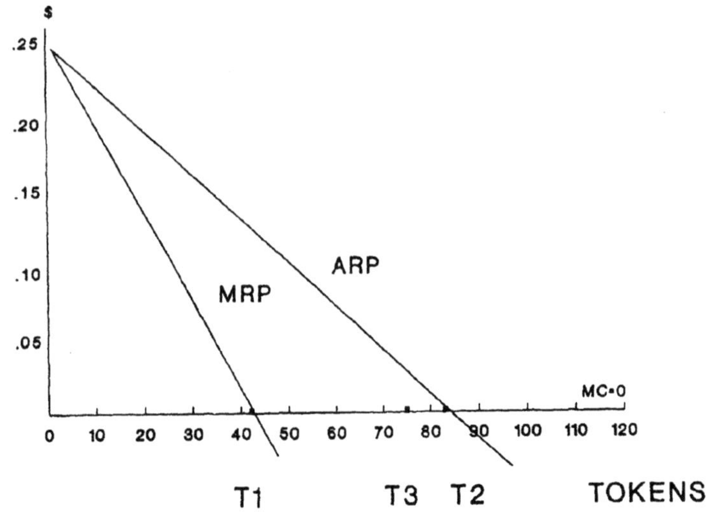

Figure 4: Theoretical Predictions: Design II

Given the equilibria for this design, one can see why we modified our design by increasing token endowments and endowing subjects with an up front capital r ward. With these payoff conditions, ARP=0 at a group investment of 83 tokens. For this reason, we increased individual token endowments to 15 (from 10) so that full rent dissipation would not be a corner solution. Further, with this design, it is possible for subjects to have negative returns for a decision period. To leave subjects with some minimal experimental earnings, we added the up front cash endowment.

Finally, consider the level of subject payoffs accrued at the Nash equilibria and at the zero rent predictions in Designs I and II. In Design I, subjects earned \$.67 per period at the Nash prediction and \$.50 per period at the zero rent prediction. In Design II, however, these payoffs change to \$.26 at the Nash prediction and \$.00 at the zero rent prediction. Thus, the magnitude in the payoff differences between the Nash pr diction and the zero rent prediction increases in Design II. Further, one might argue that subjects would be more sensitive to strategic decisions given that the z ro rent prediction actually yielded a zero payoff. We turn to the results from our Design II experiments to see if these parametric changes have a significant impact on aggregate and individual behavior.

B. Experimental Results - Design II

As with Design I, we begin our interpretation of the experimental observations with a descriptive look at the data across each experiment. In Figure 5 we present p riod by

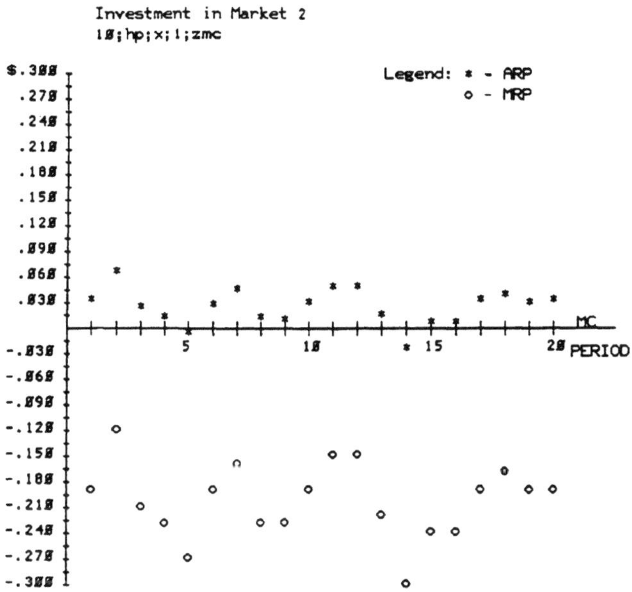

FIGURE 5a: DESIGN II - ARP AND MRP BY PERIOD - EXPERIMENT 1

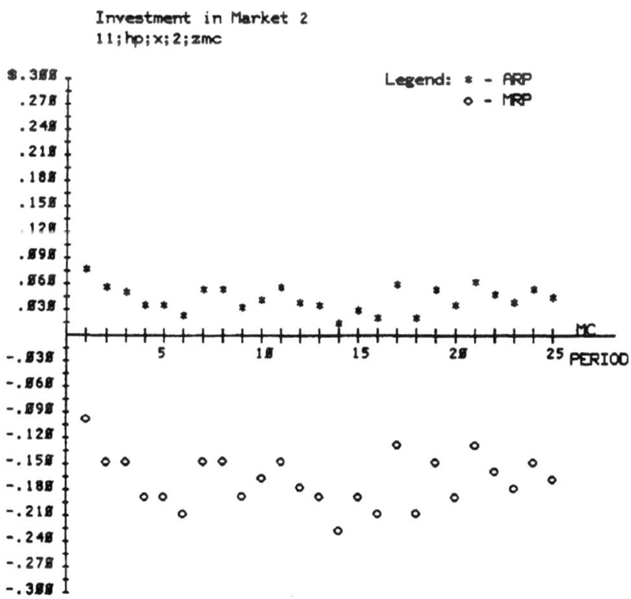

FIGURE 5b: DESIGN II - ARP AND MRP BY PERIOD - EXPERIMENT 2

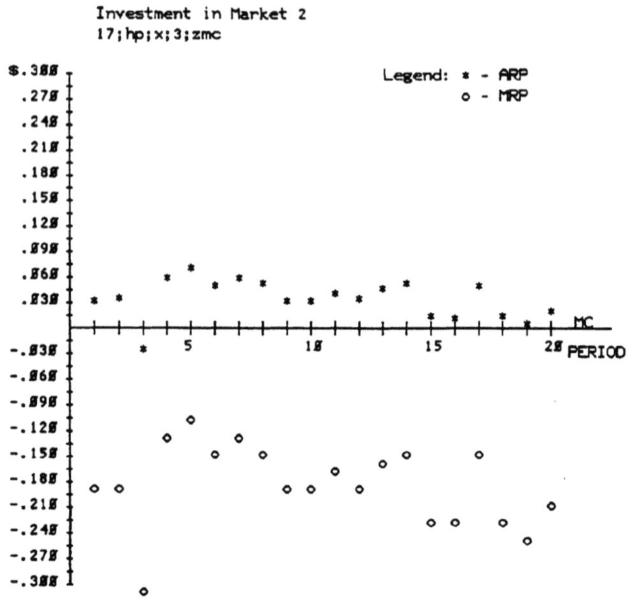

FIGURE 5c: DESIGN II - ARP AND MRP BY PERIOD - EXPERIMENT 3

period observations on the ARP and MRP across experimental decision periods. Similar to the Design I experiments and contrary to predictions of zero rents, we do not find our experimental markets stabilizing at full rent dissipation. Instead, we observe a general pattern across experiments where rents decay toward zero then rebound as subjects reduce the level of investment in the CPR. In Table 6, we present summary information with regard to the aggregate tendencies of these four experiments. On average (pooling across all experiments) we find average rents equal to 40.8% of optimum. Thus, the parametric changes in Design II appear to increase the level of rents relative to those observed in the Design I experiments. Further, the 1 vel of rents (on average) are almost identical to the Nash prediction of 40%.

DESIGN II - ZERO MARGINAL COSTS EXPERIMENTS

		Length
1-x-hp-zmc:	29.8, 28.57, -46 - 76	n = 20
2-x-hp-zmc:	50.5, 16.74, 15 - 83	n = 25
3-x-hp-zmc:	39.5, 30.47, -52 - 78	n = 20
Pooled:	40.8, 26.42, -52 - 83	

*All decision periods are used in calculating the descriptive statistics.

Table 6: Descriptive Statistics - Percentage of Rents Accrued (mean - standard deviation - range)*

How does our data conform to the individual predictions for the mixed strategy equilibrium for this environment? As with Design I, this symmetric game actually has 57 Nash equilibria. 56 of these are asymmetric and in pure strategies, with $\Sigma x_i = 74$. These are generated by having 6 players play $x_i = 9$ and 2 players play $x_i = 10$. The game also has a symmetric Nash equilibrium in mixed strategies, with $E(\Sigma x_i) = 74$. This equilibrium is generated by each player playing $x_i = 9$ with probability .74 and $x_i = 10$ with probability .26.

In Tables 7 and 8 we present summary data on individual data parallel to that which was presented for the Design I experiments. Table 7 presents frequency counts across experiments that describe the extent to which individual decisions match those predicted by the possible Nash equilibria. Since the Nash equilibria predictions now require an investment of 9 or 10 tokens, questions Q1 and Q2 are modified accordingly. The two principal research questions are:

Q1 - To what extent do period by period observations meet the criteria of 74 tokens allocated to market 2?

a) the number of periods in which 74 tokens were allocated in market 2.
b) if 74 tokens were contributed to market 2, the number of periods in which all subjects invested 9 or 10.
c) if 74 tokens were contributed to market 2, the number of periods in which there were 5 equal to 7 and 3 equal to 8.

Q2 - What is the frequency of periods in which investments of 9 or 10 were made?

a) periods in which all investments were 9 or 10.
b) periods in which all but 1 investment was 9 or 10.
c) periods in which all but 2 investments were 9 or 10.
d) periods in which all but 3 investments were 9 or 10.
e) periods in which all but 4 investments were 9 or 10.
f) periods in which all but 5 investments were 9 or 10.
g) periods in which all but 6 investments were 9 or 10.
h) periods in which all but 7 investments were 9 or 10.
i) periods in which no investments of 9 or 10 were made.

Experimental	Q1			Q2									Number of
Condition	a	b	c	a	b	c	d	e	f	g	h	i	Periods
ZMC	6	0	0	0	0	0	2	6	11	21	22	3	65

Table 7: Periods in Which Individual Investments were Consistent with Nash Design II Experiments

In summary, the r sults are not consistent with the individual Nash equilibria predictions. In only 6 of 65 periods do we see token investments in market 2 equal to the Nash prediction of 74. Further, in none of these 6 periods do we observe all subjects investing 9 or 10. With regard to the frequency in which subjects invest 9 or 10 tokens, one can see from Table 7 (columns Q2e through Q2i) that in 63 of 65 periods at least one half of the players (4) do not play the Nash strategy of investing 9 or 10 tokens in market 2.

Turning to Table 8, we focus on individual strategies across periods. Out of the 24 subjects participating in these three experiments, we see no subject consistently following the Nash strategy of 9 or 10 tokens. However, parallel to the argument given for our Design I experiments, the rational (income maximizing) strategy for an individual only matches that predicted by the Nash equilibrium concept if all individuals are following the Nash strategy. In the final periods of these experiments (similar to our Design I experiments) there does appear to be some movement toward Nash strategies. From Panel C, looking at observations confined to periods 16-20, we see that 3 of 24 players consistently invest the Nash strategy of 9 or 10 tokens. Further, 9 of 24 individuals restrict their market 2 investments to the range of 11 to 8 tokens, a one token deviation on either side of the Nash strategy.

PANEL A - ALL PERIODS

Experimental Condition	Number of Individuals Always Investing 15,14,13	12,11	10,9	8,7	6,5	4,3	2,1,0	15-12	11-8	7-0
ZMC	0	0	0	0	0	0	0	1	0	2

PANEL B - PERIODS 1-5

Experimental Condition	Number of Individuals Always Investing 15,14,13	12,11	10,9	8,7	6,5	4,3	2,1,0	15-12	11-8	7-0
ZMC	0	0	0	0	1	2	0	1	2	2

PANEL C - Periods 16-20

Experimental Condition	Number of Individuals Always Investing 15,14,13	12,11	10,9	8,7	6,5	4,3	2,1,0	15-12	11-8	7-0
ZMC	2	0	3	1	1	1	0	2	9	2

Table 8: Individual Data: Consistency in Investment Patterns Design II Experiments

In conclusion, results from our Design II experiments follow an aggregate pattern that is on average consistent with the predictions of the Nash equilibria. At the individual level, however, our results offer little support to the hypothesis that individuals will stabilize at an individually rational Nash equilibria. Is this result due to the fact that the Designs for I and II produce multiple Nash equilibria, only one of which is symmetric and that one is in mixed strategies? In Section V, we investigate this question employing a modification on Design I that allows for a unique Nash equilibrium prediction.

V. DESIGN III - EXPERIMENTS WITH A UNIQUE NASH EQUILIBRIUM

The parametric conditions for our three "Design III" experiments are shown in Table 9. These additional experiments allowed us the opportunity to investigate whether our initial results were at least partially the result of a design with multiple equilibria. We conducted a new set of three experiments (all using experienced subjects) with experimental procedures identical to Design I except the payoff for player i $h_i(x)$, in cents, was changed to:

$$h_i(x) = \begin{cases} 50 & \text{if } x_i = 0 \\ 5(10-x_i) + (x_i/\Sigma x_i)(23\ \Sigma x_i - .25(\Sigma x_i)^2) & \text{if } x_i>0 \end{cases}$$

where $x = (x_1,...,x_8)$ is the vector of strategies of all players.

Experiment Type	Design III
Number of Subjects	8
Individual Token Endowment	10
Production Function: Mkt.2*	$23(\Sigma x_i)-.25(\Sigma x_i)^2$
Market 2 Return/unit of output	\$.01
Market 1 Return/unit of output	\$.05
Earnings/Subject at Group Max.	\$.91
Earnings/Subject at Nash Equil.	\$.66
Earnings/Subject at Zero Rent	\$.50

*Σx_i = the total number of tokens invested by the group in market 2. The production function shows the number of units of output produced in market 2 for each level of tokens invested in market 2.

Table 9: Experimental Design II - Parameters for a Given Decision Period

A. Theoretical Pr dictions

As with the Design I parameterizations, Design III allows for three non-cooperative theoretical predictions. Figure 6 illustrates group behavior that would be consistent with these alternative predictions. A group investment of 36 tokens yields a level of investment at which MRP = MC and thus maximum rents (denoted T1). Conversely, a group investment of 72 tokens yields a level of investment at which ARP = MC and thus zero rents from market 2 (denoted T2). Finally, this symmetric game has a unique Nash equilibrium with each subject investing 8 tokens in market 2.[8] Note that at the Nash equilibrium prediction, subjects would earn (similar to Designs I and II) approximat ly 39.5% of maximum rents.

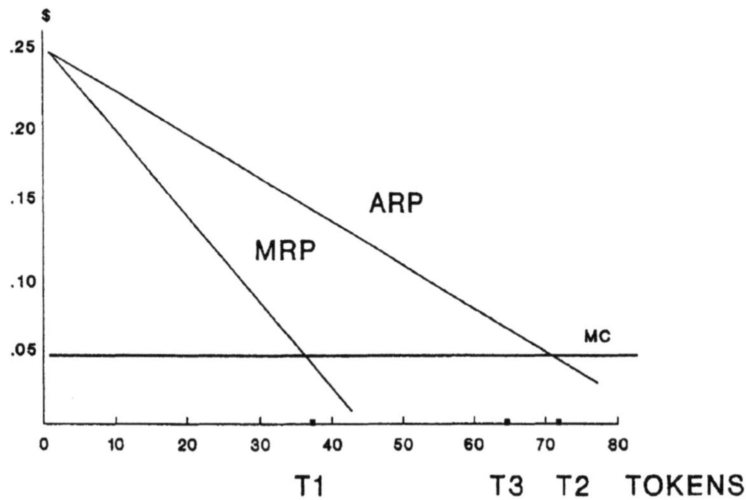

PREDICTIONS: T1=36; T2=72; T3=64

Figure 6: Theoretical Predictions: Design III

---

[8]When the one shot game has a unique Nash equilibrium, the repeated game has a unique subgame perfect equilibrium that plays Nash each period. As Benoit and Krishna (1985) show, when a one shot game has multiple Nash equilibria the finite repetition of the one shot game has subgame perfect equilibria with average payoffs close to maximin l vels. In our context, this means we could have subgame perfect equilibrium with average payoffs close to complete rent dissipation outcomes. Of course there are also subgame perfect equilibrium with average payoffs close to the one shot Nash equilibrium payoffs. Thus, in a repeated game framework, equilibrium point theory does not lead to such clear cut conclusions as in the one shot game. At the aggregate level, our mpirical results suggest that the latter equilibrium were being played, rather than the former. Moreover, this could be part of the reason that at the individual level, subjects do not appear to be playing one shot Nash equilibrium.

B. Experimental Results - Design III

As with our previous results, we begin our interpretation of the experimental observations with a descriptive look at the data across each experiment. In Figure 7 we present period by period observations on the ARP and MRP across experimental decision periods. Note that all of these experiments were run for 30 periods as compared with 20 to 25 in our previous designs. This change was made to increase the opportunity for a group to stabilize at a "behavioral equilibrium." Similar to the results from our previous designs and contrary to predictions of zero rents, rents do not stabilize at full dissipation. Instead, we observe the previously noted pattern where rents decay toward zero and then rebound as subjects reduce the level of investment in the CPR. In Table 10 we present summary information with regard to the aggregate tendencies of these three experiments. On average (pooling across all experiments), we find average rents equal to 37.2% of optimum. Thus, at least on average, this design also yields rents consistent with the Nash prediction. The analysis that follows examines whether the implementation of the unique Nash equilibrium alters our results with respect to individual behavior.

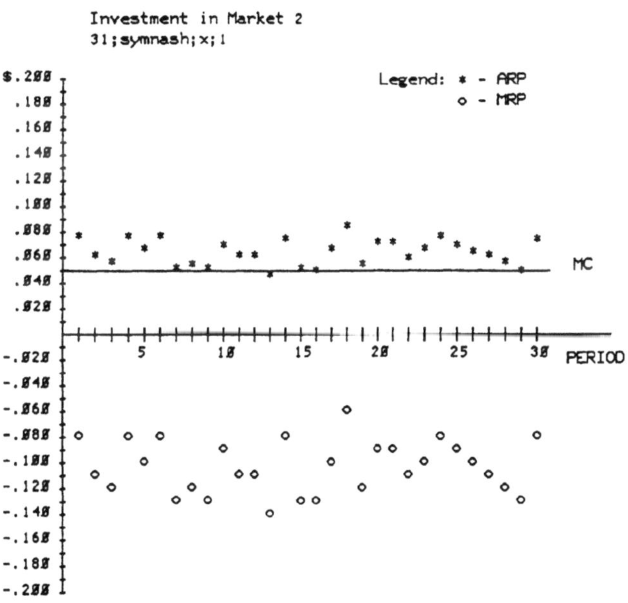

FIGURE 7a: DESIGN III - ARP AND MRP BY PERIOD - EXPERIMENT 1

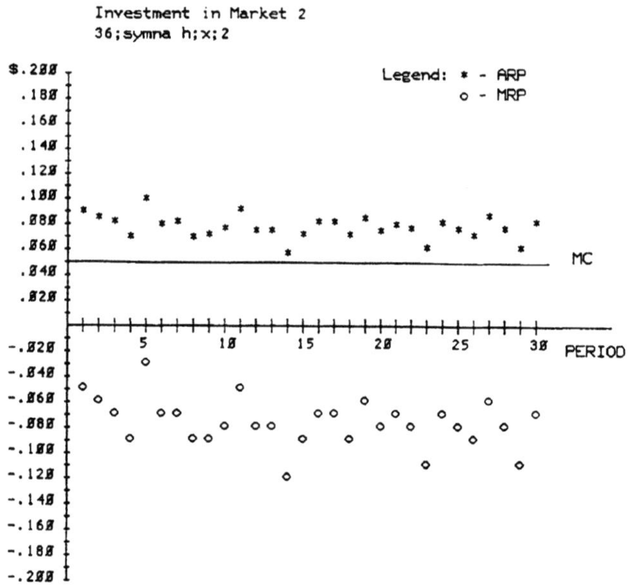

FIGURE 7b: DESIGN III - ARP AND MRP BY PERIOD - EXPERIMENT 2

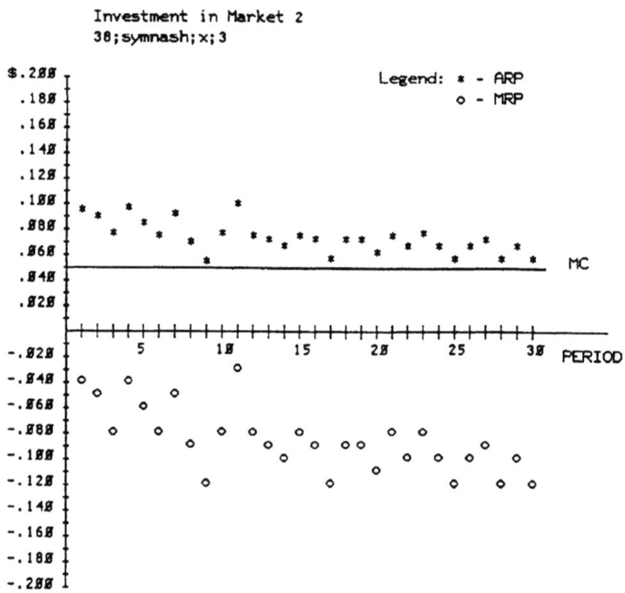

FIGURE 7c: DESIGN III - ARP AND MRP BY PERIOD - EXPERIMENT 3

DESIGN III - SYMMETRIC NASH EXPERIMENTS	
1xsn:	23.8, 19.5, (-11 to 59)
2xsn:	47.8, 14.4, (11 to 78)
3xsn:	39.8, 19.7, (06 to 78)
Pooled:	37.2, 20.4, (-11 to 78)

*All decision periods are used in calculating the descriptive statistics. All experiments contained 30 periods.

Table 10: Descriptive Statistics: Percentage of Maximum Rents Accrued (mean - standard deviation - range)*

In Table 11 we present frequency counts across experiments that describe the extent to which individual decisions match those predicted by the unique Nash equilibrium. Since the Nash equilibrium prediction requires a market 2 investment of 8 tokens by all subjects, questions Q1 and Q2 are modified accordingly. The two principal research questions are:

Q1 - To what extent do period by period observations meet the criteria of 64 tokens allocated to market 2?

a) the number of periods in which 64 tokens were allocated in market 2.

b) if 64 tokens were contributed to market 2, the number of periods in which all subjects invested 8.

Q2 - What is the frequency of periods in which investments of 8 were made?

a) periods in which all investments were 8.

b) periods in which all but 1 investment was 8.

c) periods in which all but 2 investments were 8.

d) periods in which all but 3 investments were 8.

e) periods in which all but 4 investments were 8.

f) periods in which all but 5 investments were 8.

g) periods in which all but 6 investments were 8.

h) periods in which all but 7 investments were 8.

i) periods in which no investments of 8 were made.

Experimental		Q1						Q2					Number of
Condition	a	b	c	a	b	c	d	e	f	g	h	i	Periods
Unique Nash	11	0	0	0	0	0	0	1	3	24	29	33	90

Table 11: Periods in which Individual Investments were Consistent with Nash Design III Experiments

In summary, the data provide very little support for the research hypothesis that our investment environment will stabilize at the unique Nash equilibrium. Out of 90 investment periods, we find only 11 in which aggregate investment in market 2 equals 64. In none of those 11 cases did we find a pattern of 8 tokens invested by each subject. Further, in all 90 investment periods we find 4 or less of the 8 subjects investing the Nash equilibrium prediction of 8 tokens.

Turning to Table 12, we focus on individual strategies across periods. Of the 24 subjects in these three experiments we found no subject always playing the strategy of 8 tokens invested in market 2. Further, as can be seen in Panel A, we found no subject consistently playing within one token of the Nash prediction (playing 9, 8, or 7). What happens if we analyze only the last 5 periods of these 30 period experiments (Panel C)? In these final periods, we find 3 subjects consistently playing within the one token band around the Nash prediction. We also find 19 of 24 playing in the broader range of 6 to 10 tokens invested in market 2. However, consistent with our previous designs, we find a strong pattern of players investing all 10 tokens in market 2. In fact, in the final 5 periods, 6 of the 24 players always invest 10 tokens.

PANEL A - ALL PERIODS

Experimental Condition	Number of Individuals Always Investing 10	9,8,7	6,5	4,3	2,1,0	10 - 6	5 - 0
Unique Nash	3	0	0	0	0	9	0

PANEL B - PERIODS 1-5

Experimental Condition	Number of Individuals Always Investing 10	9,8,7	6,5	4,3	2,1,0	10 - 6	5 - 0
Unique Nash	4	1	0	0	0	13	1

PANEL C - Periods 26-30

Experimental Condition	Number of Individuals Always Investing 10	9,8,7	6,5	4,3	2,1,0	10 - 6	5 - 0
Unique Nash	6	3	1	0	0	19	1

Table 12: Individual Data: Consistency in Investment Patterns Design III Experiments

In the context of a Nash equilibrium, this is the simplest of our three designs. For the interested reader, we present in Figure 8 the complete history of Market 2 investments from our three Design III experiments. This data allows a detailed look at the evolution of decision strategies within and across experiments. The most striking observation is the frequency with which numerous players will consistently make market 2 investment decisions of 10 tokens. In our view, it is this pattern of balanced deviation disequilibrium decisions that inhibits, to the greatest degree, the inability for group behavior to reach a Nash equilibrium. Given that subjects have an endowment of only 10 tokens, the strategy of 10 tokens may also be a "constrained" choice. In our future experiments, we will analyze the impact of increasing individual endowments so that individual's choices are not constrained by this parameterized constraint.

Experimental Data by Period (1-30)
Subject Number (1-8)

	1xsn								2xsn								3xsn							
	1	2	3	4	5	6	7	8	1	2	3	4	5	6	7	8	1	2	3	4	5	6	7	8
1	5	6	6	5	10	10	10	10	8	8	7	10	5	10	6	3	0	6	7	6	10	9	9	8
2	5	8	9	7	10	10	9	10	8	7	6	7	7	10	7	7	3	7	6	5	10	8	9	9
3	7	7	10	8	10	10	8	10	7	8	8	6	8	10	6	7	4	8	6	6	10	10	10	8
4	3	6	3	10	10	10	10	10	9	8	9	7	7	10	8	7	3	8	5	4	10	9	9	6
5	8	5	5	9	10	10	9	10	7	4	7	7	6	10	5	7	3	9	5	7	10	9	9	7
6	7	8	4	10	10	5	8	10	8	7	7	8	6	10	6	9	4	8	6	7	10	10	10	8
7	7	10	5	10	10	10	10	10	7	9	8	7	6	10	5	8	2	8	4	7	10	9	9	7
8	7	10	6	8	10	10	10	10	9	9	8	7	5	10	9	8	10	8	7	8	10	8	8	6
9	4	10	10	8	10	10	10	10	7	8	10	6	5	10	9	9	10	8	7	7	10	10	9	10
10	3	10	7	10	6	10	9	10	6	6	9	6	7	10	9	9	8	9	5	6	10	7	9	8
11	5	9	10	10	6	10	8	10	7	4	8	5	6	10	6	10	9	7	2	6	10	10	0	9
12	7	9	10	9	6	10	7	10	8	10	8	7	5	10	6	9	8	8	3	6	10	10	9	9
13	7	10	10	10	10	10	7	10	7	10	6	6	6	10	9	9	10	10	2	6	10	10	8	8
14	4	10	6	10	6	10	7	10	9	10	8	6	8	10	9	10	10	9	6	6	10	9	9	7
15	8	10	10	10	6	10	8	10	7	8	10	5	6	10	8	10	8	6	5	5	10	10	10	9
16	4	9	10	10	10	10	10	10	8	7	10	5	6	10	5	9	9	9	4	5	10	10	8	9
17	5	8	10	9	6	10	8	10	7	8	9	4	6	10	6	10	10	9	7	7	10	10	9	8
18	5	9	3	10	5	10	7	10	7	10	10	5	5	10	7	10	10	6	5	5	10	10	9	9
19	6	10	6	10	10	10	9	10	8	8	8	2	6	10	8	9	8	9	4	5	10	10	9	9
20	5	6	6	10	10	10	7	10	8	7	10	5	6	10	8	9	8	10	6	6	10	9	10	9
21	5	8	3	10	10	10	8	10	8	10	8	4	6	10	6	9	10	9	7	5	10	10	5	7
22	5	8	6	10	10	10	10	10	7	6	10	5	6	10	9	9	9	9	6	4	10	10	9	9
23	2	9	6	10	10	10	9	10	8	10	9	5	6	10	10	10	8	9	7	5	10	8	8	7
24	2	8	4	10	10	10	8	10	7	8	10	4	6	10	5	10	7	9	7	5	10	10	9	9
25	5	7	5	10	10	10	8	10	7	9	10	5	6	10	6	9	10	9	7	7	10	10	9	8
26	5	8	6	10	10	10	8	10	8	8	10	5	6	10	8	9	10	9	6	6	10	8	8	9
27	7	8	4	10	10	10	9	10	7	7	10	4	5	10	6	9	7	9	7	5	10	10	9	7
28	7	7	6	10	10	10	10	10	9	7	10	5	6	10	6	9	10	9	7	6	10	10	10	8
29	5	8	10	10	10	10	10	10	8	10	10	5	7	10	9	9	10	9	7	5	10	10	9	6
30	3	6	6	10	10	10	8	10	7	9	8	5	6	10	6	9	10	9	8	5	10	10	9	9

Figure 8: Individual Market 2 Investments
Symmetric Nash Equilibrium Experiments

VI. SUMMARY AND CONCLUDING COMMENTS

A generally accepted premise in the literature focusing on open access resources is that resource users, working independently, will overexploit the resource in question, leading to rent dissipation and possible resource destruction. Using experimental methods to control for subject incentives and to induce a set of institutional arrangements that capture the strategic essence of the appropriation dilemma in a CPR, this study investigates the degree to which suboptimal appropriation occurs.

In summary, our results strongly support a research hypothesis of suboptimal appropriation and thus significant rent dissipation. Across all experimental conditions, subjects earn on average only 34% of possible rents. Further, the level of rent dissipation tends to increase when subjects are experienced in the decision environment and with repetition of the decision process. Contrary to one of the standard predictions for this type of environment, we do not find our experimental markets stabilizing at a level of rents equal to full dissipation. Instead, we observe a general pattern across experiments where rents decay toward zero then rebound as subjects reduce the level of investment in the common-pool resource. Investigating across three distinct parameterizations, we find that at the aggregate level, our results lend strong support to the aggregate Nash equilibrium prediction. At the individual decision level, however, we do not find behavior consistent with the Nash prediction.

Several factors may be contributing to the balanced deviation disequilibrium results we observe in these experiments. First and foremost is the computational complexity of the task. The payoff functions are highly nonlinear and nondifferentiable, making them difficult for our subjects to process. Indeed, in post-experiment questionnaires we administered, we found that many subjects were using the rule of thumb "Invest more in market 2 whenever the rate of return is above $.05 per token." Then, when the rate of return fell below $.05, they reduce investments in market 2, giving rise to the cycle in returns we observe across all environments. A related factor is the focal point effect of investing 10 tokens in market 2 (for Designs I and III), which is indeed the modal strategic response. Here, the role of thumb seems to be "Invest <u>all</u> tokens in market 2 whenever the rate of return there is above $.05 per token in previous decision rounds." This behavior is clearly inconsistent with best response behavior in these experiments. Finally, the fact that equilibrium is never reached at the individual level means that each player is continually having to revise his or her response to the current "anticipated" situation. This strategic turbulence on top of an already complex task increases the chances that a player may not attempt a best response approach to the task, but rather invoke simple rules of thumb of the type reported above. In current work we are formally investigating the extent to which subjects appear to follow a reaction function consistent with Nash type behavior (Dudley and Walker, 1989). This work also extends our environment by increasing subject endowments. This change significantly reduces the possibility of having a player's market 2 investment constrained by their resource endowment.

The consistency of balanced deviation disequilibrium - indeed, the consistency with which we find no behavioral equilibrium whatsoever in these experiments - is the single largest unanswered question posed by our results. It is a complex issue for experimental research in general. We know that for many institutional settings the Nash prediction can be quite robust. See for example Cox, Smith, and Walker (1988) for the case of single unit sealed bid auctions. Even in this research, however, institutional changes, such as a switch to multiple unit auctions, can lead to subject behavior which is no longer consistent with a Nash model based on expected utility maximization (see Cox, Smith, and Walker, 1984).

There are four areas that we think to be especially worth exploring, given the results presented here. First, what form of parameterization might lead to a behavioral equilibrium consistent with the Nash prediction? This work could shed important light on the types of behavioral environments in which one might expect to see Nash as a strong predictor. Second, one would like to explore the effect of communication on these environments, where one continues to retain the noncooperative structure of th situation. It may be the case that, even without their ability to implement binding contracts, the availability of channels of communication among players may lead to a substantial decline in the amount of rent dissipated. Third, it is desirable to relax the stationarity assumption. Most naturally occurring CPRs are nonstationary; indeed, it is the possibility of degradation over time that makes them so important from a policy standpoint. Here, the strategic complexity grows remarkably. If players have a hard time achieving a Nash equilibrium in a simple stationary environment, they should have even a harder time reaching an equilibrium here. However, it may well be the case that aggregate behavior persists in Nash equilibrium level of rent dissipation. Finally, the possibility of destruction of the CPR - again a major concern in many naturally occurring CPR environments - should be modeled and explored experimentally. Risk averse players taking the threat of destruction seriously can be expected to reduce their investments in the risky environment and again reduce the level of rents dissipated.

References

Agnello, R. and L. Donnelly (1975). Property Rights and Efficiency in the Oyster Industry. J. Law & Econonomics, 18: 521-533.

Alexander, P. (1982). Sri Lankan Fishermen. Rural Capitalism and Peasant Society. Canberra: Australian National University.

Alger, D. (1986). Investigating Oligopolies Within the Laboratory. Staff Report, Bureau of Economics of the Federal Trade Commission.

Bell, F. (1972). Technological Externalities and Common Property Resources: An Empirical Study of the U.S. Lobster Industry. J. Pol. Econ., 80: 148-158.

Benoit, J. and V. Krishna (1985). Finitely Repeated Games. Econometrica, 53: 905-922.

Blomquist, W. and E. Ostrom (1985). Institutional Capacity and the Resolution of a Commons Dilemma. Policy Studies Rev., 5: 383-393.

Bromley, D. (1982). Land and Water Problems: An Institutional Perspective. Amer. J. Agricultural Economics, 64: 834-844.

Ciriacy-Wantrup, S. V. and R. Bishop (1975). Common Property as a Concept in Natural Resource Policy. Natural Resource J., 15: 713-727.

Clark, C. (1976). Mathematical Bioeconomics: The Optimal Management of Renewable Resources. New York: Wiley.

Cox, J., V. Smith, and J. Walker (1984). Theory and Behavior of Multiple Unit Discriminative Price Auctions. J. Finance, 34: 983-1010.

Cox, J., V. Smith, and J. Walker (1988). Theory and Individual Behavior of First Price Auctions. J. Risk & Uncertainty, 1: 61-99.

Dudley, D. and J. Walker (1989). Nash Type Behavior in Common Pool Resource Environments. Working paper. Indiana University.

Gardner, R., E. Ostrom, and J. Walker (1988). The Nature of Common Pool Resource Problems. Working paper. Bloomington, IN: Indiana University, Workshop in Political Theory and Policy Analysis.

Gordon, S. (1954). The Economic Theory of a Common Property Resource: The Fishery. J. Pol. Economy: 124-142.

Harsanyi, J. and R. Selten (1988) A General Theory of Equilibrium of Equilibrium Selection in Games. Cambridge, MASS: MIT Press.

Isaac, R. M., J. Walker, and S. Thomas (1984). Divergent Evidence on Free Riding: An Experimental Examination of Possible Explanations. Pub. Choice: 43: 113-149.

Isaac, R. M. and J. Walker (1988) Group Size Effects in Public Goods Provision: The Voluntary Contributions Mechanism. Quarterly J. Economics, February: 179-200.

Johnson, R. and G. Libecap (1982). Contracting Problems and Regulation: The Case of the Fishery. Amer. Econ. Rev., December: 1005-1023.

Jorgenson, D. and A. Papciak (1980). The Effects of Communication, Resource Feedback, and Identifiability on Behavior in a Simulated Commons. J. Experimental Social Psychology, 17: 373-385.

McCay, B. and J. Acheson, eds. (1987). The Question of the Commons. Tucson: University of Arizona Press.

Messick, D. and C. McClelland (1983). Social Traps and Temporal Traps. Personality & Social Psychology Bulletin, 9: 105-110.

Millner, E. and M. Pratt (1988). An Experimental Investigation of Efficient Rent Seeking. Public Choice, forthcoming.

Morrison, C. and H. Kamarei (1989). Some Experimental Testing of the Cournot-Nash Hypothesis in Small Group Rivalry Situations. Journal of Economic Behavior and Organization, forthcoming.

National Research Council (1986). Proceedings of the Conference on Common Property Resource Management. Washington, D.C.: National Academy Press.

Ostrom, E. (1988). Institutional Arrangements and the Commons Dilemma. In V. Ostrom, D. Feeny, and H. Picht, eds. Rethinking Institutional Analysis and Development: Issues, Alternatives, and Choices. San Francisco: Institute for Contemporary Studies Press, 103-139.

Plott, C. and R. Meyer (1975). The Technology of Public Goods, Externalities, and the Exclusion Principle. In Edwin S. Mills, ed. Economic Analysis of Environmental Problems. New York: National Bureau of Economic Research.

Samuelson, C. and D. Messick (1988). When Do People Want to Change the Rules for Allocating Shared Resources. Unpublished manuscript.

Schlager, E. and E. Ostrom (1988). Common Property, Communal Property, and Natural Resources: A Conceptual Analysis. Working paper. Bloomington, IN: Indiana University, Workshop in Political Theory and Policy Analysis.

Smith, V. (1968). Economics of Production from Natural Resources. Amer. Econ. Rev., 85: 409-431.

MIX
Papier aus verantwortungsvollen Quellen
Paper from responsible sources
FSC® C105338
www.fsc.org

If you have any concerns about our products,
you can contact us on
ProductSafety@springernature.com

In case Publisher is established outside the EU,
the EU authorized representative is:
Springer Nature Customer Service Center GmbH
Europaplatz 3, 69115 Heidelberg, Germany

Printed by Libri Plureos GmbH
in Hamburg, Germany